Basic Research on Lung Surfactant

Progress in Respiration Research

Vol. 25

Series Editor
H. Herzog, Basel

KARGER

Basel · München · Paris · London · NewYork · New Delhi · Bangkok · Singapore · Tokyo · Sydney

3rd International Symposium on Basic Research on Lung Surfactant,
Marburg, September 12–14, 1988

Basic Research on Lung Surfactant

Volume Editors
P. von Wichert, Marburg
B. Müller, Marburg

112 figures and 45 tables, 1990

Basel · München · Paris · London · New York · New Delhi · Bangkok · Singapore · Tokyo · Sydney

Progress in Respiration Research

Library of Congress Cataloging-in-Publication Data
International Symposium on Basic Research on Lung Surfactant
(3rd: 1988: Marburg, Germany)
Basic research on lung surfactant/
3rd International Symposium on Basic Research on Lung Surfactant, Marburg,
September 12–14, 1988: volume editors, P. von Wichert, B. Müller.
p.   cm. – (Progress in respiration research; vol. 25)
Includes bibliographical references.
1. Pulmonary surfactant – Congresses.   I. Wichert, P. von (Peter)
II. Müller, B. (Bernd), 1952–   .   III. Title.   IV. Series.
[DNLM: 1. Pulmonary Surfactants – congresses.]
ISBN 3–8055–5030–8

Drug Dosage

The authors and the publisher have exerted every effort to ensure that drug selection and dosage set forth in this text are in accord with current recommendations and practice at the time of publication. However, in view of ongoing research, changes in government regulations, and the constant flow of information relating to drug therapy and drug reactions, the reader is urged to check the package insert for each drug for any change in indications and dosage and for added warnings and precautions. This is particularly important when the recommended agent is a new and/or infrequently employed drug.

© Copyright 1990 by S. Karger AG, P.O. Box, CH– 4009 Basel (Switzerland)
Printed in Switzerland by Thür AG Offsetdruck, Pratteln
ISBN 3–8055–5030–8

Contents

Surfactant Synthesis and Secretion

Intraalveolar Surfactant Metabolism and Recycling

Actual Fields of Research

Contents

Application of Basic Research Data in Surfactant Therapy

Posters

Contents

Foreword

One of the most exciting developments in lung research in the last three decades has been the increasing appreciation of the metabolic activities of the lung. In particular, the link between biosynthetic and secretory processes in lung cells and respiratory function exemplified by the surfactant system has stimulated a large body of scientific work which is already finding some clinical application.

The lung that lacks surfactant does not work very well. This phenomenon was first recognized in premature infants as the respiratory distress syndrome, or hyaline membrane disease, but it is now becoming clear that other lung diseases, among them acute pulmonary failure in adults, may involve dysfunction of the surfactant system. Starting from purely scientific beginnings, surfactant research has broadened to include clinical investigations by pediatricians, internists, anesthesiologists, surgeons, and pathologists who hope to evolve new forms of therapy for their seriously ill patients. It may be hazardous, however, to attempt to translate research findings into clinical practice, if the scientific basis is not solid or if therapeutic aspirations too greatly exceed realistic expectations. This difficulty particularly affects surfactant research. Many unanswered questions plague our understanding of events in the alveolar regions of the lungs, especially in the presence of disease processes. It is reasonable, therefore, to go back to basic issues in order to strengthen the rationale for new therapies. This idea motivated the organization of the Third International Symposium on Basic Research on Lung Surfactant, whose proceedings are summarized in this volume.

At present, surfactant research is an especially exciting field. On the one hand, the surfactant system, Janus-like, links pulmonary function and lung biology just as mechanics and intermediary metabolism were long ago linked in muscle contraction. On the other hand, new knowledge about the generation, transport and regeneration of surfactant components has fundamentally changed our view of the lung and opened a field of research that leaves traditional respiratory physiology far behind, not withstanding the dominant position of gas exchange efficiency as the gold standard for lung function. Nonetheless, investigations of lung hormones, of clearance of mediators in the pulmonary circulation, and of many other functions, sprang from the belated recognition that the lung has an active metabolism. The body as a whole depends on this group of functions only slightly less acutely than on efficient gas exchange.

It is now well known that lung surfactant contains both phospholipids and specific proteins. The phospholipid components have been studied for years, and their metabolism and properties in normal and diseased lungs are fairly well described. The surfactant proteins, on the contrary, have only recently come under close scrutiny, despite evidence from the earliest experiments that surfactant function is protein dependent. Progress in the methods of protein chemistry, immunology and molecular biology has greatly accelerated study of these protein components. Information is now accumulating at a rapid rate about the detailed structure of the proteins, about their metabolism, about their interactions with surfactant lipids, about their effects on alveolar cells, and about their genetic variations. Naturally, the more detailed our understanding becomes, the more new questions arise; and we can see now the need for much further clarification of the physiological and phatological roles of the surfactant proteins. How these roles are disturbed in lung diseases is not clear at present. Although enthusiasm runs high for innovative therapy based on current perceptions of what these proteins do, we should recognize that our knowledge about them is still limited.

How many functions does lung surfactant have? Clearly, it lowers alveolar surface tension, but may it promote airway clearance of toxins or opsonize micro-organisms? How else may it modulate alveolar function and architecture? Does it play a significant role in small airways? How is surfactant turnover regulated in normal and – of special interest to clinicians – pathological states? Do surfactant components counteract, or perhaps promote, certain disease processes? Can surfactant components be used as drugs or to target other drugs to alveolar or bronchiolar cells?

These and many other questions about the surfactant system certainly lack answers at present. To speak more precisely, we are unable to answer these questions because we do not have adequate basic knowledge about the system. Therefore, fundamental research on lung surfactant is critical, not only for immediate clinical application but also for the longer range goal of more completely understanding pulmonary function in normal and pathological conditions. Towards that end this record of the proceedings of the Third International Symposium on Basic Research on Lung Surfactant not only summarizes recent progress but also points out some gaps in current knowledge that need further attention.

The organizers as well as the participants of the symposium express their thanks for generous support to many institutions and persons.

The Deutsche Forschungsgemeinschaft supports the meeting by covering most of the expenses. The help of the president of the Philipps University of Marburg as well as the Dean of the Faculty of Medicine of the Philipps University is greatfully acknowledged. Support comes also from the Ministry of Science of Hessen and international companies as Byk-Gulden, Konstanz, Calbiochem, San Francisco, Calif., Dr. Karl Thomae, Biberach, Deutsche Wellcome, Burgwedel, E. Merck, Darmstadt, Hoechst, Frankfurt and Boehringer, Ingelheim.

A particular support by Byk-Gulden, Konstanz, and Dr. Karl Thomae, Biberach, makes the publishing of the records of the symposium possible.

The editors of this volume and the authors of the many different papers wish to stimulate many scientists in a fascinating, growing and still expanding field of research.

P. v. Wichert, Marburg
J.A. Clements, San Francisco, Calif.

Structure and Polymorphism of Surfactant Related Proteins

Wichert P von, Müller B (eds): Basic Research on Lung Surfactant.
Prog Respir Res. Basel, Karger, 1990, vol 25, pp 1–7

Regulation of Surfactant Protein Expression[1]

*Jeffrey A. Whitsett, Timothy E. Weaver, Stephan W. Glasser,
Thomas R. Korfhagen*

University of Cincinnati College of Medicine, Department of Pediatrics,
Cincinnati, Ohio, USA

The study of the regulation of surfactant synthesis has taken on rapid complexity with the increasing awareness of the importance of three surfactant proteins (SP-A, SP-B and SP-C) to surfactant structures and function. Significant progress has been made in the elucidation of their primary structures, derived primarily from the analysis of the cDNAs and genes encoding the proteins. This progress has been made at a time at which there is intense interest in the mechanisms controlling gene expression in general. In the present work, factors which influence surfactant protein expression will be discussed and summarized.

The surfactant proteins SP-A, SP-B and SP-C are each under strict and distinct regulatory influences. Their expression is confined to the epithelial cells of the lung and their synthesis is controlled both developmentally and hormonally. These proteins are synthesized primarily by type II epithelial cells in the alveolus and neither pre-type II cells nor type I cells express these cell-specific proteins in significant abundance. The synthesis of the surfactant proteins is closely linked to the type II cell phenotype. Type II epithelial cells rapidly lose their phenotypic characteristics and their ability to synthesize the surfactant proteins after isolation and in vitro culture, greatly complicating the direct study of surfactant protein expression.

[1] This work was supported in part by Children's Hospital, Division of Pulmonary Biology and NIH Grants HL38859, HL28623 and HD11725.

Table 1. Levels of control of protein synthesis

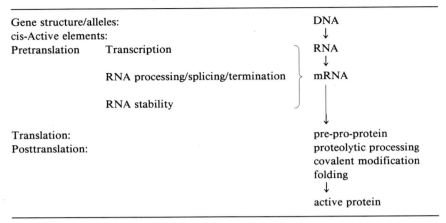

Gene structure/alleles:		DNA
cis-Active elements:		↓
Pretranslation	Transcription	RNA
		↓
	RNA processing/splicing/termination	mRNA
	RNA stability	
		↓
Translation:		pre-pro-protein
Posttranslation:		proteolytic processing
		covalent modification
		folding
		↓
		active protein

Levels of Control of Surfactant Protein Expression

The abundance of specific gene products can be influenced at various regulatory levels of biosynthetic and metabolic pathways. At a pretranslational level, the amount of mRNA can be influenced by factors altering gene transcription and RNA processing, termination and stability. Genes encoding the proteins may contain cis-active elements (intrinsic to the DNA encoding the proteins) which may directly determine gene expression by interaction with so-called trans-active factors which influence the transcriptional machinery, see table 1. Complex, cotranslational and posttranslational processing of the surfactant proteins further influences their structure, function and stability. The great diversity of control elements at these various levels of cellular organization add significantly to the amount and diversity of molecules generated from a single gene locus.

Surfactant Protein SP-A

SP-A is a group of glycoproteins migrating with $M_r = 28,000–36,000$ in SDS-PAGE. These migrate as sulfhydryl-dependent multimers in the non-reduced state. A summary of the various levels of regulation of SP-A

Table 2. Levels of regulation: SP-A

Gene structure/alleles:		multiple loci, chromosome 10
cis-Active elements:		GRE, CRE
Pretranslation	Transcription	⎱ ↑glucocorticoid, EGF, cAMP
	RNA processing	⎰ ↓insulin, TGF-β
	RNA stability	↓glucocorticoid
Translation:		$M_r = 23{,}000\text{--}24{,}000$
Posttranslation:		leader
		covalent modification
		S-S tertiary/quaternary
		$M_r > 400{,}000$

expression is seen in table 2. SP-A is synthesized primarily by type II epithelial cells in the alveolus of the lung and is generated by transcription of gene(s) [1] on human chromosome 10. This locus appears to be complex and contains at least several genes encoding SP-A-related sequences [1–3]. Cis-active elements which may influence SP-A expression include potential glucocorticoid and cAMP responsive elements located in the 5′ flanking region the human and rabbit SP-A genes [1, 4]. SP-A mRNA and protein increase during advancing gestation in a variety of species. This induction of SP-A expression has been studied in fetal lung explant cultures where SP-A mRNA and SP-A protein are, in general, coordinately modulated during culture in the absence of serum or hormones. The expression of SP-A is further influenced by the addition of a variety of hormones to the culture media. SP-A is enhanced by cAMP, β-adrenergic agents, glucocorticoids, interferon-γ, and epidermal growth factor in such cultures [5–9]. Insulin [10], TGF-β [6], and glucocorticoids [5, 6] inhibit both SP-A mRNA and SP-A protein synthesis in vitro. Inhibitory effects of glucocorticoids have been noted in explant cultures of human fetal lung occurring at higher concentrations of dexamethasone [5]. Glucocorticoids also markedly inhibit the synthesis of SP-A mRNA in a human adenocarcinoma cell line in vitro [11]. Marked induction of SP-A synthesis and mRNA are detected in lungs of rats exposed to 85% oxygen [12]. In all of these experiments, the abundance of mRNA correlated in general with the amount of SP-A protein synthesized, suggesting that control of SP-A expression is exerted primarily at a pretranslational level.

Table 3. Levels of regulation: SP-B

Gene structure/alleles:		single locus, chromosome 2, Asn_{129}
cis-Active elements:		GRE, CRE
Pretranslation	Transcription RNA processing RNA stability	↑glucocorticoid, ±cAMP
Translation:		$M_r = 40,000$
Posttranslation:		leader sequence Asn-carbohydrate proteolytic processing S-S oligomerization ↓ $M_r = 8,000 \rightarrow 18,000$

SP-B

SP-B, isolated from alveolar surfactant is a hydrophobic peptide, migrating with $M_r = 8,000$, that forms sulfhydryl-dependent oligomers. SP-B is present in lamellar bodies and in alveolar surfactant. Human SP-B is derived from a single gene encompassing approximately 9.5 kilobases of DNA and is located on human chromosome 2 [13]. The human SP-B gene has been sequenced in its entirety and contains potential cis-acting sequences compatible with regulation by both glucocorticoid and cAMP. Human SP-B is generated by proteolytic processing of a larger precursor protein of $M_r = 40,000–46,000$ [14, 15]. The mature $M_r = 8,000$ protein confers important surface-active properties to phospholipids. Allelic polymorphisms may account for the heterogeneity of cDNAs which predict the presence and absence of a glycosylation concensus sequence in the amino-terminal portion of the preproprotein. Expression of SP-B protein and mRNA increase during development in the rat lung, increasing markedly between 19.5 and 20.5 days of gestation [Weaver et al., unpubl. observation] SP-B mRNA and precursor synthesis increase during explant culture of human fetal lung in the absence of serum or hormones and expression is further enhanced by the addition of glucocorticoid [16]. SP-B mRNA and synthesis are markedly enhanced by dexmethasone in human lung adenocarcinoma cell lines [11]. Proteolytic processing of SP-B preproprotein occurs both intracellularly and extracellularly to generate the active $M_r = 8,000$ peptide; however, the precise sites of processing have not been fully clarified. cAMP did not significantly enhance SP-B content in human lung explant cultures in marked contrast to its effects on SP-A expression in this system [11]. A summary of factors altering SP-B expression is seen in table 3. As for SP-A, SP-B mRNA correlates well with SP-B protein synthesis suggesting a pretranslational control of its synthesis.

SP-C

SP-C, an extremely hydrophobic peptide (migrating with $M_r = 4,000–5,000$) is derived from precursor proteins of 22–23 kdaltons. The active peptide of approximately 35 amino acids appears to have important surfactant-enhancing properties. SP-C is generated by proteolytic processing of a precursor which is distinguished from SP-A and SP-B by the absence of a hydrophobic leader sequence. Heterogeneity of cDNAs encoding the primary polypeptide was observed in the cloning and analysis of SP-C cDNA, and is accounted for by the presence of alternative RNA splice sites in exon 5 of the SP-C gene(s). The splice variants predict a polypeptide with or without a six amino acid deletion in the proprotein [17–19]. The human SP-C gene is located on chromosome 8 in the human [18, 19]. SP-C is encoded by a relatively small gene(s) which contains potential cis-active elements responsive to cAMP and glucocorticoid, both located in the 5′ flanking region of the gene. Likewise, nucleotide sequences compatible with Z-DNA formation are located in the 3′ flanking region of the gene. Sequence analysis has been completed on two distinct but highly related human SP-C genes, both of which are expressed; however, whether these represent a gene duplication or allelic differences at the loci remains to be clarified. SP-C mRNA increases with advancing gestation in the rat increasing markedly between days 19.5 and 20.5 in a time course similar to that of SP-A and SP-B [Weaver et al., unpubl. observations]. The ontogenic changes in SP-C mRNA occur concomitantly with the enhanced SP-C protein expression. Regulation of SP-C mRNA in fetal lung explants is similar to that seen for SP-B. SP-C mRNA is enhanced during organ culture of human fetal lung in the absence of hormones and is further increased by glucocorticoids [16]. Glucocorticoids also increased SP-C mRNA in a human adenocarcinoma cell line in a similar dose-response relationship to the enhancement of SP-B mRNA [unpubl. observations]. A summary of factors influencing SP-C expression is listed in table 4.

Table 4. Levels of regulation: SP-C

Gene structure/alleles:		2 genes, chromosome 8
cis-Active elements:		GRE, CRE, Z-DNA
Pretranslation	Transcription	↑glucocorticoid, ±cAMP
	RNA processing	alternative splice sites
	RNA stability	
Translation:		$M_r = 22,000$
Posttranslation:		proteolytic processing
		↓
		$M_r = 3,000-4,000$

Summary

The biosynthesis of the surfactant proteins involves cell-specific, hormonal and developmental controls of expression. The synthesis of the active surfactant peptides appears to be modulated by systems with considerable complexity, including numerous levels of regulatory control. Certainly, we have only begun to explore the nature of these control elements. The reagents for further analysis of the mechanisms controlling expression of these three proteins are now available. It is hoped that the elucidation of the factors controlling the synthesis and metabolism of the surfactant proteins will aid in understanding the pathogenesis of hyaline membrane disease and offer new avenues for the therapy and diagnosis of other pulmonary disorders as well.

References

1   White, R.T.; Damm, D.; Miller, J.; Spratt, K.; Schilling, J.; Hawgood, S.; Benson, B.; Cordell, B.: Isolation and characterization of the human pulmonary surfactant apoprotein gene. Nature, Lond. *317:* 361–363 (1985).

2   Fisher, J.H.; Kao, F.T.; Jones, C.; White, R.T.; Benson, B.J.; Mason, R.J.: The coding sequence for the 32,000 Dalton pulmonary surfactant-associated protein A is located in the chromosome 10 and identifies two separate restriction-fragment-length polymorphisms. Am. J. hum. Genet. *40:* 503–511 (1987).

3   Bruns, G.; Stroh, H.; Veldman, G.M.; Latt, S.A.; Floros, J.: The 35 kd pulmonary surfactant-associated protein is encoded on chromosome 10. Hum. Genet. *76:* 58–62 (1987).

4   Boggaram, V.; Qing, K.; Mendelson, C.R.: Characteristic of cDNAs and the gene encoding the rabbit pulmonary surfactant apoprotein. Fed. Proc. *46:* 2038 (1987).

5   Whitsett, J.A.; Pilot, T.; Clark, J.; Weaver, T.E.: Induction of surfactant protein in fetal lung. Effects of cAMP and dexamethasone on SAP-35 RNA and synthesis. J. biol. Chem. *262:* 5256–5261 (1987).

6   Whitsett, J.A.; Weaver, T.E.; Lieberman, M.A.; Clark, J.C.; Daugherty, C.: Differential effects of epidermal growth factor and transforming growth factor-β on synthesis of $M_r = 35,000$ surfactant-associated protein in fetal lung. J. biol. Chem. *262:* 7908–7913 (1987).

7   Ballard, P.L.; Hawgood, S.; Liley, H.; Wellenstein, G.; Gonzales, L.W.; Benson, B.; Cordell, B.; White, R.T.: Regulation of pulmonary surfactant apoprotein SP-28-36 gene in fetal human lung. Proc. natn. Acad. Sci. USA *83:* 9527–9531 (1986).

8   Mendelson, C.R.; Chen, C.; Boggaram, V.: Regulation of the synthesis of the major surfactant apoprotein in fetal rabbit lung tissue. J. biol. Chem. *261:* 9938–9943 (1986).

9   Ballard, P.L.; Liley, H.G.; Gonzales, L.W.; Warr, R.G.; White, R.T.; Benson, B.; Ammann, A.J.: Interferon-γ (IFN-γ) and surfactant synthesis in human fetal lung. Pediat. Res. *23:* suppl., p. 497A (1988).

10 Snyder, J.M.; Mendelson, C.R.: Insulin inhibits the accumulation of the major lung surfactant apoprotein in human fetal lung explants maintained in vitro. Endocrinology *120:* 1250–1257 (1987).

11 O'Reilly, M.A.; Gazdar, A.F.; Morris, R.E.; Whitsett, J.A.: Differential effects of glucocorticoid on expression of surfactant proteins in a human lung adenocarcinoma cell line. Biochim. biophys. Acta *970:* 194–204 (1988).

12 Nogee, L.M.; Wispé, J.R.: Effect of hyperoxia on alveolar surfactant-associated protein (SAP-35) and pulmonary compliance. Pediat. Res. *24:* 565–573 (1988).

13 Pilot-Matias, T.J.; Kister, S.E.; Fox, J.L.; Kropp, K.; Glasser, S.W.; Whitsett, J.A.: Structure and organization of the gene encoding human pulmonary surfactant proteolipid SP-B. DNA *8:* 75–86 (1989).

14 Glasser, S.W.; Korfhagen, T.R.; Weaver, T.E.; Pilot-Matias, T.; Fox, J.L.; Whitsett, J.A.: cDNA and deduced amino acid sequence of human pulmonary surfactant-associated proteolipid SPL(Phe). Proc natn. Acad. Sci. USA *84:* 4007–4011 (1987).

15 Jacobs, K.A.; Phelps, D.S.; Steinbrink, R.; Fisch, J.; Kriz, R.; Mitsock, L.; Dougherty, J.P.; Taeusch, H.W.; Floros, J.: Isolation of a cDNA clone encoding a high molecular weight precursor to a 6-kDa pulmonary surfactant-associated protein. J. biol. Chem. *262:* 9808–9811 (1987).

16 Whitsett, J.A.; Weaver, T.E.; Clark, J.C.; Sawtell, N.; Glasser, S.W.; Korfhagen, T.R.; Hull, W.M.: Glucocorticoid enhances surfactant proteolipid Phe and pVal synthesis and RNA in fetal lung. J. biol. Chem. *262:* 15618–15623 (1987).

17 Warr, R.G.; Hawgood, S.; Buckley, D.I.; Crisp, T.M.; Schilling, J.; Benson, B.J.; Ballard, P.L.; Clements, J.A.; White, R.T.: Low molecular weight human pulmonary surfactant protein (SP-5): isolation, characterization, and cDNA and amino acid sequences. Proc. natn. Acad. Sci. USA *84:* 7915–7919 (1987).

18 Glasser, S.W.; Korfhagen, T.R.; Weaver, T.E.; Clark, J.; Pilot-Matias, T.; Meuth, J.; Fox, J.L.; Whitsett, J.A.: cDNA, deduced polypeptide structure and chromosomal assignment of human pulmonary surfactant proteolipid: SPL(pVal). J. biol. Chem. *263:* 9–12 (1988).

19 Glasser, S.W.; Korfhagen, T.R.; Perme, C.M.; Pilot-Matias, T.J.; Kister, S.E.; Whitsett, J.A.: Two genes encoding human pulmonary surfactant proteolipid SP-C. J. biol. Chem. *263:* 10326–10331 (1988).

Jeffrey A. Whitsett, MD, University of Cincinnati College of Medicine, Department of Pediatrics, 231 Bethesda Avenue, Cincinnati, OH 45267-0541 (USA)

Wichert P von, Müller B (eds): Basic Research on Lung Surfactant.
Prog Respir Res. Basel, Karger, 1990, vol 25, pp 8–14

Chromosomal Location of Human Surfactant Protein Genes and a Comparative Study of Surfactant Protein Structure[1]

James H. Fisher[a–c], *Philip A. Emrie*[a, d], *John M. Shannon*[a, d],
Robert J. Mason[a, d]

[a] Department of Medicine, University of Colorado Health Sciences Center,
Division of Pulmonary Sciences;
[b] Eleanor Roosevelt Institute for Cancer Research;
[c] Webb-Waring Lung Institute;
[d] National Jewish Center for Immunology and Respiratory Medicine,
Department of Medicine, Denver, Colo., USA

The surfactant protein SP-A, a 26- to 36-kilodalton glycoprotein, SP-B, an 18-kilodalton hydrophobic protein, and SP-C, a 5-kilodalton hydrophobic protein are believed to have crucial roles in the adsorption of surfactant to an air-liquid interface [1, 2], the regulation of surfactant secretion [3, 4], and the regulation of surfactant re-uptake and processing [5, 6]. However, little is known about the in vivo regulation of surfactant proteins. Is the synthesis of each surfactant protein co-ordinately regulated with the synthesis of phospholipid? Is each protein synthesized in a fixed ratio to the other surfactant proteins? Are there measurable physiologic consequences from alterations in the abundance of one or another of the surfactant proteins? Are there alterations in the regulation of the individual surfactant proteins in disease states such as adult respiratory distress syndrome?

To answer these similar questions we have taken two paralell approaches. First, we have used gene mapping to understand the regulation

[1] This research was supported by a Physician Scientist Award from the American College of Chest Physicians, NIH grant No. HL41320-01, and a grant from Great West Life Assurance company.

Table 1. Chromosomal location of surfactant protein genes

	Chromosome	Region
SP-A	10	long arm [7]
SP-B	2	? [8]
SP-C	8	short arm 8p22-centromere [9]

of surfactant proteins. Secondly, we have cloned and sequended cDNAs for each rat surfactant protein to determine if the rat is an appropriate model for studying surfactant protein regulation in vivo and for use as reagents.

Chromosomal Localization of Surfactant Proteins

Our goals in gene mapping included: (1) determination of chromosomal and regional location for each surfactant protein; (2) determination of the number of genes encoding each protein; (3) correlation of congenital chromosomal abnormalities involving surfactant protein genes with the presence or absence of respiratory pathology, and (4) identification of markers that can be used to determine if structural alterations in the gene or its regulatory sequences correlate with human respiratory diseases.

By using a battery of somatic cell hybrids constructed by Dr. Carol Jones and others within the Eleanor Roosevelt Institute for Cancer Research each containing a well-defined complement of whole and fragment human chromosomes and cDNA probes for each surfactant protein (kindly provided by Drs. R.T. White and Brad Benson of California Biotechnology, Emeryville, Calif.), it was possible to map SP-A, SP-B, and SP-C to chromosome locations (table 1).

Do Several Genes Encode SP-A?

During chromosomal mapping, it was apparent that for SP-A hybridization bands were observed which could not be accounted for by the known nucleotide sequence of one genomic clone encoding SP-A [17], indicating that additional sequences similar to SP-A exist on the long arm of chromosome 10 [7]. Floros et al. [10] have published two cDNA sequences for SP-A which differ within the coding region by 1.3% and in the 3′ noncoding region by 9%. Thus, SP-A may be encoded by two or more genetic loci both of which reside on the long arm of chromosome 10.

The existence of multiple genes encoding SP-A is intriguing given the observations that two differently sized messenger RNA species of 0.9 and 1.6 kb (kilobases) encode SP-A in rat [11]. These mRNA species have arisen as a result of differential polyadenylation of one transcript [12]. However, the smaller messenger RNA lacks a classic polyadenylation signal [12]. Recently, the 1.6-kb insert was cloned into a mammalian expression vector and transformed into CHO cells by Drs. McCormack and Voelker. Such transformed CHO cells produce immunologically reactive SP-A and produce two messenger RNA species, i.e. a predominant mRNA of 1.6 kb and a less abundant mRNA of 0.9 kb. Other rodent species including mouse and rabbit appear to express SP-A as multiple-sized mRNAs. One can *speculate* that the existence of multiple mRNA sequences may have regulatory significance. Since differential polyadenylation occurs in a Chinese hamster ovary cell, differential polyadenylation itself may not be the most important regulatory event. The mRNA species may be subject to differential stability or processing.

SP-B and C Are Each Encoded by Single Genes

Human SP-B has been mapped to chromosome 2. We have not been able to regionally localize SP-B more closely. However, there appears to be only one gene encoding SP-B [8].

The gene encoding SP-C appears to reside on the short arm of chromosome 8 between band 8p23 and the centromere [9]. There appears to be only one gene encoding that protein but there is some controversy. Whitsett and others have sequenced two separate genomic clones encoding SP-C which differ remarkably in the nucleotide sequence and also the pattern of DNA fragments generated when they are cut with a variety of restriction enzymes [13]. We have shown that when human DNA is cut with Eco RI and subjected to DNA blot analysis that three common restriction patterns are observed [9]. The most common is a single band at 19.5 kb which occurs 56% of the time. The second most common pattern is the existence of bands at both 6.2 and 19.5 kb which occurs approximately 37% of the time. The least common pattern is a single band which occurs at 6.2 kb approximately 6% of the time. Most often when variations such as these occur the explanation is that either an insertion or deletion of DNA or a mutation of a specific nucleotide has given rise to a restriction fragment length polymorphism (RFLP) and thus the differential banding is allelic. Only the smaller 6.2 kb band was observed when one CHO hybrid cell containing only the short arm of chromosome 8 was probed with human SP-C. It is difficult to conceive that two separate SP-C genes could be contained within the length of that DNA fragment given the size of the SP-C gene. We have also done family studies showing co-dominant transmission of alleles. One individual homozygous for the large allele had heterozygous parents supporting that hypothesis. Although these observations are most easily explained by allelic variation, an alternative explanation includes the possibility that there are two separate loci for SP-C on the short arm of chromosome 8, one generating a 6.5-kb band and one generating a 19.5-kb band when Eco RI digested DNA is probed with SP-C. We have obtained DNA from 30 unrelated individuals and have not identified any individual who was null for both loci but to exclude a null individual will require examination of about 110 unrelated individuals to reach statistical significance.

RFLP have been identified for the other surfactant proteins. The approximate frequency of each polymorphism and the enzyme which recognizes the polymorphism have been published [7–9].

Surfactant Protein Genes and Congenital Chromosomal
Abnormalities

The concept of gene dose predicts that if a gene encoding a protein is deleted in a haploid fashion its expression will be decreased by 50% from normal when the gene is present in its usual diploid gene dose. Alternatively, if the gene is present in 3 copies as in a trisomy, its expression should be increased by 50%. The current literature was reviewed for individuals that have trisomies and monosomies for the specific regions encoding SP-A and SP-C. It is apparent that individuals with trisomies and monosomies involving the SP-A loci and trisomies involving the SP-C locus can have normal respiratory systems. We have not identified individuals who are clearly monosomic for SP-C. Normal respiratory function in the presence of abnormal surfactant gene dose may occur as a result of well-regulated gene expression for the surfactant proteins or, alternatively, it may be that the exact level of surfactant proteins is not important for normal surfactant function.

Surfactant Protein Structure in Rat

We have chosen a more direct approach of studying surfactant proteins by using rat. We have cloned and sequenced cDNAs for rat SP-A [11, 12], SP-B [14], and SP-C [15]. Rat SP-A is remarkably homologous with human SP-A. The position of the cysteines, the collagen-like region, and the position of the glycosylation site are all 100% conserved across man, dog, and rat [11]. SP-B is approximately 80% homologous with human SP-B. The position of the 25 cysteine residues is conserved with 100% fidelity across 3 species [14]. As discussed by Emrie there are a number of regions with significant structural homology between human and rat SP-B. Rat SP-C is approximately 80% similar to human and dog SP-C [15]. There are a number of extremely well-conserved regions within this molecule. SP-C is posttranslationally cleaved in the amino- and carboxy-terminals to procedure a mature molecule. Within what is believed to be the mature portion of the SP-C molecule approximately 94–95% similarity exists between human and rat. Secondary structure analysis of the deduced amino acid sequences of rat, human, dog, and cow SP-C predicts structures that are nearly identical. In particular, the region between amino acids 35 and 55 is present in the mature SP-C molecule, and when analyzed by the criteria of Eisenberg et al. [16] predicts a membrane-associated monomeric alpha-helix. Additionally, there is an extremely well-conserved cationic region in the mature SP-C molecule. One can speculate that the hydrophobic portion interacts with the alkyl groups of phospholipids while the positively charged region interacts with water or the zwitterionic portions of phopholipids. The extreme conservation of sequence and structure in the rat surfactant proteins predicts that the rat will probably be a valid model for understanding surfactant function. Therefore, conclusions drawn from studying surfactant function and regulation in the rat can probably be extended to humans.

References

1 Hawgood, S.; Benson, B.J.; Schilling, J.; Damm, D.; Clemets, J.; White, R.T.: Nucleotide and amino acid sequences of pulmonary surfactant protein SP18 and evidence for cooperation between SP18 and SP28-36 in surfactant lipid adsorption. Proc. natn. Acad. Sci. USA *84:* 66–70 (1987).

2 Whitsett, J.A.; Ohning, B.L.; Ross, G.; Meuth, J.; Weaver, T.; Holm, B.A.; Shapiro, D.L.; Notter, R.H.: Hydrophobic surfactant associated protein in whole lung surfactant and its importance for biophysicial activity in lung surfactant extracts. Pediat. Res *20:* 460–467 (1986).

3 Dobbs, L.G.; Wright, J.R.; Hawgood, S.; Gonzales, R.; Venstrom, K.; Nellensorsen, J.: Pulmonary surfactant and its components inhibit secretion of phosphotidyl choline from cultured alveolar type II cells. Proc. natn. Acad. Sci. USA *84:* 1010–1014 (1987).

4 Rice, W.; Ross, G.F.; Singleton, F.M.; Dingle, S.; Whitsett, J.A.: Surfactant-associated protein inhibits phospholipid secretion from type II cells. J. appl. Physiol. *63:* 692–698 (1987).

5 Claypool, W.D.; Wang, D.L.; Chander, A.; Fisher, A.B.: An ethanol ether soluble apoprotein from rat lung surfactant augments pneumocytes. J. clin. Invest. *74:* 677–684 (1984).

6 Wright, J.R.; Wager, R.; Hawgood, S.; Dobbs, L.; Clements, J.: Surfactant apoprotein M_r = 26,000–36,000 enhances uptake of liposomes by type II cells. J. biol. Chem. *262:* 2888–2894 (1987).

7 Fisher, J.H.; Kao, F.T.; Jones, C.; White, R.T.; Benson, B.J.; Mason, R.: The coding sequence for the 32,000 dalton pulmonary surfactant associated protein is located on chromosome 10 and identifies two separate restriction fragment length polymorphisms. Am. J. hum. Genet. *40:* 503–511 (1987).

8 Emrie, P.A.; Jones, C.; Hofmann, T.; Fisher, J.H.: The coding sequence for the human 18,000 dalton hydrophobic surfactant protein is located on chromosome 2 and identifies a restriction fragment length polymorphism. Somatic Cell mol. Genet. *14:* 105–110 (1988).

9 Fisher, J.H.; Emrie, P.A.; Drabkin, H.A.; Koshnick, T.; Gerber, M.; Hofmann, T.; Jones, C.: The gene encoding the hydrophobic surfactant protein SP-C is located on 8p and identifies an Eco RI restriction fragment length polymorphism. Am. J. hum. Gent. (in press).

10 Floros, J.; Steinbrink, R.; Jacobs, K.; Phelps, D.; Kriz, R.; Recny, M.; Sultzman, L.; Jones, S.; Taeusch, H.W.; Frank, H.A.; Fritsch, E.F.: Isolation and characterization of cDNA clones for the 35-kDa pulmonary surfactant associated protein. J. biol. Chem. *261:* 9024–9033 (1986).

11 Sano, K.; Fisher, J.H.; Mason, R.J.; Kuroki, Y.; Schilling, J.; Benson, B.; Voelker, D.: Isolation and sequence of a cDNA clone for the rat pulmonary surfactant associated protein (SP-A). Biochem. biophys. Res. Commun. *144:* 367–373 (1987).

12 Fischer, J.H.; Emrie, P.A.; Shannon, J.M.; Sano, K.; Hattler, B.; Mason, R.J.: Rat pulmonary surfactant protein-A is expressed as two differently sized mRNA species which arise from differential polyadenylation of one transcript. Biochim. biophys. Acta (in press).

13   Glasser, S.W.; Korfhagen, T.R.; Perme, C.M.; Pilot-Matias, T.; Kister, S.E.; Whit-
     sett, J.A.: Two genes encoding human pulmonary surfactant protolipid. J. biol.
     Chem. *203:* 10326–10331 (1988).
14   Emrie, P.A.; Shannon, J.M.; Mason, R.J.; Fisher, J.H.: The deduced amino acid
     sequence for SP-B. A highly conserved protein. Biochim. biophys. Acta (submit-
     ted).
15   Fisher, J.H.; Shannon, J.M.; Hofmann, T.; Mason, R.J.: Nucleotide and amino acid
     sequence of the hydrophobic surfactant protein SP-C from rat. Expression in alveo-
     lar type II cells and homology with SP-C from other species. Biochim. biophys. Acta
     (submitted).
16   Eisenberg, D.; Schwarz, E.; Komdromy, M.; Wall, R.: Analysis of membrane and
     surface protein sequence with the hydrophobic moment plot. J. molec. Biol. *179:*
     125–142 (1984).

James H. Fisher, MD,
Department of Medicine, University of Colorado Health Sciences Center,
Division of Pulmonary Sciences, 4200 E. Ninth Avenue, Denver, CO 80262 (USA)

Wichert P von, Müller B (eds): Basic Research on Lung Surfactant.
Prog Respir Res. Basel, Karger, 1990, vol 25, pp 15–23

Structural and Functional Aspects of SP-A: Evidence for a SP-A Specific Receptor

Tilman Voss, Ursula Mock, Klaus Peter Schäfer, Harald Eistetter

Byk-Gulden Pharmazeutika, Abteilung für Molekularbiologie, Konstanz, BRD

Pulmonary surfactant plays an important role in maintaining the structural integrity of the alveoli by reducing the surface tension. Surfactant consists of a complex mixture of phospholipids and three genetically distinct proteins (SP-A, SP-B, SP-C). It is synthesized by alveolar type II pneumonocytes and secreted as tightly packed lamellar bodies into the alveoli [King, 1982]. The function of the two hydrophobic proteins SP-B [Hawgood et al., 1987] and SP-C [Warr et al., 1987] are clearly related to the surface tension-reducing properties of surfactant [Suzuki et al., 1986; Yu and Possmayer 1988]. Their mode of function, however, is less well understood.

The major surfactant-associated protein SP-A has been shown to enhance the spreading of phospholipids [Hawgood et al., 1985, 1987]. Furthermore, it has been described to inhibit the secretion of surfactant [Dobbs et al., 1987; Kuroki et al., 1988a; Rice et al., 1987] and enhance the uptake of liposomes by type II pneumonocytes [Wright et al., 1987], thus playing a role in the recycling of surfactant. Moreover, SP-A binds specifically to immobilized mannose indicating lectin-like properties [Haagsman et al., 1987]. Finally, the specific binding of radiolabeled SP-A to primary cultures of rat alveolar type II cells was published recently [Kuroki et al., 1988b].

In this study we compared the specific binding of radiolabeled human normal, proteinosis and recombinant SP-A to isolated alveolar type II pneumonocytes. The interaction of SP-A with its target molecule on type II cells seems to involve carbohydrate recognition and is calcium dependent.

The potential receptor for SP-A in the membrane of type II cells was characterized by affinity chromatography to be a protein with M_r of about 30 kD.

Materials and Methods

Purification of SP-A. SP-A was isolated from alveolar lavage from healthy human volunteers (normal SP-A) or proteinosis patients (proteinosis SP-A), basically as described [Hawgood et al., 1987]. The expression and purification of human recombinant SP-A has been described recently [Voss et al., 1988].

Cells and Cell-Binding Assay. L2, LTK⁻ (American Tissue Type Collection) and rat type II cells were cultivated in DMEM, Ham's F 12 (1:1, Gibco), supplemented with 10% fetal calf serum (Gibco), 31.45 µg/ml penicillin, 25 µg/ml streptomycin and 2 mM glutamine at 37 °C in a humified 10% CO_2 atmosphere.

The isolation and cultivation of rat alveolar type II pneumonocytes was carried out as described by Dobbs et al. [1986]. Freshly isolated type II cells were cultivated for 40 h on 96-multiwell plates. After adherence to the culture dish, cells were washed with PBS and new medium was added. LTK⁻ and L-2 cells were cultivated for 24 h on multiwell plates. The ligand-binding study was carried out essentially as described for the binding of high density lipoprotein to cultured fibroblasts [Oram et al., 1987], except that no albumin was added to the binding medium or wash solutions. One microgram radiolabeled normal, proteinosis or recombinant SP-A per milliliter binding medium was used. In the case of inhibition experiments the competing agents were added first. Each reaction assay was done in duplicate. To express specific binding, either each measurement was corrected by the corresponding binding of SP-A in the presence of 100 times excess unlabeled SP-A or a 'zero value' was subtracted. The 'zero value' was determined by harvesting the cells immediately after addition of radiolabeled SP-A. Unspecific binding was up to 40%.

Labeling of Proteins with Iodine. SP-A from different sources or solubilized type II membrane proteins were labeled with Na¹²⁵I (Amersham) in the presence of Iodo-Gen (Sigma). For the iodination proteins and Na¹²⁵I were added to dried Iodo-Gen aliquots (20 µl of 100 µg Iodo-Gen/ml chloroform). After 30 min incubation at room temperature the reaction wa stopped with 1 mg/ml L-tyrosine in PBS. Nonincorporated radioactivity was removed by passing the reaction mixture over a Sephadex G-25 (Pharmacia) column, equilibrated with 20 mg/ml hemoglobin (Sigma). The average yield obtained was 4,000 cpm/ng SP-A and 1,000 cpm/ng membrane proteins. To control the integrity of the radiolabeled SP-A, an aliquot was analyzed on an SDS-PAGE (data not shown).

Enzymatic Digests of SP-A. Collagenase treatment: dried protein was dissolved in 25 mM Tris-HCl, pH 7.5, 5 mM $CaCl_2$, 12.5 µg/ml N-ethylmaleimide (Sigma) and digested overnight at 37 °C with 0.8 U collagenase (CLSPA, Worthington) per microgram protein. N-glycanase digest: aspargin-linked carbohydrates were removed by N-glycanase (Genzyme) from dried protein in 0.1 M $NaPO_4$, pH 8.6, 20 mM EDTA with 0.02 U enzyme/µg protein at 37 °C overnight.

SDS-Polyacrylamide Gel Electrophoresis (SDS-PAGE). Radiolabeled proteins were separated on 10 or 12% polyacrylamide slab gels [Laemmli et al., 1970].

Membrane Preparation. Type II cell membranes were prepared using a polyethylene glycol (PEG)/dextran step gradient according to the method described by Halfter et al. [1981]. Membrane fractions were solubilized with 20 mM Tris-HCl, pH 7.5, 0.2 M NaCl, 0.02% NaN$_3$, 0.2 mM phenylmethanesulfonylfluoride (PMSF), 10 mM CaCl$_2$, 2% CHAPS (Calbiochem, solubilization buffer) at 4 °C overnight under gentle shaking. Solubilized proteins were separated from undissolved material by centrifugation at 12,000 rpm for 10 min in an Eppendorf centrifuge.

SP-A Affinity Column. Recombinant SP-A was coupled to CNBr-activated Sepharose 4B (Pharmacia) according to the manufacturer's instructions. Radiolabeled type II membrane proteins were passed three times over the SP-A affinity column. The column was washed with solubilization buffer (20 mM Tris-HCl, pH 7.5, 0.2 M NaCl, 0.02% NaN$_3$, 0.2 mM PMSF, 10 mM CaCl$_2$, 2% CHAPS) and with 20 mM Tris-HCl, pH 7.5, 0.2 M NaCl, 0.02% NaN$_3$, 0.2 mM PMSF, 0.5% CHAPS. Bound membrane proteins were eluted by the addition of 20 or 50 mM EDTA to the latter buffer. Final elution was carried out with 8 M urea. The eluted proteins wre precipitated in 10% TCA and prepared for SDS-PAGE. After electrophoresis the gel was exposed to a Du Pont Cronex 4 X-ray film.

Results

The binding of radiolabeled surfactant-associated protein A (SP-A) to isolated alveolar type II pneumonocytes was studied using human SP-A isolated from natural sources (lung lavage fluid from a healthy volunteer and from proteinosis patients) and recombinant SP-A, produced by CHO cells.

Recently the complex macromolecular structure of SP-A was resolved and compared for natural and recombinant SP-A [Voss et al., 1988]. The proteinosis SP-A was examined by electron microscopy in order to show that this molecule reveals the correct hexameric structure and was not destroyed by the extraction with organic solvents (fig. 1).

SP-A interacts specifically with alveolar type II pneumonocytes. We compared the binding of radiolabeled SP-A to the rat alveolar epithelial cell line L2, to mouse L fibroblasts (LTK$^-$) and to rat type II cells.

As shown in figure 2, binding to isolated type II cells occurs to a greater extent. Within 60 min, 3–4 times more SP-A was bound to type II cells than to L2 and LTK$^-$ cells.

Figure 3 shows the binding kinetics of labeled SP-A from natural sources to rat alveolar type II cells at 0 °C (fig. 3a) and 37 °C (fig. 3b). Saturation was not observed within 60 min. In long-time kinetics, however, the binding reached a plateau at around 6 h (data not shown). Binding of SP-A at 37 °C was higher than at 0 °C. Between 4 and 6 times less

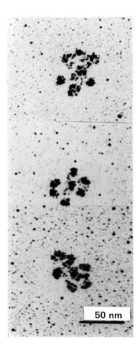

Fig. 1. Electron micrograph showing the hexameric assembly products of human SP-A isolated from lung lavage from proteinosis patients after rotary shadowing. The bar indicates 50 nm.

SP-A was bound under the latter conditions. The difference in the binding of normal and proteinosis SP-A was not significant.

In order to further characterize the mode of interaction of SP-A to its potential receptor, we examined the influence of putative inhibitors on the binding of radiolabeled proteinosis and recombinant SP-A to type II cells. SP-A from both sources interacts equally well with type II cells (Fig. 4). Various sugars like mannose and galactose were able to reduce the binding to about 30% of the control (no inhibitor). Mannan, a polymer of mannose, however, did not show such a strong inhibitory effect. The C-terminal manomeric globular domains of SP-A obtained by collagenase digestion as well as deglycosylated SP-A (glycanase digested) competed with the labeled SP-A thus inhibiting the binding as efficient as mannose or galactose.

Affinity chromatography was applied to further characterize the potential receptor of SP-A in the membrane fraction of type II cells. Solubilized and labeled membrane proteins were passed over an SP-A affinity column and eluted with EDTA. Various fractions were analyzed by SDS-

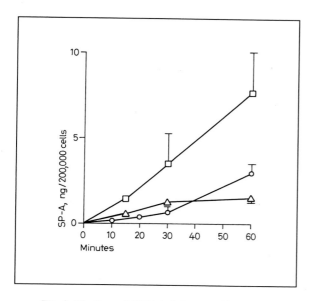

Fig. 2. Binding of ^{125}I-labeled recombinant SP-A to L2 (△) and LTK⁻ (o) cells in comparison to binding of ^{125}I-labeled normal SP-A to isolated type II cells (□) at 37 °C. Data expressed are mean ± SE from duplicate determinations.

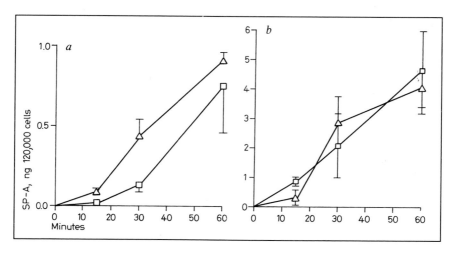

Fig. 3. Binding of 1 µg/ml ^{125}I-labeled normal (□) and proteinosis (△) SP-A to 120,000 type II cells: (*a*) binding at 0 °C; (*b*) binding at 37 °C. Data are expressed as mean ± SE from duplicate determinations.

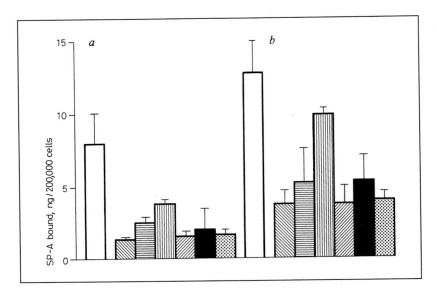

Fig. 4. Binding of 1 µg/ml ^{125}I-labeled recombinant (*a*) and proteinosis (*b*) SP-A to 200,000 type II cells in the presence or absence of different inhibitors after 30 min incubation at 37 °C. Control (☐): binding without competitor; unlabeled SP-A (▨): +100 × excess unlabeled recombinant or proteinosis SP-A; mannose (▤): +0.1 *M* mannose; mannan (▥): +10 µg/ml mannan; galactose (▨): +0.1 *M* galactose; collagenase (■): +100 × excess unlabeled collagenase-resistant SP-A fragments; glycanase (▦): +100 × excess unlabeled N-glycanase digested SP-A. Data are expressed as mean ± SE from duplicate determinations.

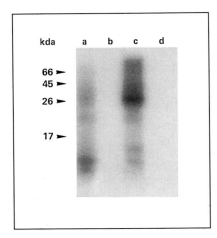

Fig. 5. SDS gel electrophoresis of type II cell membrane proteins separated by SP-A affinity column. a = Flow through and wash; b = elution without calcium; c = elution with 20 m*M* EDTA; d = elution with 8 *M* urea.

PAGE (fig. 5). As the major component, a distinct protein band of about 30 kD could be eluted with 20 mM EDTA from the column. Also, some minor bands can be seen.

Discussion

Several studies suggest that a major route of clearance of surfactant is the uptake of surfactant lipids by the alveolar type II pneumonocytes. The large surfactant apoprotein SP-A seems to play an important role in this process. It enhances the uptake of the surfactant lipids [Wright et al., 1987]. Moreover, SP-A has been found to inhibit the secretion of surfactant lipids from alveolar type II cells [Dobbs et al., 1986; Kuroki et al., 1988a; Rice et al., 1987]. Recently, evidence was presented for a high-affinity receptor for SP-A on alveolar type II cells [Kuroki et al., 1988b].

In our study we compared the binding of SP-A from various sources to alveolar type II cells. In order to assess the structural integrity of the SP-A preparations the material was examined by electron microscopy. The correct hexameric structures described by Voss et al. [1988] were observed for both normal and proteinosis SP-A. The isolation procedure obviously had no effect on the structure.

Human recombinant SP-A binds as well to type II cells as SP-A isolated from normal lung lavage and proteinosis lavage. The complex oligomeric structure of SP-A [Voss et al., 1988] seems to be necessary for the binding. Reductive breakdown of the structure abolishes the binding to type II cells (data not shown). This corresponds well with the findings that the secretion of surfactant is not any longer inhibited by reduced SP-A [Kuroki et al., 1988b].

Since SP-A is involved in the regulation of secretion and uptake of surfactant, the similar binding behavior of normal and proteinosis SP-A to alveolar type II cells might be interesting in regard to the still unknown regulatory defects of the proteinosis disease. The data presented here do not provide any evidence that proteinosis SP-A is malfunctioning as far as its binding to type II cells is concerned.

The binding mechanism of SP-A to its putative receptor molecule seems to be related to the lectin-like properties of SP-A which are calcium dependent [Haagsman et al., 1987]. In contrast to other reports we found that the binding of the protein to alveolar type II cells could be competed by mannose and galactose, but also by unlabeled deglycosylated SP-A. This indicates the involvement of carbohydrate structures on the target molecule and not on SP-A itself. High mannose structures are unlikely to be involved since mannan, a polymer of mannose, did not inhibit the binding efficiently. The interacting site is located at the C-terminal globular domain of SP-A (collagenase-resistant fragment) because this monomeric domain competes with the intact molecule at the target structure.

As demonstrated by affinity chromatography, SP-A seems to bind to a 30-kD protein in the type II cell membrane. The successful elution of this protein with EDTA confirms the calcium-dependent, carbohydrate-recognizing properties of SP-A involved in the interaction with the receptor. The receptor of SP-A in type II cells actually seems to be a molecule of larger size. Proteolytic events leading to a degradation might have occurred during the preparation, thus yielding in the 30-kD molecule described here (data not shown).

Further investigations will elucidate the structure of the SP-A specific receptor and its interaction with SP-A. The knowledge of this interaction and the fate of the bound protein should help to clear the role of SP-A in the recycling process of pulmonary surfactant.

References

Dobbs, L.G.; Gonzales, R.; Williams, M.C.: An improved method for isolating type II cells in high yield and purity. Am. Rev. resp. Dis. *143:* 141–145 (1986).
Dobbs, L.G.; Wright, J.R.; Hawgood, S.; Gonzalez, R.; Venstrom, K.; Nellenbogen, J.: Pulmonary surfactant and its components inhibit secretion of phophaditylcholine from cultured rat alveolar type II cells. Proc. natn. Acad. Sci. USA *84:* 1010–1014 (1987).
Haagsman, H.P.; Hawgood, S.; Sargeant, T.; Buckley, D.; White, R.T.; Drickamer, K.; Benson, B.J.: The major lung surfactant protein SP 28–36 is a calcium-dependent carbohydrate-binding protein. J. biol. Chem. *262:* 13877–13880 (1987).
Halfter, W.; Claviez, M.; Schwarz, U.: Preferential adhesion of tectal membranes to anterior embryonic chick retina neurites. Nature *292:* 67–70 (1981).
Hawgood, S.; Benson, B.J.; Hamilton, R.C.: Effects of a surfactant associated protein and calcium ions on structure and surface activity of lung surfactant lipids. Biochemistry *24:* 184–190 (1985).
Hawgood, S.; Benson, B.J.; Schilling, J.; Damm, D.; Clements, J.A.; White, R.T.: Nucleotide and amino acid sequences of pulmonary surfactant protein SP 18 and evidence for cooperation between SP 18 and SP 28–36 in surfactant lipid adsorption. Proc. natn. Acad. Sci. USA *84:* 66–70 (1987).
King, R.J.: Pulmonary surfactant. J. appl. Physiol. *53:* 1–8 (1982).

Kuroki, Y.; Mason, R.J.; Voelker, D.R.: Alveolar type II cells express a high affinity receptor for pulmonary surfactant. Proc. natn. Acad. Sci. USA *85:* 5566–5570 (1988a).

Kuroki, Y.; Mason, R.J.; Voelker, D.R.: Pulmonary surfactant apoprotein A structure and modulation of surfactant secretion by rat alveolar type II cells. J. biol. Chem. *263:* 3388–3394 (1988b).

Laemmli, U.K.: Cleavage of structural proteins during assembly of the head of bacteriophage T4. Nature *227:* 680–685 (1970).

Oram, J.F.; Johnson, C.J.; Brown, T.A.: Interaction of high density lipoprotein with its receptor on cultured fibroblasts and macrophages. J. biol. Chem. *262:* 2405–2410 (1987).

Rice, W.R.; Ross, G.F.; Singleton, F.M.; Dingle, S.; Whitsett, J.A.: Surfactant-associated protein inhibits phospholipid secretion from type II cells. J. appl. Physiol. *63:* 692–698 (1987).

Suzuki, Y.; Curstedt, T.; Grossmann, G.; Kobayashi, T.; Nilsson, R.; Nohara, K.; Robertson, B.: The role of the low-molecular-weight ($<$ 15,000 daltons) apoproteins of pulmonary surfactant. Eur. J. resp. Dis. *69:* 336–345 (1986).

Voss, T.; Eistetter, H.; Schäfer, K.P.; Engel, J.: Macromolecular organization of natural and recombinant lung surfactant protein SP 28–36. J. molec. Biol. *201:* 219–227 (1988).

Warr, R.G.; Hawgood, S.; Buckely, D.I.; Crisp, T.M.; Schilling, J.; Benson, B.J.; Ballard, P.L.; Clements, J.A.; White, R.T.: Low molecular weight human pulmonary surfactant protein (SP5): Isolation, characterization, and cDNA and amino acid sequences. Proc. nat. Acad. Sci. USA *84:* 7915–7919 (1987).

White, R.T.; Damm, D.; Miller, J.; Sprat, K.; Schilling, J.; Hawgood, S.; Benson, B.; Cordell, B.: Isolation and characterization of the human pulmonary surfactant apoprotein gene. Nature *317:* 361–363 (1985).

Wright, J.R.; Wager, R.E.; Hawgood, S.; Dobbs, L.; Clements, J.A.: Surfactant apoprotein M_r = 26,000–36,000 enhances uptake of liposomes by type II cells. J. biol. Chem. *262:* 2888–2894 (1987).

Yu, S.H.; Possmayer, F.: Comparative studies on the biophysical activities of the low-molecular-weight hydrophobic proteins purified from bovine pulmonary surfactant. Biochim. biophys. Acta *961:* 337–350 (1988).

Tilman Voss, PhD, Byk-Gulden Pharmazeutika, Abteilung für Molekularbiologie, D–7750 Konstanz (FRG)

Wichert P von, Müller B (eds): Basic Research on Lung Surfactant.
Prog Respir Res. Basel, Karger, 1990, vol 25, pp 24–28

Intracellular Localization of Surfactant-Associated 35-Kilodalton Glycoproteins in Human Alveolar Type II Cells

Toyoaki Akino[a], *Hiroki Takahashi*[a], *Yoshio Kuroki*[a], *Kimimaro Dempo*[b]

Departments of [a]Biochemistry and [b]Pathology, Sapporo Medical College, Sapporo, Japan

Main surfactant-associated protein, 35-kdalton protein (SP-A), is well known to have structural heterogeneity due to differences in the carbohydrate chains [1–3]. However, the intracellular metabolism of the SP-A subspecies remains to be clarified. In the present study, we intended to study the intracellular localization of SP-A subspecies in alveolar type II cells. For this purpose, a monoclonal antibody, PE10, to human SP-A was used to find SP-A in whole lung tissue, which we prepared previously [4]. The antibody recognizes specifically SP-A in alveolar type II cells [5]. Therefore, proteins reacted with the antibody in lung tissue are thought to be intracellular protein components of alveolar type II cells or extracellular surfactant proteins derived from the cells. Human lung tissue was subfractionated, and the proteins in each subfraction were examined using the antibody. In this way, the intracellular localization of the SP-A subspecies could be revealed.

Materials and Methods

Human lung tissues obtained at operation from lung cancer patients were immediately minced, homogenized, and briefly centrifuged. Then the supernatant was centrifuged at $18,000\,g$ for 10 min twice. From the $18,000\,g$ pellet, five lung fractions were isolated by discontinuous sucrose density gradient centrifugation according to the method reported by Gil and Reiss [6], which has been designed to separate surfactant subfractions. Each subfraction was identified by electron microscopy, marker enzyme assay and lipid analysis. Activities of acid phosphatase, α-glucosidase without binding affinity with concanavalin A [7] and NADPH cytochrome c reductase [8] were determined. Sodium dodecyl sulfate-polyacrylamide gel (13%) electrophoresis (SDS-PAGE) was performed according to the method of Laemmli [9]. Western blotting using a monoclonal antibody, PE10, was carried out by a modification of the procedure of Towbin et al. [10] as described previously [4]. The molecular species of phosphatidylcholine were determined by high-performance liquid chromatography of 3,5-dinitrobenzoyl derivatives of diacylglycerol derived from the phospholipid [11]. The N-glycosylated carbohydrate chain in SP-A was cleavaged by treatment of N-glycosidase F (Genzyme Co., Boston, Mass.) [12].

Results and Discussion

By electron-microscopic examination, fractions 1 and 2 which were recovered between 0.32 and 0.45 M, and 0.45 and 0.55 M sucrose, respectively, contained predominantly lamellar bodies. Fractions 3 and 4 which were recovered between 0.55 and 0.7 M, and 0.7 and 0.8 M sucrose, respectively, contained mainly two kinds of cell structure. One was the tubular myelin-like structure and the other the amorphous or small vesicular structure. Although microsomes were not expected to exist appreciably in the 18,000 g pellet, fraction 5, recovered between 0.8 and 0.9 M sucrose, contained heterogenous, vesicular membrane structures.

Acid phosphatase activity was very high in fraction 1. Fractions 2 and 5 also had higher activities of acid phosphatase. The relative proportion of α-glucosidase activity without binding affinity with concanavalin A to total α-glucosidase activity was about 75% in fractions 1 and 2. This relative activity of α-glucosidase has been shown as a marker of lamellar bodies [7]. NADPH cytochrome c reductase activity was highest in fraction 5. The ratio of phospholipid to protein in fractions 1 and 2 was 7.6 and 6.2, respectively. These higher ratios are almost equal to the values reported for purified lamellar bodies. Phospholipid composition of fractions 1–4 showed characteristic features for surfactant fraction, that is, phosphatidylcholine and phosphatidylglycerol accounted for about 70 and 10%, respectively. In contrast, the phospholipid composition of fraction 5 was different from those in other fractions, which was very similar to that of membrane fractions reported previously. Dipalmitoyl species of phosphatidylcholine, known as a characteristic molecular species of surfactant phospholipids [13], was about 50% in fractions 1–4. However, in fraction 5 it was only about 30%. From these findings, the lung subfractions iso-

Table 1. Localization of SP-A subspecies in human lung subfractions

Fraction No.	Designation	SP-A subspecies	
		37-kdalton protein	34-kdalton protein
1	lamellar body rich	23[1] ($\pm$)[2]	77 (++)
2	lamellar body rich	24 ($\pm$)	76 (++)
3	extracellular surfactant and intracellular small vesicles	35 (+)	65 (++)
4	extracellular surfactant and intracellular small vesicles	35 (+)	65 (++)
5	cellular membranes	61 (++)	39 (+)

The lung subfractions were isolated by discontinuous sucrose density gradient centrifugation according to the method reported by Gil and Reiss [6].
[1] The values represent percent distribution between SP-A subspecies, which were calculated from peak area obtained by densitometric tracing of Coomassie staining.
[2] The grade in parenthesis represents the intensity of immunoreaction with a monoclonal antibody, PE10, to human SP-A.

lated in this study were identified as follows: In order of increasing density, fractions 1 and 2 were lamellar body-rich fractions; fractions 3 and 4 were probably made up of extracellular surfactant and also intracellular small vesicles. Fraction 5 could not be identified, but it seemed to be composed of a mixture of cellular membranes such as plasma membranes, microsomes and Golgi membranes.

Proteins in each fraction were then separated, and detected by Coomassie brilliant blue and immunoblotting using a monoclonal antibody, PE10. SP-A was clearly separated into two bands on SDS-PAGE, i.e. 37- and 34-kdalton proteins. Marked differences in the distribution of the two protein bands among the lung subfractions were observed by both Coomassie staining and immunoblotting (table 1). Fractions 1 and 2 contained the 34-kdalton protein, while fraction 5 contained the 37-kdalton protein as the major protein band. In fractions 1 and 2, the 34-kdalton protein was about threefold the 37-kdalton protein. In contrast, fraction 5 contained more 37- than 34-kdalton protein. The intensity of immunoreaction with PE10 was in good agreement with the profiles of Coomassie staining. These results indicate that the main SP-A subspecies is different between the lamellar bodies and cellular membranes.

Recently, we found that the epitope of the monoclonal antibody, PE10, is located at the C-terminal side of the peptide portion of SP-A. When the N-glycosylated oligosaccharide chain of SP-A in the lung subfractions was cleaved by the treatment with N-glycosidase F, a 30-kdalton protein band was newly detected in each fraction by both Coomassie staining and immunoblotting. This finding indicates that the 34- and 37-kdalton glycoproteins have a common core protein, the 30-kdalton protein as the nonglycosylated form.

The present results led us to an important suggestion concerning the intracellular metabolism of SP-A subspecies in alveolar type II cells. The primary product of SP-A translated from mRNA may be the 30-kdalton protein [14] observed as the nonglycosylated protein in this study. This 30-kdalton protein may be immediately N-glycosylated [15] and grow into the 37-kdalton glycoprotein in some cellular membranes other than lamellar bodies. The 37-kdalton glycoprotein may then be partially deglycosylated outside of the membranes. The processed form, the 34-kdalton glycoprotein, may be selectively incorporated with specific phospholipids into the lamellar bodies and secreted into the alveolar space. These two distinct forms of human SP-A may be derived from the corresponding high-mannose precursors of SP-A as reported in rat alveolar type II cells [16].

References

1   Katyal, S.L.; Singh, G.: Analysis of pulmonary surfactant apoproteins by isoelectric focusing. Biochim. biophys. Acta 794: 411–418 (1984).

2   Weaver, T.E.; Hull, W.M.; Ross, G.F.; Whitsett, J.A.: Intracellular and oligomeric form of surfactant-associated apolipoprotein(s) A in the rat. Biochim. biophys. Acta 827: 260–267 (1985).

3   Whitsett, J.A.; Ross, G.; Weaver, T.; Rice, W.; Dion, C.; Hull, W.: Glycosylation and secretion of surfactant-associated glycoprotein(s) A. J. biol. Chem. 260: 15273–15279 (1985).

4   Kuroki, Y.; Fukada, Y.; Takahashi, H.; Akino, T.: Monoclonal antibodies against human pulmonary surfactant apoproteins. Specificity and application in immunoassay. Biochim. biophys. Acta 836: 201–209 (1985).

5   Kuroki, Y.; Dempo, K.; Akino, T.: Immunohistochemical study of human pulmonary surfactant apoproteins with monoclonal antibodies. Pathologic application for hyaline membrane disease. Am. J. Path. 124: 25–33 (1986).

6   Gil, J.; Reiss, O.K.: Isolation and characterization of lamellar bodies and tubular myelin from rat lung homogenates. J. Cell Biol. 58: 152–171 (1973).

7   Vries, A.C.J. de; Schram, A.W.; Tager, J.M.; Batenburg, J.J.; Golde, L.M.G. van: A specific acid α-glucosidase in lamellar bodies of the human lung. Biochim. biophys. Acta 837: 230–238 (1985).

8   Phillips, A.H.; Langdon, R.G.: Hepatic triphosphopyridine nucleotide-cytochrome c reductase. Isolation, characterization and kinetic studies. J. biol. Chem. *237:* 2652–2660 (1962).
9   Laemmli, U.K.: Cleavage of structural proteins during assembly of the head of bacteriophage T. Nature, Lond. *227:* 680–685 (1970).
10   Towbin, H.; Staehelin, T.; Gordon, J.: Electrophoretic transfer of proteins from polyacrylamide gels to nitrocellulose sheets. Proc. natn. Acad. Sci. USA *76:* 4350–4352 (1979).
11   Kito, M.; Takamura, H.; Narita, H.; Urade, R.: A sensitive method for quantitative analysis of phospholipid molecular species by high-performance liquid chromatography. J. Biochem. *98:* 327–331 (1985).
12   Tarentino, A.L.; Gomez, C.M.; Plummer, T.H., Jr.: Deglycosylation of asparagine-linked glycans by peptide: N-glycosidase F. Biochemistry *24:* 4665–4671 (1985).
13   Akino, T.; Ohno, K.: Phospholipids of the lung in normal, toxic and diseased states. CRC crit. Rev. Toxicol. *9:* 204–274 (1981).
14   Floros, J.; Phelps, D.S.; Taeusch, H.W.: Biosynthesis and in vitro translation of the major surfactant-associated protein from human lung. J. Biol. Chem. *260:* 495–500 (1985).
15   Phelps, D.S.; Floros, J.; Taeusch, H.W.: Post-translational modification of the major human surfactant-associated proteins. Biochem. J. *237:* 373–377 (1986).
16   O'Reilly, M.A.; Nogee, L.; Whitsett, J.A.: Requirement of the collagenous domain for carbohydrate processing and secretion of a surfactant protein, SP-A. Biochim. biophys. Acta *969:* 176–184 (1988).

Toyoaki Akino, MD, PhD, Department of Biochemistry, Sapporo Medical College, South-1, West-17, Chuo-Ku, Sapporo 060 (Japan)

Wichert P von, Müller B (eds): Basic Research on Lung Surfactant.
Prog Respir Res. Basel, Karger, 1990, vol 25, pp 29–35

Clara Cell Secretory (10 kdaltons) Protein: Amino Acid and cDNA Nucleotide Sequences, and Developmental Expression[1]

S.L. Katyal, Gurmukh Singh, William E. Brown, Amy L. Kennedy, N. Squeglia, Mari-Lou Wong Chong

Department of Pathology and Laboratory Service, V.A. Medical Center, University of Pittsburgh School of Medicine, and Department of Biological Sciences, Carnegie-Mellon University, Pittsburgh, Pa., USA

We have identified two proteins of molecular weights 55,000 and 10,000 from rat lung lavage fluid ([1–5]. Both proteins are found in the secretory granules of Clara cells by immunocytochemistry [1, 3]. The Clara cell 10-kdalton protein was purified from the lavage fluid by a combination of gel filtration, ion-exchange chromatography on Mono Q column, column chromatofocusing on Mono P, and reverse-phase chromatography [5]. Three isotypes (I, II and III) of the 10-kdalton Clara cell protein were thus isolated [5]. Upon chemical reduction, the 10-kdalton protein yields two chains of 5 kdaltons each [2, 5]. Isotypes I and III are homodimers, whereas isotype II is a heterodimer and contains chains of different pI but of similar sizes [Singh et al., unpubl. observation]. The amino acid composition of the three isotypes is similar [5]. In this report, we present the nucleotide sequence of the cDNA for the rat Clara cell 10-kdalton protein (RtCC10) and deduced amino acid sequence of the protein. The amino acid sequence of RtCC10 shows about 55% homology with the published amino acid sequence of rabbit uteroglobin [6]. In addition, we present evidence that the protein is developmentally regulated.

[1] This work was supported by the Veterans Administration, and grants from the NIH (HL 37739 and 28193), and by the Pathology Education and Research Foundation, Department of Pathology, University of Pittsburgh.

Material and Methods

Preparation of Antiserum to RtCC10

RtCC10 was isolated from bronchoalveolar lavage by the methods described in Singh et al. [5]. Rabbit antiserum to the purified protein was prepared and its reactivity and specificity tested by immunohistochemistry, immunoelectron microscopy and Western blotting technique [2, 3, 5].

Amino Acid Sequencing

Purified, reduced and alkylated protein (isotype III) was subjected to partial N-terminal sequencing on an Applied Biosystems 470 A Gas-Phase sequencer using the standard sequencing program of Hunkapiller et al. [7, 8]. Twenty cycles were performed on 1 nmol of the protein.

Immunoscreening of Rat Lung cDNA Expression Library

A rat lung cDNA expression library in lambda gt11 phage was purchased from Clontech (Palo Alto, Calif.). An antiserum to RtCC10 was used to screen the cDNA library according to the instructions provided by Clontech.

Sequencing of RtCC10 cDNA

The cDNA clones isolated by the immunoscreening were grown in bulk cultures and further purified by CsCl gradient centrifugation [9]. Phage DNA was extracted and digested with EcoRI to release the cDNA insert. The insert DNA was subcloned into M13 in both orientations and sequenced by the dideoxy chain termination method [10], using a Sequanase kit purchased from United States Biochemical Corp. (Cleveland, Ohio).

Tissue Distribution and Developmental Expression of RtCC10 mRNA

Total RNA from salivary gland, intestine, liver, prostate and seminal vesicles, testes and epididymis, uterus at day 5 of pregnancy, and lungs of adult rats was isolated by extraction with guanidine isothiocyanate, and further purified by CsCl gradient centrifugation [11]. RNA was also extracted from the lungs of fetal (16–21 days of gestation), 1-day-old and adult rats (Sprague-Dawley strain obtained from Zivic Miller, Allison Park, Pa.). RNA from above sources was dissolved in 20 × SSC = 150 mM NaCl and 15 mM sodium citrate, pH 7) applied as slots on nitrocellulose paper. The slot blots were baked at 80 °C in vacuo, and incubated at 60 °C with hybridization buffer (50% fromamide, 5 × SSC, 1 × Denhardt's solution and 50 mM potassium phosphate buffer, pH 7.4) containing 100 µg/ml each of tRNA and sonicated salmon sperm DNA. An antisense RNA probe was prepared from RtCC10 cDNA subcloned in pGEM (Promega, Madison, Wisc.) and linearized with SmaI. The protocol used to label the RNA probe with [alpha-^{32}P] CTP was obtained from Promega (Madison). After hybridization the blots were first washed in 2 × SSC containing 0.1% SDS at room temperature, and then in 0.1 × SSC with 0.1% SDS at 60 °C.

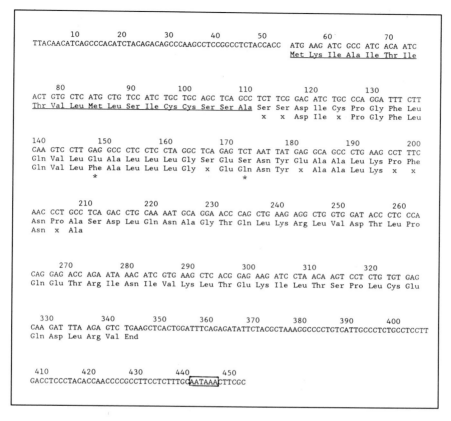

Fig. 1. Nucleotide sequence of RtCC10 cDNA. The cDNA consists of 452 base pairs. The amino acid sequence of the purified protein is presented below the deduced amino acid sequence. Differences in the two sequences are shown by *. A 19-amino acid long signal peptide (underlined) is cleaved from the nascent protein. The poly(A) addition signal is boxed.

Results and Discussion

RtCC10 cDNA Analysis

Screening of a cDNA lambda gt11 expression library with a polyclonal antiserum raised against purified RtCC10 led to the isolation of several positive clones. The nucleotide sequence of one of these clones is presented in figure 1. The cDNA consists of 452 base pairs. An ATG translation start codon is found at nucleotide 54, and lies within the sequence CCACCAT-

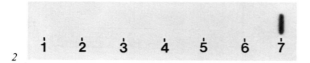

2

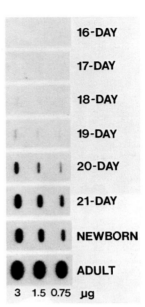

Fig. 2. Slot blot hybridization of total RNA from various rat tissues. Total RNA (4 μg) from various tissues was applied as slots. Slot blot was probed with an antisense strand labeled RNA probe prepared from RtCC10 cDNA. Slot blot after hybridization and washing was subjected to autoradiography. RtCC10 mRNA is not detectable in any tissue other than the lung. 1 = Salivary gland; 2 = intestine; 3 = liver; 4 = prostate gland; 5 = testis; 6 = uterus; 7 = lung.

Fig. 3. Expression of RtCC10 mRNA in fetal, postnatal and adult rat lungs. RNA was isolated from the lungs of fetal (16–21 days of gestation), 1-day-old and adult rats, and applied as slots (3, 1.5 and 0.75 μg). The slot blot was probed with antisense RNA probe prepared from RtCC10 cDNA. After hybridization and washings, the blot was subjected to autoradiography.

3

GA, similar to the consensus start sequence proposed by Kozak [12]. The cDNA has an open reading frame of 96 amino acid residues. Comparison of the amino acid sequence of a purified RtCC10 isotype (presented below the deduced sequence) and the deduced amino acid sequence shows several matches and two exceptions (Glu and Ser instead of Phe and Gln; positions indicated by *). The exceptions may be due to the comparison of amino acid sequences of two different isotypes. The comparison also reveals that a 19-amino acid long signal peptide (underlined) is cleaved from the nascent protein. The cDNA is nearly full-length, since the 3′ noncoding portion of the cDNA contains a poly(A) addition signal, AATAAA (boxed), but no poly(A) tail.

Sequence Homology

A computer search for RtCC10 sequence homology was performed with the Fast P program, using database from the NIH Protein Identification Resource [13]. A search of the protein data bank revealed about 55% identity between the open reading frame of RtCC10 and rabbit uteroglobin precursor [6]. A similar protein (HuCC10) has been isolated from human lung lavage, and its deduced amino acid sequence and its similarity to

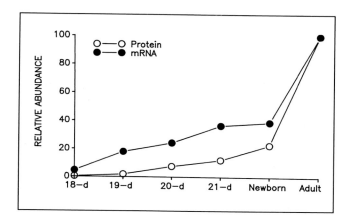

Fig. 4. Developmental expression of RtCC10 and its mRNA. Results obtained from the densitometry of the autoradiogram in figure 4 are compared with those from quantitation of RtCC10 [4]. A good correlation is seen in the amounts of RtCC10 and its mRNA in fetal and postnatal lungs.

rabbit uteroglobin have been reported [14, 15]. One could argue that RtCC10 is rat uteroglobin, and the observed differences are due to species differences. Tissue distribution of RtCC10 and its mRNA was therefore studied.

Analysis of Various Tissues for RtCC10 Expression

By immunocytochemistry, RtCC10 is exclusively seen in the nonciliated cells (Clara cells) in the airways [3]. This contrasts the distribution of rabbit uteroglobin in the lung, which is shown to be present in both alveolar and airway cells [16]. We further investigated whether RtCC10 mRNA is found in tissues other than lung. RtCC10 is not detectable in salivary gland, intestine, liver, prostate gland, testis, or gravid uterus (fig. 2). Rabbit uteroglobin, on the other hand, is detectable in the lung, endometrium, prostate and epididymis [16]. These results show that the distribution of the two proteins is clearly different.

Developmental Expression of RtCC10

RtCC10 mRNA is detectable in fetal lungs at gestational age of 17 days (fig. 3). RtCC10, on the other hand, is first detectable in fetal lung at a gestational age of 18 days (fig. 4). Their contents increase thereafter, but most of the increase is observed postnatally (fig. 3, 4). This is consistent with the development of the airway epithelium in the rat, which occurs postnatally [17]. The protein is also secreted into amniotic fluid by the fetal lung [4]. Whether the concentration of the Clara cell protein in human amniotic fluid can be used to predict fetal lung maturation and the risk of neonatal respiratory distress syndrome remains to be determined.

Secondary Structure of RtCC10

Comparison of the predicted secondary structures of RtCC10 and HuCC10 with the secondary structure of rabbit uteroglobin derived from the crystal structure shows striking similarity [15, 18]. The marked similarity in the amino acid sequence and the predicted secondary structure of RtCC10, HuCC10 and rabbit uteroglobin suggests that the two proteins belong to the same family and have similar functions. Rabbit uteroglobin has been reported to bind to progesterone, to inhibit phospholipase A2, and to have immunosuppressive, antiinflammatory and antiproteinase activities [19–21]. Whether the Clara cell proteins demonstrate one or more of these properties remains to be determined.

References

1 Singh, G.; Katyal, S.L.: An immunologic study of the secretory products of rat Clara cells. J. Histochem. Cytochem. *32:* 49–54 (1984).

2 Singh, G.; Katyal, S.L.; Gottron, S.A.: Antigenic, molecular and functional heterogeneity of Clara cell secretory proteins in the rat. Biochim. biophys. Acta *829:* 156–163 (1985).

3 Bedetti, C.D.; Singh, J.; Singh, G.; Katyal, S.L.; Wong-Chong, M.L.: Ultrastructural localization of rat Clara cell 10 KDa secretory protein by the immunogold technique using polyclonal and monoclonal antibodies. J. Histochem. Cytochem. *35:* 789–794 (1987).

4 Singh, G.; Katyal, S.L.; Wong-Chong, M.L.: A quantitative assay for a Clara cell-specific protein and its application in the study of development of pulmonary airways in the rat. Pediat. Res. *20:* 802–805 (1986).

5 Singh, G.; Singal, S.; Katyal, S.L.; Brown, W.E.; Gottron, S.A.: Isolation and amino acid composition of the isotypes of a rat Clara cell specific protein. Exp. Lung. Res. *13:* 299–309 (1987).

6 Menne, C.; Suske, G.; Arnemann, J.; Wenz, M.; Cato, A.C.; Beato, M.: Isolation and structure of the gene for the progesterone-inducible protein uteroglobin. Proc. natn. Acad. Sci. USA *79:* 4853–4857 (1982).

7 Hunkapiller, M.W.; Hewick, R.M.; Dreyer, W.J.; Hood, L.E.: High sensitivity sequencing with a gas-phase sequenator; in Hirs, Timasheff, Methods in enzymology, vol. 91, pp. 399–413 (Academic Press, New York 1983).

8   Hunkapiller, M.W.; Hood, L.E.: Analysis of phenylthiohydantoins by ultrasensitive gradient high performance liquid chromatography; in Hirs, Timasheff, Methods in enzymology, vol. 91, pp. 486–493 (Academic Press, New York 1983).

9   Maniatis, T.; Fritsch, E.F.; Sambrook, J.: Molecular cloning (Cold Spring Harbor Laboratory, New York 1982).

10 Sanger, F.; Nicklen, S.; Coulson, A.R.: DNA sequencing with chain-terminating inhibitors. Proc. natn. Acad. Sci. USA *74:* 5463–5467 (1977).

11 Chirgwin, J.M.; Przybyla, A.E.; MacDonald, R.J.; Rutter, W.T.: Isolation of biologically active ribonucleic acid from sources enriched in ribonuclease. Biochemistry *18:* 5295–5299 (1979).

12 Kozak, M.: Compilation and analysis of sequences upstream from the translational start site in eukaryotic mRNA. Nucl. Acids Res. *12:* 857–872 (1984).

13 Lipman, D.J.; Pearson, W.R.: Rapid and sensitive protein similarity searches. Science *227:* 1435–1441 (1985).

14 Singh, G.; Singh, J.; Katyal, S.L.; Brown, W.E.; Kramps, J.A.; Paradis, I.L.; Dauber, J.H.; Macpherson, T.A.; Squeglia, N.: Identification, cellular localization, isolation, and characterization of human Clara cell-specific 10 kDa protein. J. Histochem. Cytochem. *36:* 73–80 (1988).

15 Singh, G.; Katyal, S.L.; Brown, W.E.; Phillips, S.; Kennedy A.L.; Anthony, J.; Squeglia, N.: Amino-acid and cDNA nucleotide sequences of human Clara cell 10 kDa protein. Biochim. biophys. Acta *950:* 329–337 (1988).

16 Warembourg, M.; Tranchant, O.; Atger, M.; Milgrom, E.: Uteroglobin messenger ribonucleic acid: localization in rabbit uterus and lung by in situ hybridization. Endocrinology *119:* 1632–1640 (1986).

17 Massaro, G.D.; Davis, L.; Massaro, D.: Postnatal development of the bronchiolar Clara cell in rats. Am. J. Physiol. *247:* C197–C203 (1984).

18 Mornon, J.P.; Fridlansky, F.; Bally, R.; Milgrom, E.: X-ray crystallographic analysis of a progesterone-binding protein. The C2221 crystal form of oxidized uteroglobin at 2.2 Å resolution. J. molec. Biol. *137:* 415–429 (1980).

19 Levin, S.W.; Butler, J.D.; Schumacher, U.K.; Wightman, P.D.; Mukherjee, A.B.: Uteroglobin inhibits phospholipase A2 activity. Life Sci *38:* 1813–1819 (1986).

20 Gupta, R.P.; Patton, S.E.; Jetten, A.M.; Hook, G.E.: Purification, characterization and proteinase-inhibitory activity of a Clara-cell secretory protein from the pulmonary extracellular lining of rabbits. Biochem. J. *248:* 337–344 (1987).

21 Miele, L.; Cordella-Miele, E.; Facchiano, A.; Mukherjee, A.B.: Novel anti-inflammatory peptides from the region of highest similarity between uteroglobin and lipocortin I. Nature, Lond *335:* 726–730 (1988).

S.L. Katyal, PhD, Department of Pathology, University of Pittsburgh, School of Medicine, Pittsburgh, PA 15261 (USA)

Wichert P von, Müller B (eds): Basic Research on Lung Surfactant.
Prog Respir Res. Basel, Karger, 1990, vol 25, pp 36–47

Comparative Properties of Reassembly Lipoproteins Formed with Surfactant Proteins A and C[1]

Richard J. King

Department of Physiology, University of Texas Health Science Center,
San Antonio, Tex., USA

The physiological importance of reducing the alveolar surface tension to very low values is now a fundamental aspect of respiratory physiology, and is accepted by most scientists without hesitation. Pulmonary surfactant, a lipoproteinaceous material, is synthesized by alveolar epithelial type II cells and appears to be largely, if not solely, responsible for this function. There are three important properties of surfactant which enable it to be effective. First, it contains sufficiently large amounts of DPPC to form a tightly packed film adsorbed at the interface between the alveolar liquid and air. Bangham [1] points out that such a film would act as an intervening solid phase, with a solid/liquid interfacial tension close to zero. Second, it rapidly adsorbs from the alveolar liquid subphase to the surface. Third, within the subphase it forms a unique, largely lamellar structure termed tubular myelin, which is likely to be the supramolecular construct taken by this material. These latter two properties of pulmonary surfactant appear to be dependent upon the incorporation of several unique proteins within the largely lipid surfactant matrix.

Three proteins have, to date, been found in surfactant, and their major properties are shown in table 1. The largest and probably the most abundant of the proteins is SP-A, having a molecular weight of about 35,000, which is processed from a smaller precursor by glycosylation and other posttranslation modifications [2] to produce the dominant forms found in extracellular surfactant. The protein is notable in having an amino acid

[1] This work was supported by grant HL 19676 awarded by the National Heart, Lung and Blood Institute.

Table 1. Unique proteins in surfactant

Protein	Approximate molecular weight (–SH reduced)	Precursor	Special properties
SP-A	30–36	28	collagen sequence
SP-B	7	40	hydrophobic
SP-C	5	20	hydrophobic

sequence characteristic of collagen-like domain [2–4] and a carbohydrate-binding region, similar to certain lectins [5]. SP-B and SP-C are two highly hydrophobic proteins synthesized intracellularly as larger molecular weight precursors which are then cleaved intracellularly or extracellularly to form the low molecular weight hydrophobic entities found in pulmonary surfactant in alveolar fluid [6–9]. To date, all studies indicate that these proteins are unique to surfactant, perhaps implying that they contribute specific structure-function contributions. Other proteins in high abundance in the body, such as plasma albumin or other plasma proteins, either are not especially effective in changing the surface properties of surfactant, that is, in inducing rapid adsorption or in directing the formation of tubular myelin, or in some cases are inhibitory of these functions, such as the protein studied by Ikegami et al. [10] found in certain conditions of neonatal injury. The surface properties of the reassembly products formed with these proteins, especially their rates of surface adsorption, have been studied and are shown in figure 1. The lipid that is used is a mixture of DPPC with PG. What is apparent, and has been known for some number of years, is that the lipids by themselves adsorb quite slowly. When you mix these lipids with SP-A, the larger protein, there is an acceleration of

Abbreviations

CD	circular dichroism	PE	phosphatidylethanolamine
DMPC	dimyristoylphosphatidylcholine	PG	phosphatidylglycerol
DO	dioleoylglycerol	PI	phosphatidylinositol
DOPG	dioleoylphosphatidylglycerol	PS	phosphatidylserine
DPPC	dipalmitoylphosphatidylcholine	SP-A	surfactant protein A
DPPG	dipalmitoylphosphatidylglycerol	SP-B	surfactant protein B
PA	phosphatidic acid	SP-C	surfactant protein C
PC	phosphatidylcholine		

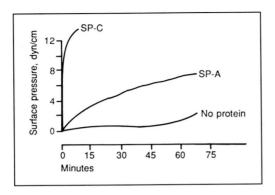

Fig. 1. The adsorption of DPPC/DPPG (85:15 weight %), a combinant lipoprotein formed with SP-A and this lipid, or a recombinant formed with SP-C and DPPC/DOPG (70:30 weight %). The SP-A recombinant was formed with a ratio of lipid to protein of 20:1; that of the SP-C recombinant was 10:1. Adsorption was at 37 °C in a buffer of pH 7.4 containing 5 mM Ca^{2+}. The concentration of lipid alone or in combination with SP-A was 10 µg/ml; that in combination with SP-C was 15 µg/ml.

adsorption; however, the rate is still relatively slow compared to what would be found with normal surfactant. In contrast, after mixing these phospholipids with SP-B or SP-C, the adsorption rate is markedly stimulated. The experiments that are reported here were stimulated by these unexplained findings, that is, we attempted to determine why the physical properties of the recombinants formed with these two proteins were so markedly different, at least as reflected by adsorption rate. The data are organized into two sections: (1) the reassembly of SP-A with surfactant phospholipids, summarizing observations made previously, and (2) the reassembly of SP-C with phospholipids, with study of the surface properties of the recombinant lipoprotein.

Reassembly of SP-A with Surfactant Lipids

SP-A was purified from canine surfactant by the procedures of King and MacBeth [11] or Hawgood et al. [12]. The protein was greater than 90% pure, as judged by polyacrylamide gel chromatography. The protein formed an isotropic solution in low ionic strength buffer which was stable in storage at 5 °C for over 2 weeks.

Table 2. Properties of SP-A/phospholipid recombinants

Binding	Binds liquid-crystalline lipids poorly
	Maximum binding to nonhomogeneous lamellar domains
	Ca^{2+} not required
Physical state	Ca^{2+} and SP-A induce rapidly aggregating complex
Thermodynamics	$\Delta G°$ similar to that of other lipoprotein recombinants
	$\Delta H°$ and $\Delta S°$ are relatively large and negative
	Only two-thirds of lipid in recombinant contribute to
	P_β-L_α phase transition
	10–20 mol PL/mol SP-A are immobilized upon binding

We have carried out binding studies between SP-A and different combinations of lipids, generally using those lipids found in high concentration in pulmonary surfactant. These experiments have been described in detail in previous studies [11, 13, 14], and therefore we will only summarize the methods and results here.

SP-A, isolated from canine surfactant and labeled with ^{125}I, was mixed with ^{3}H-labeled lipids to form reassembly particles, which were separated from the unbound constituents by density gradient centrifugation. The experiments included the investigation of certain physicochemical properties of the complexes: surface properties, transition between gel and liquid crystalline phases, and state of aggregation. Because we were using radioactively labeled constituents we were able to measure the composition of the recombinants, and the ratio of free and bound components. When the interaction was studied using unilamellar vesicles at a temperature which was above the phase transition of the phospholipid, we found that the binding was reversible and was concentration dependent. Thus we felt we could estimate thermodynamic parameters. The results of some of these experiments are given in table 2.

SP-A binds lipids that are in a liquid-crystalline phase very poorly [14]. As one goes through the transition between the gel and liquid-crystalline phases the amount of bound protein decreases by almost 10-fold. The maximum binding occurs using lipid mixtures which would be expected to form nonhomogeneous lamellar arrays in which domains of gel and liquid-crystalline lipids would coexist [13]. We feel that it is likely that these may be the principal binding sites of the protein.

Calcium is not required for binding, but it does markedly affect the state of aggregation [13]. In the presence of both Ca and SP-A, a phospholipid mixture comprised of 85% DPPC/15% DPPG aggregates instantaneously. In contrast, this same lipid in 10 mM EDTA solution binds comparable amounts of protein, but will aggregate only slowly.

The $\Delta G°$ for the interaction with DPPC, measured at 42 °C, is about -8 kcal/mol protein [14]. This is comparable to that found for other lipophilic proteins. For example, A-II of serum high density lipoprotein binds to DMPC with a free energy of -7 kcal/mol [15]; mellitin and DMPC have a free energy of interaction of -9 kcal/mol [16]; cyt b_5 and egg PC about -11 kcal/mol [17]; glucagon and DMPC about -7 kcal/mol [18]. The interaction between SP-A and phospholipids is highly temperature dependent, resulting in a calculated $\Delta H°$ which is large and negative. The measured entropy, therefore, is also large and negative.

By differential scanning calorimetry, we have found that only about two-thirds of the heat expected to be adsorbed in the gel-liquid-crystalline transition is actually seen [14]. Interpreting these results as due to an immobilization of lipid around protein-binding sites, or assuming that the $\Delta H°$ calculated from binding studies is entirely due to the formation of a gel phase in binding, we estimate that 10–20 mol phospholipids/mol SP-A are bound with sufficient tightness as to affect these thermodynamic measurements. Ross et al. [19] have carried out binding studies of fragments of SP-A, and find a region just distal to the collagen-like portion that would have a predicted hydrophobic region and could form an amphipathic α helix. A fragment containing this region, but not the collagen portion, binds phospholipids weakly, but not to the same extent as does the entire protein. This suggests that the collagen-like fragment directly binds lipids, or more likely, that the rigidity imparted by this portion of the protein markedly affects the conformation of the complete protein, and this, in turn, alters binding.

Reassembly of SP-C with Surfactant Lipids

SP-C was prepared by standard t-Boc solid phase peptide synthesis methodology using an Applied Biosystems 430A peptide synthesizer. The sequence used was that of Warr et al. [8]. The peptide was solubilized in chloroform/methanol (1:1, v/v) after cleavage from the resin with anhydrous hydrofluoric acid. The solubilized peptide was purified by gel per-

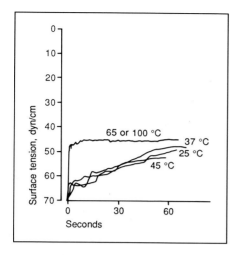

Fig. 2. The adsorption of recombinant lipoproteins formed with SP-C and DPPC/ DOPG (9:1 weight ratio), using a 10:1 ratio of lipid to protein, in which the recombinant, dispersed in 0.01 M Tris, pH 7.4, containing 5 mM Ca^{2+}, is heated for 15 min at the temperatures indicated. The samples are incubated for about 5 min at 37 °C, and then injected into a subphase of the same buffer at 37 °C.

meation chromatography over Sephadex LH-20 in chloroform/methanol (1:1).

The preparation of the reassembly lipoproteins was carried out by the following procedure: The two constituents, both in chloroform methanol 1:1 (v/v), were dried under nitrogen in a Dounce homogenizer until we saw evidence of a powder. We then added buffer (10 mM Tris, pH 7.4, containing either 5 mM Ca^{2+} or 10 mM EDTA), and incubated at 65 °C for 15 min. We have varied this temperature and found that it is important, as discussed below. Surface studies were carried out at 37 °C. All experiments used a 10:1 (w/w) ratio of lipid to protein.

We investigated the importance of the temperature at which the reassembly lipoprotein was formed in aqueous solution, and the results are shown in figure 2. We used DPPC/DOPG (9:1 w/w) incubated for 15 min at different temperatures in 0.01 M Tris, pH 7.4, with 5 mM calcium. Recombinants formed at 25, 37 and 45 °C have relatively sluggish rates of adsorption that seem to be characterized by a multiplicity of species present, as compared with recombinants formed at 65–100 °C, where the

Table 3. Adsorption: effects of lipid composition

Lipid	Concentration μg/ml	Rate dyn/cm/min	II dyn/cm
DPPC/PG 7:3 (w/w)	5	92	37
	15	> 3,000	35
DPPC/PI (PI from soybean)	11	> 3,000	39
DPPC/PS (PS from beef brain)	25	1,800	27
DPPC/PE (PE from egg)	55	350	40
DPPC/PC (PC from egg)	31	88 (1 min)	26
		9 (until max. II)	
DPPC/PA (PA from egg)	27	> 3,000	36
DPPC/DO	39	> 3,000	34
100% DPPC	26	820	19
	50	> 3,000	20

adsorption is extremely rapid. We might expect that the recombinants formed at 25 and 37 °C might be incompletely hydrated, and this might affect the rate of adsorption. This would not hold, however, for the reassembly material formed at 45 °C, which is above the P_β–L_α phase transition of both lipid components. There are at least two possibilities to explain temperature dependence: (1) the amount of lipid bound by the protein was increased with higher temperatures, or (2) structural changes occurred in the recombinant with the heating, resulting in a different lipid phase. We think it likely that reassembly particles formed with the highly lipophilic SP-C and phospholipids, mixed isotropically in chloroform/methanol and then abruptly transferred to an aqueous solution, would be complete and irreversible, making the first possibility less likely than the second, but this prediction must be confirmed by the measurement of actual lipid-protein stoichiometries.

When we use adsorption rate as a means of evaluating interactions between lipids of different composition and this protein, we generally find no major differences, with some notable exceptions. The results are shown in table 3. We used lipids of 7:3 w/w composition with DPPC and other selected unsaturated phospholipids with different phosphoester head groups. We find that the most rapid adsorption is with the acidic lipids, PG, PI, and PS. However, PE, which is charged by neutral, is only slightly slower. Phosphatidic acid, having no phosphoester alcoholic function, and

Table 4. Comparison of II (adsorption) with II (surf. compressibility change) at 37 °C

Lipid	II (adsorption)	II (compressibility change)
DPPC/PG 7:3 (w/w)	37	42
DPPC/PI	35	32
DPPC/PS	27	31
DPPC/PE	40	35
DPPC/PC	26	37
DPPC/PA	36	27
DPPC/DO	34	39
100% DPPC	20	20

diolein, which has no phosphate group at all and is neutrally charged, are both fast. Thus, acidity is not an absolute requirement. Of these combinations, only the DPPC/egg PC combination is slow. Recombinants of SP-C with 100% DPPC adsorb nearly as fast as do those with lipid mixtures.

The quasi-equilibrium surface pressure of adsorption differs markedly among the lipid mixtures, which may reflect differences in the surface pressure exerted by the respective lipids at their equilibrium adsorption packing. We carried out surface tension-area isotherms at 37 °C with each of the mixtures to investigate whether the surface pressure of adsorption corresponds to some property of the surface film. The results are shown in table 4. There is a reasonable correlation between the surface pressure of the first obvious change in the compressibility of the surface film with the quasi-equilibrium surface pressure of adsorption.

We studied the effects of Ca^{2+} on the adsorption of reassembly lipoproteins formed with 100% DPPC and DPPC/DOPG (7:3 w/w). The results are seen in table 5. Substituting EDTA for calcium in the recombinant buffer induced some change in the adsorption rate of the DPPC/PG recombinant, in that adsorption rate went from greater than 3,000 dyn/cm/min to a measurable 1,000 dyn/cm/min. In both solutions, however, adsorption to a quasi-equilibrium surface pressure required only seconds. In contrast, the adsorption of the DPPC recombinant was markedly affected by the deletion of calcium, in that the adsorption from buffers without calcium was relatively slow. After greater than 60 min the surface pressure of the adsorbed film was less than half of that expected for near equilibrium adsorption.

Table 5. Adsorption: effects of calcium

Lipid	5 mM Ca^{2+}			10 mM EDTA		
	rate dyn/cm/min	II dyn/cm	conc. μg/ml	rate dyn/cm/min	II dyn/cm	conc. μg/ml
DPPC/PG						
7:3 (w/w)	> 3,000	35	15	1,160	31	50
DPPC	820	19	26	16	13	27

Table 6. Adsorption of unsaturated lipids

Lipid	No protein		SP-C	
	rate dyn/cm/min	II dyn/cm	rate dyn/cm/min	II dyn/cm
PC	2	20	3	20
PI	10	32	7	32
PS	340	34	235	28
PE	> 3,000	42	> 3,000	41

Buffer: 5 mM Ca^{2+}; 10 mM Tris, pH 7.4. Nominal conc: 50 μg.

Returning to our original question – What makes SP-C different from SP-A in the way it forms recombinants that are highly surface active? Both SP-A and SP-C interact with DPPC and with DPPC-lipid mixtures, and this interaction is likely to involve hydrophobic bonds. Both induce significant rearrangements of the hydrocarbon chains on the phospholipid, modifying its phase transition [14, 20]. Thus, no obvious differences are apparent from these data. However, additional experiments studying interactions between SP-C and certain unsaturated phospholipids have provided us a possible hypothesis, even though it could not yet be construed as a meaningful explanation. These results are given in table 6.

PC and PI do not form an inverted micellar phase, such as the hexagonal II phase [21]. They do not rapidly adsorb, with or without protein. PS and PE do form nonlamellar phases [21]. They adsorb rapidly, with or without protein. If this generalization were to extend to the lipid mixtures containing DPPC, which do not adsorb rapidly unless combined with SP-C, then this would imply that hexagonal II or some other transitional non-lamellar phase may be formed. This possibility is consistent with the temperature data, since the hexagonal II phase cannot be directly formed from a gel phase, but must be heated at least above its phase transition temperature [21]. Other small molecular weight proteins have been shown to induce the formation of a hexagonal II phase [21]. Experiments using ^{31}P-NMR, which are now underway, may be able to verify or refute this prediction, but the results of these experiments are not available at present.

In conclusion, we stress that the data we have obtained with SP-C, and consequently our conclusions, are based solely upon assumptions made from changes in the surface properties of recombinants. The definitive data looking at compositional dependence on SP-C binding will likely not be possible until we are able to carry out binding studies where bound and free constituents are separated and quantitated. Obviously, conclusions regarding possible phase changes will await the results of NMR or possibly fluorescence measurements.

Acknowledgements

SP-C was the generous gift of Dr. Bradley Benson, California Biotechnology Corporation. The author wishes to acknowledge the very helpful discussions with Dr. Benson and Dr. Michael C. Phillips, Medical College of Pennsylvania, Throughout the course of these investigations. Mr. Shawn Shahraini provided excellent technical assistance. This work was supported in part by a grant from the National Heart, Lung and Blood Institute, HL 19676.

References

1 Bangham, A.D.: Lung surfactant: How it does and does not work. Lung *165:* 17–25 (1987).
2 Benson, B.; Hawgood, S.; Schilling, J.; Clements, J.; Damm, D.; Cordell, B.; White, R.T.: Structure of canine pulmonary surfactant apoprotein: cDNA and complete amino acid sequence. Proc. natn Acad. Sci. USA *82:* 6379–6383 (1985).
3 White, R.T.; Damm, D.; Miller, J.; Spratt, K.; Schilling, J.; Hawgood, S.; Benson, B.; Cordell, B.: Isolation and characterization of the human pulmonary surfactant apoprotein gene. Nature *317:* 361–363 (1985).
4 Floros, J.; Steinbrink, R.; Jacobs, K.; Phelps, D.; Kriz, R.; Recny, M.; Sultzman, L.;

Jones, S.; Taeusch, H.W.; Frank, H.A.; Fritsch, E.F.: Isolation and characterization of cDNA clones for the 35-kDa pulmonary surfactant-associated protein. J. biol. Chem. *261:* 9029–9033 (1986).

5   Haagsman, H.P.; Hawgood, S.; Sargeant, T.; Buckley, D.; White, R.T.; Drickamer, K.; Benson, B.J.: The major lung surfactant protein, SP28–36, is a calcium-dependent, carbohydrate-binding protein, J. biol. Chem. *262:* 13877–13880 (1987).

6   Glasser, S.W.; Korfhagen, T.R.; Weaver, T.; Pilot-Matias, T.; Fox, J.L.; Whitsett, J.A.: cDNA and deduced amino acid sequence of human pulmonary surfactant-associated proteolipid SPL (Phe). Proc. natn Acad. Sci. USA *84:* 4007–4011 (1987).

7   Glasser, S.W.; Korfhagen, T.R.; Weaver, T.E.; Clark, J.C.; Pilot-Matias, T.; Meuth, J.; Fox, J.L.; Whitstett, J.A.: cDNA, deduced polypeptide structure and chromosomal assignment of human pulmonary surfactant proteolipid, SPL (pVal). J. biol. Chem. *263:* 9–12 (1988).

8   Warr, R.G.; Hawgood, S.; Buckley, D.I.; Crisp, T.M.; Schilling, J.; Benson, B.J.; Ballard, P.L.; Clements, J.A.; White, R.T.: Low molecular weight human pulmonary surfactant protein (SP5): Isolation, characterization, and cDNA and amino acid sequences. Proc. natn Acad. Sci. USA *84:* 7915–7919 (1987).

9   Hawgood, S.; Benson, B.J.; Schilling, J.; Damm, D.; Clements, J.A.; White, R.T.: Nucleotide and amino acid sequences of pulmonary surfactant protein SP 18 and evidence for cooperativity between SP 18 and SP 28–36 in surfactant lipid adsorption. Proc. natn Acad. Sci. USA *84:* 66–70 (1987).

10 Ikegami, M.; Jobe, A.; Jacobs, H.; Lam, R.: A protein from airways of premature lambs that inhibits surfactant function. J. appl. Physiol. *57:* 1134–1142 (1984).

11 King, R.J.; MacBeth, M.C.: Interaction of the lipid and protein components of pulmonary surfactant. Role of phosphatidylglycerol and calcium. Biochim. biophys. Acta *647:* 159–168 (1981).

12 Hawgood, S.; Benson, B.; Hamilton, R.L., Jr.: Effects of surfactant-associated protein and calcium ions on the structure and surface activity of lung surfactant lipids. Biochemistry *24:* 184–190 (1985).

13 King, R.J.; Carmichael, M.S.; Horowitz, P.M.: Recombination of lipid-protein complexes of pulmonary surfactant: proposed mechanism of interaction. J. biol. Chem. *258:* 10672–10680 (1983).

14 King, R.J.; Phillips, M.C.; Horowitz, P.M.; Dang, S.C.: Interaction between the 35-kDa apolipoprotein of pulmonary surfactant and saturated phosphatidylcholines. Effects of temperature. Biochim. biophys. Acta *879:* 1–13 (1986).

15 Pownall, H.J.; Hickson, D.; Gotto, A.M., Jr.: Thermodynamics of lipid-protein association. The free energy of association of lecithin with reduced and carboxymethylated apolipoprotein A-II from human plasma high density lipoprotein. J. biol. Chem. *256:* 9849–9854 (1981).

16 Vogel, H.: Incorporation of melittin into phosphatidylcholine bilayers. FEBS Lett. *134:* 37–42 (1981).

17 Leto, T.L.; Holloway, P.W.: Mechanism of cytochrome b_5 binding to phosphatidylcholine vesicles. J. biol. Chem. *254:* 5015–5019 (1979).

18 Epand, R.M.; Sturtevant, J.M.: A calorimetric study of peptide-phospholipid interactions: the glucagon dimyristoylphosphatidylcholine complex. Biochemistry *20:* 4603–4606 (1981).

19   Ross, G.F.; Notter, R.H.; Meuth, J.; Whitsett, J.A.: Phospholipid binding and bio-
     physical activity of pulmonary surfactant-associated protein (SAP)-35 and its non-
     collagenous COOH-terminal domains. J. biol. Chem. *261:* 14283–14291 (1986).
20   Shiffer, K.; Hawgood, S.; Düzgünes, N.; Goerke, J.: The effects of the low molecular
     weight lung surfactant-associated proteins, SP-B and SP-C, on the thermotropic
     properties of various synthetic phospholipid membranes. FASEB J. *2:* A319
     (1988).
21   Cullis, P.R.; DeKruijff, B.: Lipid polymorphism and the functional roles of lipids in
     biological membranes. Biochim. biophys. Acta *559:* 399–420 (1979).

Richard J. King, MD, Department of Physiology, University of Texas Health
Science Center, 7703 Floyd Curl Drive, San Antonio, TX 78284-7756 (USA)

Wichert P von, Müller B (eds): Basic Research on Lung Surfactant.
Prog Respir Res. Basel, Karger, 1990, vol 25, pp 48–53

Pulmonary Surfactant Protein in Bilayers of Dimyristoyl Phosphatidylcholine

Gary A. Simatos, Michael R. Morrow, Kenneth B. Forward, Kevin M.W. Keough

Department of Biochemistry and Department of Physics, and Discipline of Pediatrics, Memorial University of Newfoundland, St. John's, Nfld., Canada

The ability to rapidly reduce surface tension is critical to normal surfactant function. Studies from many laboratories support the notion that the small hydrophobic proteins present in natural surfactant facilitate normal surfactant function [Hawgood et al., 1987; Takahashi and Fujiwara, 1986; Whitsett et al., 1986; Yu and Possmayer, 1986]. These observations suggest that the proteins unique to surfactant may alter the organization of the lipid surrounding them and facilitate monolayer formation at the air-water interface.

In this study we examined the ability of hydrophobic surfactant protein(s) to perturb lipid bilayers of perdeuterated dimyristoyl phosphatidylcholine (DMPC-d54) using differential scanning calorimetry (DSC) and nuclear magnetic resonance spectroscopy (NMR).

Materials and Methods

Deuterated DMPC was synthesized using the method of Gupta et al. [1977] using perdeuterated fatty acid prepared according to the method of Hsiao et al. [1974].

Isolation of Surfactant Protein

Surface-active material was recovered from the pooled lung lavage fluid as described by Keough et al. [1988]. Lipid and hydrophobic surfactant proteins were extracted according to the method of Bligh and Dyer [1959]. The material in the organic phase was chromatographed on a silicic acid column using step gradient elution with chloroform and methanol (C-M). Fractions were collected and assayed for phosphorus [Bartlett, 1959] and protein using the fluorescamine method [Udenfriend et al., 1972]. Two protein-rich pools were detected. Protein from the first peak (in C-M, 3:2, v/v) was electrophoresed on SDS-PAGE with and without mercaptoethanol and showed primarily a single band using silver staining (New England Nuclear Research Products) with an M_r of approximately 5 kdaltons and a minor band of M_r 12 kdaltons. This material in C-M, 2:1, was applied to a Sephadex LH-20 and eluted with the same solvents. The protein-rich fractions were dialyzed against chloroform-methanol (2:1, v/v) for 48 h. This procedure yielded a protein preparation with less than 1% residual phospholipid based on phosphorus and fluorescamine determinations.

Perdeuterated dimyristoyl phosphatidylcholine (DMPC-d54) or lipid plus protein were prepared in C-M (2:1, v/v), the solvents removed and the samples evacuated overnight. Samples were prepared for NMR and DSC by redispersion in 0.3 ml of 0.05 M phosphate-buffered saline at pH 7.

^{2}H-NMR experiments were conducted in a superconductive solenoid at 23.2 MHz using a spectrometer built by M.M. Samples were held at 32 °C before the series of NMR spectra were collected. At a given temperature, the sample was allowed to equilibrate for 1 h before the accumulation of the transients was initiated. Spectra were collected using a phase-cycled quadrupole echo sequence [Davis et al., 1976] with a $\pi/2$ pulse length of between 2.5 and 3.5 µs and a quadropole echo delay time of 35 µs. DSC samples were analyzed using an MC-2 differential scanning calorimeter (Microcal Inc. Northampton, Mass.) with buffer as reference. Scans were carried out three times at a scan rate of 30°/h from 5 to 40 °C. Samples were then extracted and reevaluated for phosphorus and protein content. Protein concentrations were found to be variable using the fluorescamine method and were reevaluated using a modification of the method of Bradford [1977] using 50 µl of 2-chloroethanol before addition of the Bradford reagent. To check for possible lipid breakdown, 0.5 mg phospholipid was applied to thin-layer chromatography plates and visualized by charring with 70% H_2SO_4. Aliquots were also electrophoresed on SDS-PAGE to demonstrate the presence of the 5-kdalton protein.

Results

Protein applied to SDS-PAGE gels at a concentration of 1–2 µg migrated primarily as a single band with an M_r of approximately 5 kdaltons under reducing or nonreducing conditions. When the protein was electrophoresed at a 20-fold higher load a major band at 5 kdaltons M_r was observed and a faint band at 12 kdaltons M_r was also apparent. While SDS-PAGE suggests that the protein preparation used in this study consists mainly of one protein species, likely SP-C, preliminary amino acid and N-terminal sequence analysis indicates the possible presence of an additional protein or peptide in the preparation.

Figure 1 shows the temperature dependence of the ^{2}H-NMR lineshape for DMPC-d54 in the region of the phase transition for the pure lipid and lipid plus 2.2% protein (w/w). During cooling of the pure lipid, there was a

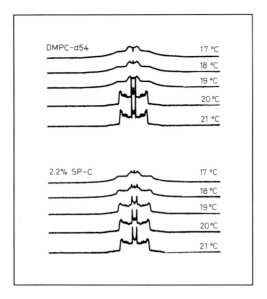

Fig. 1. The temperature dependence of the ^{2}H-NMR spectrum of DMPC-d54 near the pure lipid transition for the pure lipid and DMPC-d54 containing 2.2% (w/w) surfactant protein.

sharp transition from spectra characteristic of fast, axially symmetric motion in the liquid-crystalline phase (above 20 °C) to characteristic gel-phase spectra at 19 °C and lower. In the samples with 2.2% protein present, there was a small temperature range (18–19 °C) over which the two phases coexist with slow enough exchange between domains of the two phases to give rise to a superposition of the gel and liquid-crystalline spectra.

Heating thermograms for pure DMPC-d54 and DMPC-d54 plus 2.2% protein are presented in figure 2. In comparison to the narrow transition seen for pure DMPC-d54, a low temperature broadening was evident in the protein-supplemented dispersion and there was an increase in the half height width. The onset of transition was depressed in the protein-lipid dispersion to 15.7 °C compared to 18.8 °C seen for the pure lipid. The temperature of maximum excess heat capacity (T_m) for the pure lipid was found to be 19.5 for the pure lipid while the T_m of the dispersion containing protein was reduced to 18.9 °C. No significant difference in the enthalpy change of the transition of the two preparations was observed.

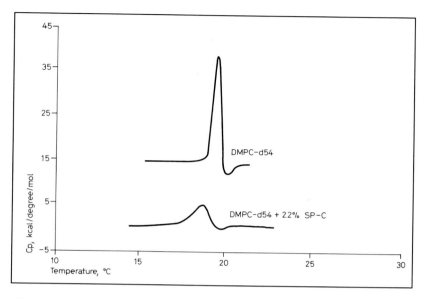

Fig. 2. DCS thermograms for pure DMPC-d54 and DMPC-d54 and 2.2% (w/w) surfactant protein.

Discussion

The hydrophobic nature of the protein and its resistance to delipidation during isolation indicates a high potential for interaction with the fatty acyl chains within the nonpolar interior of the phospholipid bilayer. The potential for SP-C to disrupt lipid packing has been investigated by Shiffer et al. [1988] using DSC. At a protein concentration of 10% (w/w), these investigators observed a 50% reduction in the enthalpy change of transition of DPPG from 6.7 kcal/mol for the pure lipid to 3.1 kcal/mol for that containing 10% SP-C. We have not observed a substantial effect of the protein on the enthalpy change of transition but we employed only 2.2% protein. However, the appearance of the low temperature shoulder, the reduction in T_m and the increase in the half-height width of the transition are consistent with the presence of the protein altering the packing arrangement of the DMPC-d54 in the gel state.

The ^{2}H-NMR spectra are also consistent with a perturbation in the lipid packing by the protein. We observed in the spectra a region in which two phases coexist over a range of about 2° below the normal transition temperature for the pure lipid.

Both methods suggest that the protein can perturb the lipid packing. If the protein had been all SP-C it would have constituted about 1 mol% of the lipid-protein sample. SP-C contains a hydrophilic N-terminal region of 12 residues and the remaining 23 residues are all hydrophobic [Johansson et al., 1988]. The hydrophobic portion would be sufficient to produce a complete transmembrane alpha-helix. The hydrophilic portion would probably orient between the lipid head groups or protrude out in the aqueous environment. The perturbation in the lipid packing caused by the protein is not insignificant in light of the relatively small protein concentration in our protein-lipid preparation.

Acknowledgements

The authors thank Prof. J.H. Davis, K.R. Jeffrey and M. Bloom for helpful comments and suggestions during construction of the NMR spectrometer. The authors also gratefully acknowledge the provision, by Prof. Davis, of facilities, material and information necessary to synthesize the DMPC-d54 used in this study.

Supported by MRC and NSERC, Canada.

References

Bartlett, G.R.: Calorimetric assay method for free and phosphorylated glyceric acids. J. biol. Chem. *234:* 466–468 (1959).

Bligh, E.G.; Dyer, W.J.: A rapid method of total lipid extraction and purification. Can. J. Biochem. Physiol. *37:* 911–917 (1959).

Bradford, M.M.: A rapid and sensitive method for the quantitization of microgram quantities of protein utilizing the principle of protein-dye binding. Anayl. Biochem. *72:* 248–254 (1976).

Davis, J.H.; Jeffrey, K.R.; Bloom, M.; Valic, M.I.; Higgs, T.P.: Quadrupolar echo deuteron magnetic-resonance spectroscopy in ordered hydrocarbon chains. Chem. Phys. Lett. *42(2):* 390–394 (1976).

Gupta, C.M.; Radhakrishnan, R.; Khorana, H.G.: Glycerophospholipid synthesis: Improved general method and new analogs containing photoactivable groups. Proc. natn. Acad. Sci. USA *74:* 4315–4319 (1977).

Hawgood, S.; Benson, B.J.; Schilling, J.; Damm, D.; Clements, J.A.; White, R.T.: Nucleotide and amino acid sequence of pulmonary surfactant SP18 and evidence for cooperation between SP18 and SP28-36 in surfactant lipid adsorption. Proc. natn. Acad. Sci. USA *84:* 66–70 (1987).

Hsiao, C.Y.Y.; Ottaway, C.A.; Wetlaufer, D.B.: Preparation of fully deuterated fatty acids by simple method. Lipids *9:* 813–915 (1974).

Johansson, J.; Curstedt, T.; Robertson, B.; Jornvall, H.: Size and structure of the hydro-phobic low molecular weight surfactant-associated polypeptide. Biochemistry *27:* 3544–3547 (1988).

Keough, K.M.W.; Parsons, C.S.; Phang, P.T.; Tweedale, M.G.: Interactions between plasma proteins and pulmonary surfactant: surface balance studies. Can. J. Physiol. Pharmacol. *66:* 1166–1173 (1988).

Shiffer, K.; Hawgood, S.; Duzgunes, N.; Goerke, J.: The effect of the low molecular weight lung surfactant-associated proteins, SP-B and SP-C, on the thermotrophic properties of various synthetic phospholipid membranes. Proc. Fed. Am. Soc. exp. Biol. *2:* A319 (1988).

Takahashi, A.; Fijiwara, T.: Proteolipid in bovine lung surfactant: Its role in surfactant function. Biochem. biophys. Res. Commun. *135:* 527–532 (1986).

Udenfriend, S.; Stein, S.; Bohlen, P.; Dairman, W.; Loimgruber, W.; Weigele, M.: Fluor-escamine: A reagent for assay of amino acids, peptides and primary amines in the picomole range. Science *178:* 871–872 (1972).

Whitsett, J.A.; Ohning, B.L.; Ross, J.; Meuth, J.; Weaver, T.; Holm, B.A.; Shapiro, D.L.; Notter, R.H.: Hydrophobic surfactant-associated protein in whole lung surfactant and its importance for biophysical activity in lung surfactant extracts used for replacement therapy. Pediat. Res. *20:* 460–467 (1986).

Yu, S.; Possmayer, F.: Reconstitution of surfactant activity by using the 6 kDa apoprotein associated with pulmonary surfactant. Biochem. J. *236:* 85–89 (1986).

Gary A. Simatos, PhD, Department of Biochemistry,
Memorial University of Newfoundland, St. John's, NF A1B 3X7 (Canada)

Wichert P von, Müller B (eds): Basic Research on Lung Surfactant.
Prog Respir Res. Basel, Karger, 1990, vol 25, pp 54–63

Role of the Low Molecular Weight Proteins in Pulmonary Surfactant

Fred Possmayer[a, b]*, Shou-Hwa Yu*[a, 1]

Departments of [a]Obstetrics and Gynaecology, and [b]Biochemistry,
University of Western Ontario, London, Ont., Canada

Pulmonary surfactant stabilizes the lung by reducing the surface tension at the air-liquid interface of the alveolar lining layer [1–5]. Lack of sufficient surfactant to stabilize the terminal airways is a major factor contributing to the establishment of the respiratory distress syndrome (RDS) of the newborn. Pulmonary surfactant obtained through bronchoalveolar lavage contains approximately 90% lipid and 10% protein. The major lipid component, dipalmitylphosphatidylcholine (DPPC), represents the principle surface tension reducing agent. However, DPPC does not adsorb and spread at the air-liquid interface at a sufficient rate to support normal lung function [1–5]. Hence, surfactant-associated proteins have been implicated in the formation of the surface active monolayer.

Pulmonary surfactant contains at least three associated proteins which have been designated surfactant-associated proteins-A (SP-A), SP-B and SP-C [6]. These proteins are discussed at length in other articles in this volume. Briefly, SP-A refers to the major surfactant protein which is a highly modified sialoglycoprotein with nominal molecular masses of 26–36 kdaltons. SP-B and SP-C are low molecular weight hydrophobic proteins which remain associated with the lipids in organic solvents.

[1] The authors would like to express their appreciation to Mrs. M.A. Ormseth for preparing the bovine surfactant and to Mr. J. Chung for technical assistance. Ms. B. McDougall provided excellent editorial services.

The present report will attempt to summarize our present view on the roles of the surfactant-associated proteins, in particular the low molecular weight hydrophobic proteins in the formation of the surface-active phospholipid monolayer that is responsible for stabilizing the alveolus. The discussion will rely primarily on studies conducted in our laboratory with bovine natural surfactant and lipid extracts of the natural product.

Materials and Methods

Bovine pulmonary surfactant and its lipid extract were obtained by slight modifications of previously described methods [7, 8]. Biophysical activities were estimated with the pulsating bubble surfactometer designed by Enhorning [9]. With this apparatus, a bubble in contact with the atmosphere is pulsated within a dispersion of the material to be tested in saline, 1.5 mM CaCl$_2$. Surface tensions at the maximum bubble radius (R_{max}) and the minimum bubble radius (R_{min}) are presented. Unless indicated, the samples were examined at a total phospholipid concentration of 10 mg/ml.

The methods used to isolate and analyse SP-A and the low molecular weight hydrophobic proteins SP-B and SP-C have been described in previous publications [10, 11]. Phospholipid content was determined by measuring total phosphorus [12] and protein content by the method of Lowry et al. [13] in the presence of 0.1% sodium dodecylsulfate. Extracts of bovine surfactant were separated on Whatman LK6D plates using chloroform:ethanol:water:triethylamine (30:34:8:35). Sodium dodecylsulfate-urea-polyacrylamide gel electrophoresis (SDS-urea-PAGE) of lipid extract surfactant and N-terminal analysis were conducted as previously described [14].

Results and Discussion

Thin-layer chromatography of lipid extract surfactant prepared from bovine surfactant [7] revealed approximately 80% phosphatidylcholine (PC), 10% phosphatidylglycerol (PG), 2% phosphatidylinositol (PI), 4% phosphatidylethanolamine (PE), 2.5% sphingomyelin and 1.5% of an acidic phospholipid tentatively identified as lyso-bis-phosphatidic acid (fig. 1). Bovine surfactant contains ~3% neutral lipids, including cholesterol, cholesterol esters, and di- and triacylglycerol [7]. Early studies conducted in collaboration with Dr. G. Enhorning revealed that both natural surfactant and chloroform:methanol extracts of natural surfactant possessed the ability to reduce the surface tension of a pulsating bubble to near 0 mN/m [15]. These studies indicated that the bulk of the protein is not necessary for reducing the surface tension of a pulsating bubble to near 0 mN/m. However, with a few notable exceptions [2], mixtures of pure

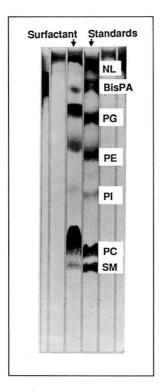

Fig. 1. Thin-layer chromatography of the phospholipids in lipid extracts of bovine pulmonary surfactant. BisPA = Bis-phosphatidic acid; NL = neutral lipid; SM = sphingomyelin.

phospholipids are not capable of duplicating these characteristic properties. Although initially unsuccessful [15], we were eventually able to confirm the presence of hydrophobic proteins with a nominal molecular mass of ～6 kdaltons in lipid extracts of bovine surfactant [16]. N-terminal Edman degradation revealed the presence of two distinct polypeptides in lipid extracts (fig. 2). One of these polypeptides, initiating with an N-terminal Phe, has proven to be a member of the SP-B family. A full-length 79 amino acid member of the SP-B family has been reported for bovine surfactant [17]. Homologous 79 amino acid sequences have been reported for porcine surfactant [18]. The second sequence, which was followed for ～30 amino acids, is a member of the SP-C family. This polypeptide possessed an N-terminal Leu. Another sequence, which could only be followed for 9 residues and possessed an N-Ile, appeared to be identical to amino acids 2–10 of the N-Leu sequence. The observation that SP-C possessed a

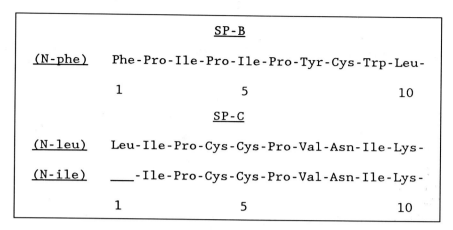

Fig. 2. Partial N-terminal amino acid sequences of the hydrophobic proteins present in lipid extracts of bovine pulmonary surfactant.

'ragged' N-terminal indicated that the alveolar form of this protein might arise via N-terminal proteolytic processing.

The realization that lipid extracts of pulmonary surfactant contained at least two distinct hydrophobic proteins prompted us to develop a SDS-urea-PAGE system to examine the intact extracts [14]. This system revealed that bovine lipid extracts contain three proteins with nominal molecular masses of ~3.5, 7 and 15 kdaltons (fig. 3). After reduction, the 15-kdalton band disappeared and was replaced by a band with a nominal molecular mass of 5 kdaltons. The intensity of the 7-kdalton band was also greatly diminished by reduction. Since these studies were conducted with intact lipid extracts without fractionation, it appears likely that all of the major protein species have been identified.

The protein bands observed on SDS-urea-PAGE were further characterized by N-terminal dansylation analysis [14]. This approach revealed that the 3.5- and 7.0-kdalton bands possessed N-terminal Leu and Ile, indicating that these proteins represented monomers and dimers of SP-C. The 15-kdalton:nonreduced and the 5-kdalton bands possessed N-terminal Phe, Leu and Ile. These observations suggested that the 15-kdalton nonreduced protein and its reduced 5-kdalton counterpart were oligomers containing both SP-B and SP-C. It should be cautioned that this conclusion must be considered tentative. The presence of N-terminal Leu and Ile in

the 15- and 5-kdalton bands could arise from variable proteolytic processing of the N-terminal end of SP-B as has been observed with SP-C. Evidence for the presence of SP-C in the 15-kdalton protein has been reported
previously by Phelps et al. [19] and by Taeusch and co-workers at the
present meeting. However, since this issue remains unresolved, we will
refer to the 15-kdalton protein as SP-B (15 kdaltons:nonreduced).

Figure 4 shows SDS-urea-PAGE of preparations of bovine SP-B (15
kdaltons:nonreduced), SP-B (5 kdaltons) and SP-C (3.5 kdaltons). The biophysical activities of these proteins were examined after combination with
a phospholipid mixture [11]. Samples containing 0.5% SP-B relative to
phospholipid can reduce the surface tension at R_{min} to near 0 mN/m
(fig. 5). Reconstituted samples containing 0.5% SP-C cause a rapid reduction in surface tension to values of $\sim$40 mN/m at R_{max} and 20 mN/m at
R_{min} but little change thereafter. Interestingly, addition of increasing
amounts of SP-C to samples containing SP-B resulted in a marked decrease
in the surface tensions, particularly at R_{max} (fig. 6).

We have previously reported that natural surfactant is approximately
four times as effective as lipid extract surfactant in reducing the surface
tension of a pulsating bubble to near 0 mN/m at R_{min} [8]. Hawgood et al.
[20] have recently reported a synergistic effect between SP-A and the low
molecular weight hydrophobic proteins in reducing the surface tension to
the equilibrium surface tension of $\sim$25 mN/m. Figure 7 demonstrates that
addition of various amounts of SP-A to lipid extract surfactant containing
SP-B and SP-C also results in a marked decrease in the surface tension of a
pulsating bubble under dynamic compression at R_{min}. Note these experiments were conducted at a phospholipid concentration of 1 mg/ml which
is one-tenth the concentration used for the other studies in this report.
These results serve to explain some of the differences noted between the
properties of natural and lipid extract surfactant when assayed with the
Wilhelmy plate and the pulsating bubble surfactometer [2, 3].

Fig. 3. SDS-urea-polyacrylamide gel electrophoresis of lipid extracts of bovine pulmonary surfactant. Lane 1 = Lipid extract; lane 2 = lipid extract reduced; PL = phospholipids. The numbers correspond to molecular weights relative to standards. Modified
from Yu et al. [14] with permission.

Fig. 4. SDS-urea-PAGE of SP-B and SP-C isolated from bovine lipid extract surfactant.

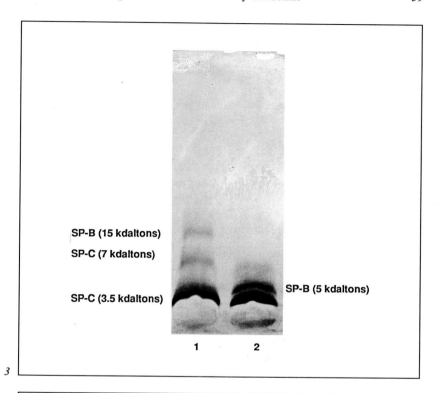

SP-B (15 kdaltons)

SP-C (7 kdaltons)

SP-C (3.5 kdaltons)

SP-B (5 kdaltons)

1   2

3

SP-B (15 kdaltons)

SP-B (5kdaltons)
SP-C (3.5 kdaltons)

1   2   3   4   5   6

4

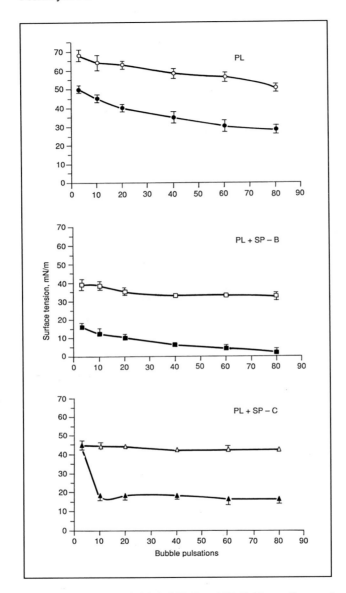

Fig. 5. Biophysical activities of SP-B and SP-C. The surface tensions at R_{max} (empty symbols) and R_{min} (filled symbols) with a pulsating bubble surfactometer are depicted. Top: Phospholipids (PL), 60% DPPC, 20% egg PC, and 20% 1-16: 0,2-18: 1 PG. Middle: phospholipids plus 1.0% SP-B. Bottom: PL plus 1.0% SP-C. See Yu and Possmayer [11] for further details.

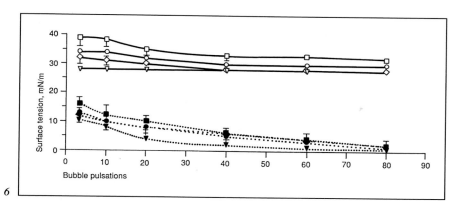

6

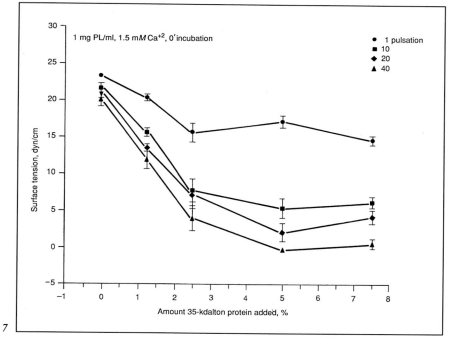

7

Fig. 6. Effect of SP-C on the biophysical activity of reconstituted surfactant containing SP-B. Reconstituted samples contained 0.5% SP-B alone (□, ■) or in the presence of 0.25% (○, ●), 0.5% (◇, ◆) or 1% (▽, ▼) SP-C. See Yu and Possmayer [11] for further details.

Fig. 7. Effect of various amounts of SP-A on the reduction of the surface tension of a pulsating bubble at R_{min} by bovine lipid extract surfactant at a concentration of 1 mg/ml.

References

1 Goerke, J.: Lung surfactant. Biochim. biophys. Acta *344:* 241–261 (1974).

2 Possmayer, F.; Yu, S.-H.; Weber, J.M.; Harding, P.G.R.: Pulmonary surfactant. Can. J. Biochem. Cell Biol. *62:* 1121–1131 (1984).

3 Goerke, J.; Clements, J.A.: Alveolar surface tension and lung surfactant; in Macklem, Mead, Handbook of physiology sect. III, vol. 3, pp. 247–261 (American Physiological Society, Washington 1986).

4 Notter, R.H.; Finkelstein, J.N.: Pulmonary surfactant. An interdisciplinary approach. J. appl. Physiol. *57:* 1613–1624 (1984).

5 van Golde, L.M.G.; Batenburg, J.J.; Robertson, B.: The pulmonary surfactant system: Biochemical aspects and functional significance. Physiol. Rev. *68:* 374–455 (1988).

6 Possmayer, F.: A proposed nomenclature for pulmonary surfactant-associated proteins. Am. Rev. resp. Dis. *138:* 990–998 (1988).

7 Yu, S.; Smith, N.; Harding, P.G.R.; Possmayer, F.: Bovine pulmonary surfactant: Chemical composition and physical properties. Lipids *18:* 522–529 (1983).

8 Weber, M.J.; Possmayer, F.: Calcium interactions in pulmonary surfactant. Biochim. biophys. Acta *796:* 83–91 (1984).

9 Enhorning, G.; Pulsating bubble technique for evaluating pulmonary surfactant. J. appl. Physiol *43:* 198–203 (1977).

10 Ross, G.F.; Notter, R.H.; Meuth, J.; Whitsett, J.A.: Phospholipid binding and biophysical activity of pulmonary surfactant-associated protein (SAP)-35 and its noncollagenous COOH-terminal domains. J. biol. Chem. *261:* 14283–14291 (1986).

11 Yu, S.-H.; Possmayer, F.: Comparative studies on the biophysical activities of the low molecular weight hydrophobic proteins purified from bovine pulmonary surfactant. Biochim. biophys. Acta *961:* 337–350 (1988).

12 Rouser, G.; Siakotos, A.N.; Fleischer, S.: Quantitative analysis of phospholipids by thin-layer chromatography and phosphorus analysis. Lipids *1:* 85–86 (1960).

13 Lowry, O.H.; Rosebrough, N.J.; Farr, A.L.; Randall, R.J.: Protein measurement with the Folin phenol reagent. J. biol. Chem. *193:* 265–275 (1951).

14 Yu, S.-H.; Chung, W.; Olafson, R.W.; Harding, P.G.R.; Possmayer, F.: Characterization of the small hydrophobic proteins associated with pulmonary surfactant. Biochim. biophys. Acta *921:* 437–448 (1987).

15 Metcalfe, I.L.; Enhorning, G.; Possmayer, F.: Pulmonary surfactant-associated proteins: their role in the expression of surface activity. J. appl. Physiol. *49:* 34–41 (1980).

16 Yu, S.-H.; Possmayer, F.: Reconstitution of surfactant activity using the 6 kDa apoprotein associated with pulmonary surfactant. Biochem. J. *236:* 85–89 (1986).

17 Olafson, R.W.; Rink, U.; Kielland, S.; Yu, S.-H.; Chung, J.; Harding, P.G.R.; Possmayer, F.: Protein sequence analysis studies on the low molecular weight hydrophobic proteins associated with bovine pulmonary surfactant. Biochem. biophys. Res. Commun. *148:* 1406–1411 (1987).

18 Curstedt, T.; Johansson, J.; Barros-Soderling, J.; Robertson, B.; Nilsson, G.; Westberg, M.; Jornvall, H.: Low molecular weight surfactant protein type 1. The primary structure of a hydrophobic 8 kDa polypeptide with 8 half-cystine residues. Eur. J. Biochem. *172:* 521–525 (1988).

19 Phelps, D.S.; Smith, L.M.; Taeusch, H.W.: Characterization and partial amino acid sequence of a low molecular weight surfactant protein. Am. Rev. resp. Dis. *135:* 1112–1117 (1987).

20 Hawgood, S.; Benson, B.J.; Schilling, J.; Damm, D.; Clements, J.A.; White, R.T.: Nucleotide and amino acid sequences of pulmonary surfactant SP18 and evidence for cooperation between SP18 and SP28-36 in surfactant lipid adsorption. Proc. natn. Acad. Sci. USA *84:* 66–70 (1987).

Dr. Fred Possmayer, Department of Obstetrics and Gynaecology, University of Western Ontario, London, ON. N6A 5A5 (Canada)

Wichert P von, Müller B (eds): Basic Research on Lung Surfactant.
Prog Respir Res. Basel, Karger, 1990, vol 25, pp 64–71

Special Qualities of Surfactant Apoproteins[1]

Henk P. Haagsman, Samuel Hawgood

Laboratory of Veterinary Biochemistry, Utrecht University, The Netherlands;
Cardiovascular Research Institute and Department of Pediatrics,
University of California, San Francisco, Calif., USA

Several forms of surfactant have been demonstrated by electron microscopy. Three classes of proteins have been described that are associated with surfactant. Fractionation of surfactant yields lipoprotein fractions that not only differ in morphology but also in protein composition, biophysical properties and metabolism. This indicates that surfactant proteins have both a role in the structure of the macromolecular complexes as well as in the biophysical properties and metabolic fate of these complexes.

In this article we want to focus on surfactant protein A (SP-A) and correlate structure with some properties of this protein.

Results and Discussion

Structure of SP-A

SP-A is a glycoprotein of pulmonary surfactant with a monomeric molecular mass of 36 kdaltons in the fully glycosylated form. It is a dimer under nonreducing, denaturing conditions. Gly-X-Y repeats in the NH_2-terminal portion suggest the presence of a collagen-like triple helix [1, 12]. To meet both requirements, disulfide-linked dimers and a triple helix, the

[1] This work is supported by a C. and C. Huygens stipend from the Netherlands Organization for Scientific Research (NWO) and by a grant (HL-24075) from the National Heart, Lung and Blood Institute.

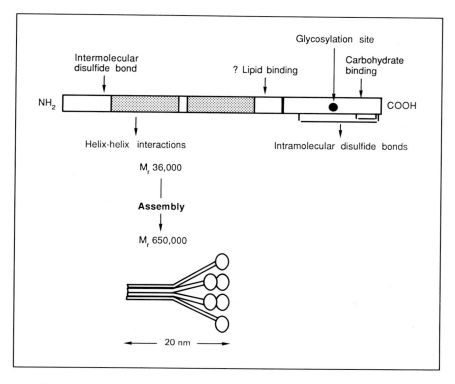

Fig. 1. Model of SP-A monomer and assembly into the native oligomeric form.

structure of native SP-A must be a hexamer or a multifold of six. The molecular mass of human recombinant SP-A as determined by Sepharose 6B chromatography is 650 kdaltons [8]. This suggests that SP-A may be an octadecamer, a structure of 6 triple helices, analogous to complement factor C1q (fig. 1). Direct support for this model has recently been provided by the elegant electron-microscopic studies of Voss et al. [11].

Heat treatment of SP-A at t > 50 °C results in the formation of particles with an apparent molecular mass of 130 kdaltons (fig. 2). These particles are further dissociated into monomers by reduction. Thus, the macromolecular structure of SP-A is determined by both noncovalent and covalent interactions.

Some aspects of the secondary structure of human and human recombinant SP-A have been studied by circular dichroism. The spectra are similar to those of C1q [2], although the small positive extremum at 230 nm was not observed with SP-A (fig. 3). The extremum at 205 nm changes as a function of temperature. The midpoint transition temperature of recombinant SP-A is $42.0 \pm 2.0\,°C$ [8], comparable with the value of $44\,°C$ recently published by Voss et al. [11]. The transition temperature of the native protein is $52.3 \pm 1.7\,°C$. Canine SP-A has a transition temperature of $51.5 \pm 1.5\,°C$ [8]. The 'melting' temperature for C1q and its collagen-like pepsin fragment is $48\,°C$; a value of $38\,°C$ was reported for lathyritic rat skin collagen [2].

Figure 4 shows that the spectrum of a fragment of human SP-A, obtained after digestion with collagenase, is comparable to the spectrum of SP-A after heat treatment. This suggests that the negative extremum at 205 nm is indeed caused by a collagen-like triple helix. No significant Ca^{2+}-induced spectral changes were observed.

The temperature-dependent structural changes, as monitored by circular dichroism, are in line with the generation of smaller fragments by heat treatment at $t > 50\,°C$. The lower transition temperature of recombinant SP-A suggests that the triple helices are less stable. A possible reason for the different properties of recombinant SP-A could be the fact that this assembled protein contains only one gene product. The collagen-like triple helices in native SP-A may consist of heterotrimers as was pointed out by Voss et al. [11].

The COOH-terminal region of SP-A is the putative carbohydrate-recognition domain of SP-A [3]. It contains four cysteines that are conserved in all lectins of this class. The intrachain disulfide bridges in human SP-A are formed between residues 135–226 and 204–218. Thus, native SP-A is assembled into an octadecamer held together at the NH_2-terminal by six

---

Fig. 2. Dissociation of oligomeric recombinant SP-A by mild heat treatment. The effect of mild heat treatment on the size of SP-A was analyzed by Sepharose 6B chromatography. Sepharose 6B chromatography of recombinant SP-A was carried out in 10 mM Tris-Cl, pH 7.4, 100 mM NaCl and 2 mM EDTA. 300 μg protein was applied. SP-A was incubated for 10 min at $50\,°C$. Most SP-A had a molecular weight of 130,000 when incubated for 10 min at $90\,°C$.

Fig. 3. Circular dichroism spectrum of human SP-A. Spectra were recorded on a JASCO J-500 A spectropolarimeter. Four scans were accumulated and averaged with the aid of a JASCO DP-500 data processor. The scan speed was 10 nm/min. SP-A was dissolved in 5 mM Tris-Cl, pH 7.4 (1 mg/ml). A cell was used with a pathlength of 0.1 mm. SP-A was purified as described previously [6, 9].

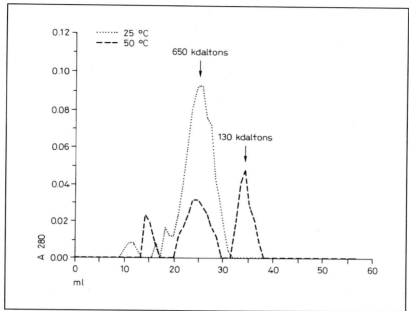

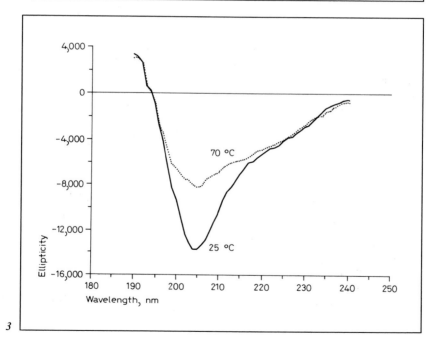

collagen-like triple helices and intermolecular disulfide bridges. A more globular structure is predicted for the COOH-terminal domain by the intramolecular disulfide pairing.

Properties of SP-A

SP-A causes phospholipid aggregation [5] and promotes adsorption of phospholipids to an air-fluid interface in the presence of surfactant proteins SP-B and SP-C [10]. These properties of SP-A are Ca^{2+} dependent. SP-A may have an important role in generating and stabilizing the structure of tubular myelin, a subfraction of surfactant that probably forms the surface film [13]. SP-A aggregates in the presence of Ca^{2+} at 37 °C [7]. Ca^{2+}-dependent aggregation of SP-A is impaired by mild heat treatment or reduction (fig. 5). This suggests that aggregation of SP-A requires the self-association of the oligomeric structure of 6 linked collagen-like triple helices.

Many properties of SP-A are Ca^{2+} dependent. We could demonstrate that SP-A is a Ca^{2+}-binding protein. It contains 2 binding sites/mol monomer as determined by equilibrium dialysis. The dissociation constants are 20 and 300 μM, respectively.

The COOH-terminal region has a remarkable sequence homology with a number of Ca^{2+}-dependent lectins [4]. In particular, the four cysteines in this domain are conserved in these lectins. The carbohydrate-binding properties, predicted by sequence analysis, have been demonstrated experimentally [6]. Binding to immobilized monosaccharides is Ca^{2+} dependent and is impaired by reduction or mild heat treatment. This suggests that also for binding to monosaccharides the protein should be oligomeric. Although binding to immobilized monosaccharides seems to require cooperativity of the binding sites, it should be emphasized that this does not imply that the cooperativity is necessary for binding to complex oligosaccharides.

Fig. 4. Circular dichroism spectrum of a collagenase-resistant fragment of human SP-A. Spectra of a collagenase-resistant fragment of SP-A (0.2 mg/ml) were recorded in 5 mM Tris-Cl, pH 7.4, 100 mM NaCl and either 2 mM CaCl$_2$ or 2 mM EDTA. Spectra were compared with a spectrum of heat-denatured human SP-A.

Fig. 5. Effects of heat treatment and reduction on the calcium-induced aggregation of canine SP-A. Divalent cation-dependent self-association of SP-A was monitored by turbidity at 300 nm. The experiment was done in 5 mM Tris-HCl, pH 7.0, 50 mM NaCl at 37 °C. Heat treatment or reduction of SP-A prevents calcium-dependent self-association.

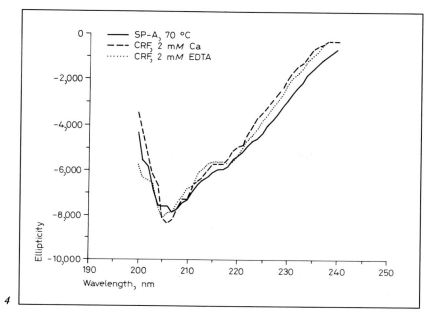

4

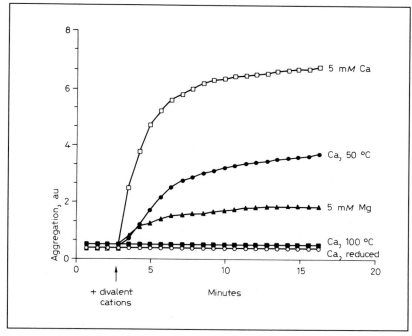

5

Summary

The molecular weight of SP-A, the major surfactant-associated protein, is 28,000–36,000 under denaturing and reducing conditions. Gel filtration shows that the molecular weight of SP-A is 650,000, consistent with a structure of 6 linked collagen-like triple helices, a structure analogous with complement factor C1q. The collagenous domain of native SP-A denatures at 52 °C, as monitored by circular dichroism. Interestingly, the transition temperature of recombinant SP-A is 10 °C lower. Heat treatment of SP-A at $t > 45$ °C decreases the molecular mass of the oligomer from 650 to 130 kdaltons. SP-A aggregates in the presence of 0.5–1mM Ca^{2+} at 37 °C. Ca^{2+}-dependent aggregation of SP-A requires the self-association of the oligomeric structure of 6 linked collagen-like triple helices.

SP-A is a Ca^{2+}-binding protein. It binds 2 mol Ca^{2+}/mol monomer. The COOH-terminal region has a remarkable sequence homology with a number of calcium-dependent lectins. In particular, the four cysteines in this domain are conserved in these lectins. The carbohydrate-binding properties predicted by sequence analysis have been demonstrated experimentally. Carbohydrate binding to immobilized monosaccharides is Ca^{2+} dependent and requires the oligomeric state of the protein.

References

1   Benson, B.; Hawgood, S.; Schilling, J.; Clements, J.; Damm, D.; Cordell, B.; White, R.T.: Structure of canine pulmonary surfactant apoprotein: cDNA and complete amino acid sequence. Proc. natn. Acad. Sci. USA 82: 6379–6383 (1985).

2   Brodsky-Doyle, B.; Leonard, K.R.; Reid, B.M.: Circular-dichroism and electron-microscopy studies of human subcomponent C1q before and after limited proteolysis by pepsin. Biochem. J. 159: 279–286 (1976).

3   Drickamer, K.; Dordal, M.S.; Reynolds, L.: Mannose-binding proteins isolated from rat liver contain carbohydrate-recognition domains linked to collagenous tails. Complete primary structures and homology with pulmonary surfactant apoprotein. J. biol. Chem. 261: 6878–6887 (1986).

4   Drickamer, K.: Two distinct classes of carbohydrate-recognition domains in animal lectins. J. biol. Chem. 263: 9557–9560 (1988).

5   Efrati, H.; Hawgood, S.; Williams, M.C.; Hong, K.; Benson, B.J.: Divalent cation and hydrogen ion effects on the structure and surface activity of pulmonary surfactant. Biochemistry 26: 7986–7993 (1987).

6   Haagsman, H.P.; Hawgood, S.; Sargeant, T.; Buckley, D.; White, R.T.; Drickamer, K.; Benson, B.J.: The major lung surfactant protein, SP 28-36, is a calcium-dependent, carbohydrate-binding protein. J. biol. Chem. 262: 13877–13880 (1987).

7   Haagsman, H.P.; Hawgood, S.: Ca^{2+}-dependent self-association of SP-A, the major lung surfactant protein. FASEB J. 2: A1044 (1988).

8   Haagsman, H.P.; Hawgood, S.; Benson, B.J.: The major lung surfactant protein, SP-A, is a 650 kDa Ca^{2+}-binding protein. Am. Rev. resp. Dis. 137: 277 (1988).

9   Hawgood, S.; Benson, B.J.; Hamilton, R.L.: Effects of a surfactant-associated protein and calcium ions on the structure and surface activity of lung surfactant lipids. Biochemistry 24: 184–190 (1985).

10 Hawgood, S.; Benson, B.J.; Schilling, J.; Damm, D.; Clements, J.A.; White, R.T.: Nucleotide and amino acid sequences of pulmonary surfactant protein SP 18 and evidence for cooperation between SP 18 and SP 28-36 in surfactant lipid adsorption. Proc. natn. Acad. Sci. USA *84:* 66–70 (1987).

11 Voss, T.; Eistetter, H.; Schäfer, K.P.; Engel, J.: Macromolecular organization of natural and recombinant lung surfactant protein SP 28-36. Structural homology with the complement factor C1q. J. molec. Biol. *201:* 219–227 (1988).

12 White, R.T.; Damm, D.; Miller, J.; Spratt, K.; Schilling, J.; Hawgood, S.; Benson, B.; Cordell, B.: Isolation and characterization of the human pulmonary surfactant apoprotein gene. Nature, Lond. *317:* 361–363 (1985).

13 Wright, J.R.; Clements, J.: Metabolism and turnover of lung surfactant. Am. Rev. resp. Dis. *135:* 426–444 (1987).

H.P. Haagsman, PhD, Laboratory of Veterinary Biochemistry, Utrecht University, PO Box 80.176, NL–3508 TD Utrecht (The Netherlands)

Wichert P von, Müller B (eds): Basic Research on Lung Surfactant.
Prog Respir Res. Basel, Karger, 1990, vol 25, pp 72–80

Structure Function:
Correlation in Reassembled Surfactant[1]

Samuel Hawgood

Cardiovascular Research Institute and Department of Pediatrics, University of
California, San Francisco, Calif., USA

The study of complex biological systems including lipoprotein membranes in an unperturbed state has obvious advantages. Establishing the
function of individual components in membranes or the nature of the
interactions between components is, however, very difficult under these
conditions and many investigators have turned to more experimentally
accessible reconstituted systems to study the role of purified membrane
components. In recent years the methodology of lipid-protein reconstitution has been applied to the surfactant system by several investigators.

The major focus of my paper is to discuss (1) the importance of solvent conditions, particularly the ionic environment, on lipid-protein interactions in both native and reconstituted surfactant, and (2) the role of the
surfactant proteins in determining the structue of lipoprotein complexes
and the effect of these proteins separately or in combination on the speed
of phospholipid surface film formation. The use of reconstituted systems
to study the potential role of the surfactant proteins in other aspects of
surfactant metabolism such as regulation of surfactant lipid clearance and
secretion are not addressed.

To follow the evolution of thinking that has occurred in our laboratory
about the role of the surfactant proteins in the adsorption process I will
first describe a limited number of experiments aimed at establishing the
conditions which regulate native surfactant structure and function and

[1] This work was supported by a grant (HL-24075) from Heart, Lung and Blood
Institutes of the National Institutes of Health, USA.

then describe a series of related reconstitution studies using increasingly homogeneous and well-characterized components. For the purposes of this paper the term native surfactant is used to describe material isolated from the bronchoalveolar washings of adult dogs by the methods of King and Clements [1]. The nomenclature for the surfactant proteins proposed by Possmayer [2] is used throughout.

Effect of Ionic Environment on Native Surfactant

Benson et al. [3] established that tubular myelin in isolated surfactant could be dispersed to vesicular forms in the divalent cation chelator EDTA but then reformed by adding back Ca^{2+}. They also noted a relationship between conditions, in particular the Ca^{2+} concentration, required for tubular myelin formation and conditions required for the maximal activity or adsorption speed of the sample. Immunohistochemical studies [4] suggest that tubular myelin is enriched in the collagen-like glycoprotein surfactant protein-A (SP-A) relative to other surfactant lipid fractions. We have therefore undertaken a series of studies to further define the relationship between the ionic conditions and SP-A content in the regulation of tubular-myelin formation and surface activity. The surface activity of native surfactant is clearly influenced by the presence of divalent cations. At a physiological pH of 6.9 Ca^{2+} but not Mg^{2+} markedly increases the surface activity of native surfactant. Electron micrographs of samples of native surfactant in Ca^{2+} or EDTA also show marked differences. Although tubular myelin is only present with Ca^{2+} there is still considerable heterogeneity with much of the native surfactant adopting structural forms other than tubular myelin [5]. A second apparently regular arrangement of membranes is seen in native surfactant consisting of roughly parallel membranes with a spacing of approximately 25 nm or about half the spacing seen in tubular myelin. The membranes are separated by a fuzzy amorphous material, presumably protein, and are often in continuity with tubular-myelin forms. Many other variable vesicular forms are also seen. The two regular membrane arrangements, tubular myelin and parallel fuzzy membranes, are lost in a reversible fashion when the Ca^{2+} is chelated. As a further correlation of structure and function in native surfactant we have determined the influence of pH on both structure and surface activity. The effects of pH and Ca^{2+} ions on surface activity and structure appear to be related. As the pH is lowered the effect of Ca^{2+} on the surface activity of the sample is less marked. At a pH of 5.0 or less native surfactant has a surface activiy in the absence of free Ca^{2+} that is comparable to the surface activity seen at pH 6.9 only in the presence of millimolar Ca^{2+}. The structures formed by native surfactant are also influenced by the pH. At pH 5.0 in the absence of free Ca^{2+} structures very similar to tubular myelin and the fuzzy parallel membranes are formed. These studies suggest that the structural rearrangements to tubular myelin and the property of rapid surface film formation may be linked mechanistically. Unfortunately, the marked heterogeneity in these samples does not allow me to exclude the possibility that the two phenomena are unrelated apart from being both regulated by the Ca^{2+} or H^+ ion concentration. Certainly many other lipid or lipoprotein structures are able to facilitate rapid film formation.

Effect of Ionic Environment on Reconstituted Surfactant

In a series of studies King et al. [6] have shown that SP-A associates with a variety of phospholipids in a Ca^{2+}-independent fashion but that the degree of lipoprotein aggregation, at least in the presence of phosphatidylglycerol (PG), is markedly affected by Ca^{2+} [6]. We have performed similar studies to those of King with a major and apparently important difference being that the lipids used by us were extracted into chloroform-methanol from native surfactant and therefore contained roughly equal amounts of SP-B and SP-C [5, 7]. The presence of SP-A markedly reduces the concentration of Ca^{2+} required for extracted surfactant lipid aggregation from greater than 10 mM to approximately 0.5 mM, well within the free Ca^{2+} concentration of the alveolar subphase. Significant structural rearrangement of the extracted lipid membranes, including the formation of appropriately spaced parallel fuzzy membranes and small patches of tubular-myelin-like structures correlates with this lipoprotein aggregation. While the membrane aggregation mediated by SP-A is very rapid, occurring in seconds after the addition of Ca^{2+} (fig. 1), the rearrangement needed for tubular myelin formation seems to be very much slower. It can be seen in figure 1 that the lipoprotein aggregation is not completely reversed by EDTA, suggesting that several processes with different kinetics, such as rapid aggregation and slow fusion, may be occurring in the presence of SP-A and Ca^{2+}. At pH less than 5.0 aggregation occurs in the absence of Ca^{2+}.

The effect of SP-A on the surface activity of the extracted surfactant lipids is also pH and Ca^{2+} dependent. As shown in table 1 the behavior of

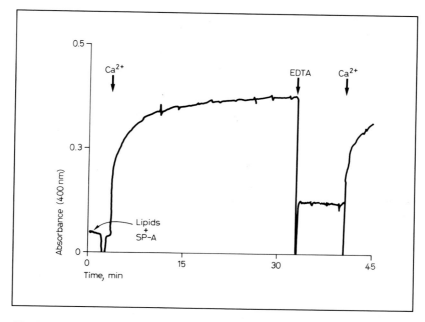

Fig. 1. Aggregation of extracted surfactant liposomes induced by SP-A and Ca^{2+}. Aggregation was assessed by measuring sample turbidity in a spectrophotometer. The concentrations of the components were phospholipid 200 $\mu g \cdot ml^{-1}$, SP-A 20 $\mu g \cdot ml^{-1}$ and Ca^{2+} 0.5 mM.

Table 1. Surface pressure

pH	SAM				Recombinant	
	7.5 min	15 min	30 min	7.5 min	15 min	30 min
7.0	7.9 ± 2.4	14.8 ± 3.5	17.0 ± 4.9	13.2 ± 0.9	15.8 ± 1.0	18.3 ± 1.0
7.0 + Ca^{2+}	39.5 ± 2.6	42.9 ± 1.2	44.3 ± 0.4	34.0 ± 6.8	40.9 ± 2.3	44.0 ± 0.0
5.0	20.0 ± 5.4	26.5 ± 3.5	33.9 ± 2.5	15.5 ± 1.1	23.0 ± 0.8	31.0 ± 2.8
4.4	41.5 ± 1.6	42.6 ± 0.3	44.3 ± 0.3	35.5 ± 3.5	41.5 ± 1.5	44.0 ± 0.0

Values (mN/m) are mean ± SD of 5–8 experiments. Samples containing lung surfactant-SAM (40 μg phospholipids) or lipoprotein-recombinant (40 μg phospholipids and 20 μg protein) were incubated for 1 h at 37 °C in Hepes buffer at various pH in the presence or absence of 2 mM Ca^{2+}. The samples were then placed below a clean surface and the change in surface pressure recorded. The subphase phospholipid concentration was 13.3 μg/ml. Reprinted with permission from Efrati et al. [5].

the reconstituted system is similar to the behavior of native surfactant in each set of ionic conditions. Again the biochemical complexity, structural heterogeneity and limitations in the assay system limit the conclusions that can be drawn from experiments of this type but the results do suggest that SP-A is one of the key components required for tubular myelin formation and that either as a consequence of the structural rearrangement of the lipids or through some other mechanism SP-A also has a role in the surface activity displayed by native surfactant.

We have recently shown that SP-A does bind Ca^{2+} directly in the presence or absence of lipid and in the presence of an excess of Mg^{2+} [Haagsman, this vol.]. We have also found that Ca^{2+} causes SP-A to extensively self-associate at pH 6.9 and 37 °C. This suggests that the initial event involved in the SP-A-induced bilayer rearrangement may be due to Ca^{2+}-mediated bridging of protein molecules on each of the interacting membranes. Clearly, this may be just the first step in a cascade of events initiated by Ca^{2+} or a change in pH. Components such as anionic phospholipids, cholesterol, SP-B or SP-C may also be critical participants in the events leading to tubular myelin formation and a surface-active complex. The role of these other components can be established in appropriate reconstitution systems.

Effects of SP-B and SP-C on the Surface Activity of Reconstituted Complexes

The presence of a hydrophobic protein fraction in surfactant was established in one of the earliest studies of surfactant proteins [8]. Phizackerley et al. [9] later demonstrated that small proteins were extracted along with the lipids of lamellar bodies into organic solvents. Numerous investigators have since shown that such organic solvent extracts of either surfactant fractions from lung homogenates or bronchoalveolar washings are surface active when tested at relatively high phospholipid concentrations.

The surface activity of these extracts does not, however, display the same Ca^{2+} or pH dependence seen with native surfactant [5]. In this paper I have summarized evidence that suggests that the striking ion-regulated phenomena observed in native surfactant are related to SP-A, a protein not found in the organic solvent lipid-rich extracts. However, complexes reconstituted from SP-A and phospholipid mixtures not containing the

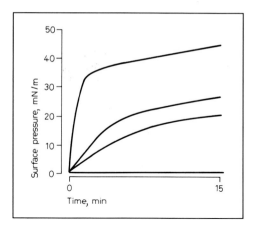

Fig. 2. Effect of SP-B on phospholipid adsorption. Experiments were performed in a Teflon trough containing 3 ml 5 mM Tris, 100 mM NaCl, 2.5 mM CaCl$_2$, pH 7.4 at 37 °C. Adsorption was assessed by the change in surface pressure. The lipids were a mixture of DPPC and egg PG (7:3) at a concentration of 33 μg·ml^{-1} in the trough in each case. The samples contain increasing amounts of SP-B (0 to 3.3:1 BW). Protein was determined by reacting the sample with fluorescamine in the presence of SDS.

hydrophobic surfactant proteins have minimal surface activity [6, 10]. There is obviously a clear difference between the effects of SP-A on lipid-rich fraction extracted from native surfactant and the effects of SP-A on protein-free lipid mixtures.

These differences support the hypothesis that either SP-B or SP-C or both have a role in the formation of a surface-active fraction and possibly the tubular myelin structure. Several studies, again employing reconstituted systems, support this hypothesis [2]. Both SP-B (fig. 2) and SP-C [11] enhance the absorption rate of phospholipid mixtures of DPPC and unsaturated PG in a dose-dependent fashion. Perhaps not surprisingly the characteristics of the adsorption isotherms are quite path dependent varying markedly with details of the reconstitution method. Formulation variables such as temperature, lipid concentration, and vesicle size markedly effect the absorption characteristics.

The adsorption characteristics of complexes of lipids and both SP-B and SP-C are also dependent on the precise lipid composition [see also Pison, this vol.]. Both proteins clearly interact with membranes of pure DPPC as evidenced by a small increase in the thermal transition temperature and a marked reduction in the enthalpy of lipid transition [12]. However, for maximal surface activity at 37 °C and physiological ionic conditions the addition of both unsaturated and anionic phospholipids to DPPC seems to be important. We have shown that the permeability of preformed large unilamellar vesicles induced by mixtures of SP-B and SP-C (both cationic proteins) added from methanol is also markedly dependent on the presence of even small amounts (10%) of anionic phospholipid in the membrane, suggesting that interactions between basic residues in the proteins and the acidic phospholipid headgroups may play a role in the effect of these proteins on membrane organization [13]. Unfortunately, these observations on the effects of SP-B and SP-C on phospholipid adsorption are descriptive and the mechanisms involved in the adsorption process are still unknown. The primary sequences of SP-B and SP-C [see Whitsett, this vol., for references] have been recently established but information regarding the orientation of these proteins in a lipid membrane, the secondary structure of these proteins, and their possible assembly into protein complexes under various conditions will be required before models can be constructed and put to the test. Eventually, such models will have to take into account the probability that in addition to modifying the behavior of the lipids in surfactant membranes the surfactant proteins may interact directly with one another in determining surfactant membrane structure and function.

In support of this last speculation we have recently found that SP-A added to lipoprotein complexes reconstituted from SP-B and relatively simple binary phospholipid mixtures further promotes the rate of surface film formation provided both an anionic phospholipid and Ca^{2+} are present (fig. 3) [10]. It is likely therefore that the presence of SP-B in the extracted lipids in our earlier studies explains the marked effect that SP-A has on the surface activity of these lipids. In our experiments to date we have not seen any effect of SP-A on the adsorption characteristics of complexes containing SP-C and phospholipids but further experiments utilizing different methods of lipid-protein assembly are required before interactions between these two proteins can be excluded.

The studies I have described in this paper address just one step in the complex surfactant cycle, the formation of the surface-active tubular myelin structure. The interactions involved in creating the impressive long-range order in this structure are still unknown but it appears that the inventory of components required is complex and includes at least a mixture of saturated and unsaturated phospholipids, anionic phospholipids,

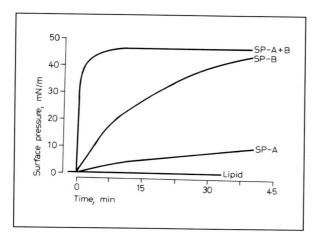

Fig. 3. Combined effect of SP-A and SP-B on phospholipid adsorption. The samples each contained μg phospholipid (DPPG:egg PG, 4:1 BW) and were placed below a clean surface of buffer containing Ca²⁺ in a Teflon trough. Adsorption was assessed by the change in surface pressure. The phospholipid:protein ratio was 5:1 for SP-A and 30:1 for SP-B. Protein was determined by the Lowry reaction. Modified with permission of the author from Hawgood et al. [10].

SP-A and at least one of the hydrophobic surfactant proteins. It also appears that alveolar conditions, particularly the Ca²⁺ concentration at physiological pH, may regulate or modify the interactions that occur between these components. Perhaps when these interactions are understood it will be apparent why this structure appears to be such an efficient vehicle for the rapid transport of phospholipid molecules into the surface film.

References

1   King, R.J.; Clements, J.A.: Surface active materials from dog lung. I. Method of isolation. Am. J. Phys. *223:* 707–714 (1972).
2   Possmayer, F.: A proposed nomenclature for pulmonary surfactant-associated proteins. Am. Rev. resp. Dis. *138:* 990–998 (1988).
3   Benson, B.J.; Williams, M.C.; Sueishi, K.; Goerke, J.; Sargeant, T.: Role of calcium ions in the structure and function of pulmonary surfactant. Biochim. biophys. Acta *793:* 18–27 (1984).

4   Williams, M.C.; Benson, B.J.: Immunocytochemical localization and identification of the major surfactant protein in adult rat lung. J. Histochem. Cytochem. *29:* 291–305 (1981).

5   Efrati, H.; Hawgood, S.; Williams, M.C.; Hong, K.; Benson, B.J.: Divalent cation and hydrogen ion effects on the structure and surface activity of pulmonary surfactant. Biochemistry *26:* 7986–7993 (1987).

6   King, R.J.; Carmichael, M.C.; Horowitz, P.M.: Reassembly of lipid-protein complexes of pulmonary surfactant. J. Biol. Chem. *258:* 10672–10680 (1983).

7   Hawgood, S.; Benson, B.J.; Hamilton, R.J.J.: Effects of a surfactant-associated protein and calcium ions on the structure and surface activity of lung surfactant lipids. Biochemistry *24:* 184–190 (1985).

8   King, R.J.; Klass, D.J.; Gikas, E.G.; Clements, J.A.: Isolation of apoproteins from canine surface active material. Am. J. Phys. *224:* 788–795 (1973).

9   Phizackerley, J.R.; Town, M.; Newman, G.E.: Hydrophobic proteins of lamellated osmiophilic bodies isolated from pig lung. J. Biochem. *183:* 731–736 (1979).

10 Hawgood, S.; Benson, B.J.; Schilling, J.; Damm, D.; Clements, J.A.: Nucleotide and amino acid sequences of pulmonary surfactant protein SP 18 and evidence for cooperation between SP 18 and SP 28-36 in surfactant lipid adsorption. Proc. natn. Acad. Sci. USA *84:* 66-70 (1987).

11 Warr, R.G.; Hawgood, S.; Buckley, D.I.; Crisp, T.M.; Schilling, J.; Benson, B.J.; Ballard, P.L.; Clements, J.A.; White, R.T.: Low molecular weight human pulmonary surfactant protein (SP5). Isolation, characterization, and cDNA and amino sequences. Proc. natn. Acad. Sci. USA *84:* 7915–7919 (1987).

12 Shiffer, K.; Hawgood, S.; Duzgunes, N.; Goerke, J.: The effects of the low molecular weight lung surfactant-associated proteins SP-B and SP-C on the thermotropic properties of various synthetic phospholipid membranes. FASEB *2:* A319 (1988).

13 Shiffer, K.; Hawgood, S.; Duzgunes, N.; Goerke, J.: Interactions of the low molecular weight group of surfactant-associated protein (SP 5–18) with pulmonary surfactant lipids. Biochemistry *27:* 2689–2695 (1988).

Samuel Hawgood, MB, BS, Cardiovascular Research Institute and Department of Pediatrics, University of California, San Francisco, CA 94143 (USA)

Wichert P von, Müller B (eds): Basic Research on Lung Surfactant.
Prog Respir Res. Basel, Karger, 1990, vol 25, pp 81–90

Endocytosis in Alveolar Type II Cells in vivo and in vitro[1]

Mary C. Williams, Leland G. Dobbs

Departments of Anatomy and Medicine, School of Medicine, University of
California, San Francisco, Calif., USA

The cellular plasma membrane participates in various types of functions including many which regulate cellular behavior. These functions include endocytosis and transcytosis, secretion, signal transduction, ion transport, receptor-ligand interactions and others. Numerous studies indicate that these processes are carried out by type II cells and have characteristics which are generally similar to those of other cell types (e.g. Na^+ transport, signal transduction systems) [1–5]. Of these, endocytosis and secretion in type II cells are two cellular processes which have been the focus of investigation because of their importance in the metabolism of surfactant lipids and proteins [6–14]. As described in various publications and in this volume, the secretory and endocytic functions of type II cells appear to have general characteristics similar to those of other cell types but have, in addition, some adaptations which may be type II cell specific [see Dobbs et al., this vol.]. The following discussion will describe some characteristics of endocytic pathways in type II cells in situ and how these appear to be altered over time by monolayer culture.

Methods and Results

Type II Cells in situ

The following methods were used to study endocytic processes in intact lungs: Adult male rats were anesthetized with 6 mg/100 g body weight sodium methohexital i.p. A Teflon cannula was inserted into the

[1] Supported by Program Project Grant HL-24075 from the National Heart, Lung, and Blood Institute, National Institutes of Health.

trachea via the oral cavity with the aid of fiberoptic illumination. A small polyethylene catheter (PE 10), threaded through the larger cannula, was inserted 2–3 cm until it met gentle resistance. A tracer solution (0.05–0.10 ml) was instilled into the lungs after which both catheters were immediately removed. These procedures required about 1–2 min; animals recovered from anesthesia in 5–10 min. The following tracers were studied: cationic ferritin (CF) and anionic ferritin (11 nm diameter), and the ferritin-labelled lectins *Maclura pomifera* (MPA), concanavalin A (Con A), and *Wisteria floribunda* (WFA). In addition, each instillate contained 0.01–0.02 ml biological grade colloidal carbon (25 nm diameter) to mark the site of tracer deposition. These tracers were selected for specific reasons. By virtue of its high positive charge cationic ferritin may bind to many different membrane proteins, particularly those with an abundance of sialic acid residues. CF is likely therefore to be endocytosed in abundance but may in fact enter more than one pathway if several exist. Native (anionic) ferritin serves as an excellent nonbinding (i.e. soluble) control protein for CF because of its similar size and composition. Of the three lectins only MPA appears to bind specifically to type II cell membranes as shown by Brandt and co-workers [15, 16] (fig. 1a). Binding can be inhibited with the addition of the hapten sugar for MPA, alpha-galactose. Colloidal carbon, 25 nm diameter, was used to assess the upper limit of particle size capable of entering the endocytic pathway.

At 5 min to 18 h after instillation of tracers, animals were reanesthetized. The lungs were fixed for electron microscopy by instillation with buffered 2% glutaraldehyde, 1% paraformaldehyde. As detailed elsewhere, standard procedures of dehydration, embedding, etc. were used [17]. The time course of uptake and participating organelles were studied for each tracer, and the amount of each internalized tracer was assessed qualitatively.

Of the tracers studied, only CF and MPA entered type II cells in significant amounts. Occasional molecules of the other tracers, except carbon, were present in some type II cells, but most cells contained none after 1- to 2-hour exposures. The selectivity of uptake of tracers by type II cells contrasts markedly with endocytic characteristics of nearby macrophages. All tracers, including colloidal carbon, were ingested rapidly and in abundance by macrophages.

To determine whether uptake of MPA was dependent on its specific interaction with alpha-galactose-containing membrane components, both MPA and methyl-alpha-*D*-galactoside were instilled into the lungs of 3

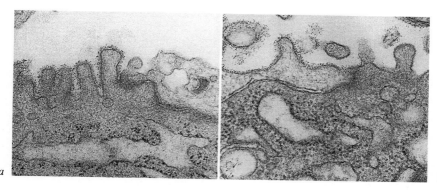

Fig. 1. The specificity of lectin binding by type II and type I cells can be readily illustrated by incubating lung tissue from adult rats in ferritin-labelled lectins (electron-dense particles). *a* MPA binds to type II cell membranes up to the intercellular junction which connects adjacent type II-type I cells. *b* The pattern of labelling is reversed with RCA-ferritin which shows heavy labelling on type I cells and little on type II cells.

animals. For 30–60 min the presence of the hapten sugar appeared to block the uptake of MPA-F into type II cells but not into nearby macrophages. After 60 min, small amounts of MPA were present in type II cells, presumably because the inhibiting hapten sugar diffused away from the instillation site and was no longer effective.

The type II cell appears to have a minimum of two pathways through which endocytosed materials can move as evidenced by the uptake of MPA and CF. In terms of the amount of material transported, the principal pathway utilizes those intracellular organelles known to participate in receptor-mediated endocytosis of ligands such as asialoglycoproteins (in liver) and low-density lipoproteins (in fibroblasts) [18, 19]. The organelles which participate in type II cell endocytosis are, in sequence: (1) small pinocytic vesicles, labelled first at 2–10 min; (2) large electron-lucent endosomes and electron-lucent multivesicular bodies (MVBs), labelled first at 5–10 min; (3) electron-dense MVBs, labelled first at 10–30 min, and (4) composite MVB-lamellar bodies (LBs) and lamellar bodies, labelled first at 30 min (fig. 2a, b). At 1 h after instillation most LBs in any labelled cell contained tracer which was preferentially localized within the nonlamellar protein content of the LB [13] (fig. 2b). At later times when the amount of tracer in an LB increased, tracer molecules sometimes intercalated between adjacent phospholipid lamellae.

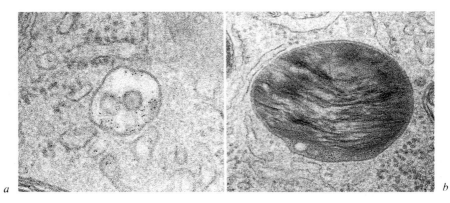

Fig. 2. When type II cells are exposed to MPA-F in intact animals, they rapidly internalize it into endocytic organelles such as MVBs (*a*). *b* Within 30 min MPA-F is present in lamellar bodies where it is preferentially localized to the nonlamellar content of the granule.

CF appeared to label a second transcytotic pathway which ended with the release of CF into the interstitium at the base of type II cells [12]. No similar pathway was observed in type I cells for any tracer studied. Particles first appeared in the interstitium about 30 min after tracer instillation. Details of the organelles, specificity, and time course of the second pathway have not been studied carefully although the transfer of substances across the epithelial layers by vesicular transcytosis has been described for many organs and is likely to be a common epithelial function.

Type II Cells in Culture

To determine whether cultured type II cells display endocytic pathways with similar characteristics of time course and specificity, some uptake studies were carried out on cells in vitro. Type II cells were isolated from adult rat lungs by standard methods [20, 21], cultured on extracellular matrix [16], and incubated with lectins with or without hapten sugars for 30–90 minutes. Cells were studied on days 1, 2, 5, or 7 in culture. The lectins MPA-F, and *Ricinus communis* I-ferritin (RCA-I-F) were selected as tracers. RCA-I was selected as an additional tracer because it binds specifically to type I alveolar epithelial cells in intact lung but not to type II cells (fig. 1b) [15, 16]. Because early studies indicated that endocytic functions of cultured cells were different from those of type

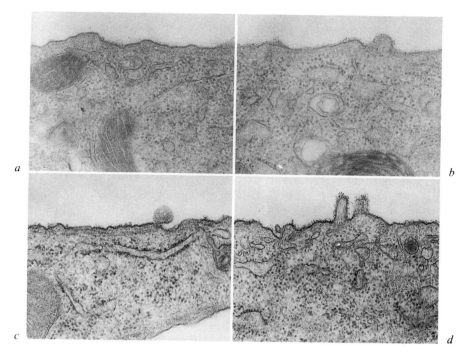

Fig. 3. By day 2 after plating isolated type II cells in monolayer culture, the cells have lost most of their MPA binding (*a*) and have increased binding of RCA-I (*b*). This trend continues with extended time in culture as shown here on day 7. Little if any MPA is bound (*c*) while RCA binding is further increased (*d*).

II cells in situ not all tracers were studied at all time points or at all days in culture.

The binding of lectins to the apical plasma membranes of cultured type II cells differed markedly from that which characterizes these cells in the intact lung. By day 2 in culture, little MPA binds to the cellular membrane (fig. 3a) although MPA binding is abundant on day 0 cells at the termination of the isolation procedure (not shown). Less MPA binding is observed on day 7 than on day 2 (fig. 3c). As the cells lose the ability to bind MPA with time in culture, they gain binding sites for RCA (fig. 3b, d). RCA binding is readily demonstrable by day 2 and increases until, by day 7, the lectin-ferritin complex forms a quasi-continuous layer at the surface of the membrane (fig. 3d). If, as proposed earlier, binding is required for uptake of tracers in amounts greater than that of soluble phase markers, it

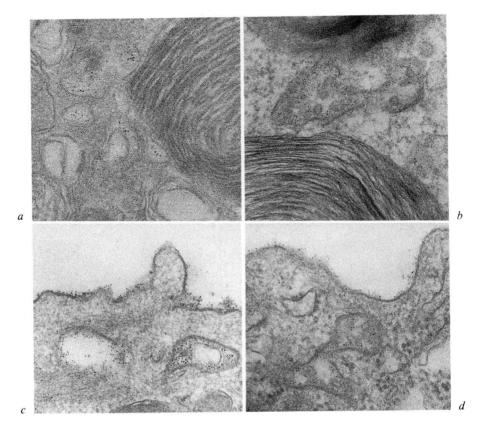

Fig. 4. Cells in culture appear to endocytose different lectins at different rates from those in intact lungs. *a* Small amounts of MPA are present in endocytic vesicles of cells on day 1 in culture after 3 h exposure to the lectin. Although vesicles are adjacent to LBs, little MPA was deposited into the granules in this time period. *b, d* By culture day 2, Con A-F is present in abundance in MVBs which, as in (*a*), do not appear to empty the tracer into nearby LBs, presumably by fusion. *c* By day 7 in culture, large amounts of RCA-I are taken up by the cells into pinocytic vesicles and MVBs but little is present in LBs.

can be hypothesized that abundant RCA but little MPA would enter cultured type II cells. Observations at various times in culture support this suggestion.

In contrast to type II cells in situ cultured type II cells endocytosed abundant Con A-F (fig. 4b, d) and RCA-I-F (fig. 4c) but little (fig. 4a) or no MPA-F. The specificity of materials taken up by the cells therefore appeared to have been influenced by the isolation and/or culture conditions. The participating organelles for the early steps in the endocytic pathway (pinocytic vesicles, endosomes, and MVBs), however, were the same as those of cells in situ. The failure of most of the ingested tracer to enter LBs within the 90 min studied was striking because tracer-loaded MBs were often observed at short distances from intracellular LBs (fig. 4b).

Discussion

These studies suggest several important characteristics of endocytic functions of type II cells. First, they indicate that the amount of a particular tracer which is internalized is greatly enhanced if the tracer binds to the plasma membrane; this is a general characteristic of endocytic processes in other cells [22]. This property was demonstrated by comparing the uptake of MPA and CF to that of nonbinding lectins or anionic or uncharged molecules. Because binding occurs extracellularly, the specificity of endocytic uptake is likely to be due to specific chemical or physical characteristics of molecular constituents of the plasma membrane. Little is known about specific constituents of type II cell membranes except that they include several MPA-binding proteins [23], a 143-kdalton cell-specific antigen [24], and a high concentration of molecules with alkaline phosphatase activity [25], etc. Biochemical analyses of the apical plasma membrane are likely to reveal other cell-specific molecules but these have not yet been carried out in detail.

Like other epithelia, a second characteristic of type II cells is that ingested materials can be transported through at least two pathways which have different endpoints. One pathway terminates in LBs (from which tracers are presumed to be resecreted and/or degraded) and another pathway terminates by release of material into the interstitial space. Of the tracers studied, only CF enters the transcytotic pathway; most CF which enters type II cells is transported to LBs rather than across the cell. The deposition of tracers into secretory granules, in this case LBs, does not appear to be a general characteristic of endocytic pathways in other epithelial cells and may be specialized for some unique function of type II cells. It has been shown that membrane recycling, after induced secretion, may result in the presence of CF in secretion granules after an intermediate association with Golgi membranes. The direct route to LBs in type II cells is therefore unusual.

Last, the time of course of uptake is similar and the participating organelles identical (up to LBs) to those in cells with well-characterized processes of receptor-mediated endocytosis.

Cell isolation and culture appear to alter the specificity of uptake, the time course, and perhaps the final targetting to LBs by 2 days after plating the cells in monolayer culture. The lectins Con A and RCA-I are preferentially taken up by cultured type II cells; little MPA is taken up on day 2 and none was detected at day 7. These changes in specificity are likely to be the consequence of changes in the chemical composition of the plasma membrane which occur during culture. We have recently shown that type II cells cultured for 2 days and longer express a 42-kdalton (reduced) protein which is specific for type I cells in intact lungs [26]. Type II cells cultured for similar periods lose the ability to bind MPA and bind instead RCA-I [16]. The changes in lectin binding are presumed to be the result in changes in glycoprotein species in the plasma membrane. Together these characteristics suggest that the cells are shifting towards a type I cell-like phenotype as they are maintained in monolayer culture.

It is not clear why very little tracer enters LBs even after a 3-hour exposure but this could be the result of a general sluggishness of endocytic or other metabolic processes, or to the loss of specific molecules required for the intracellular transit, targetting, or fusion events which mediate interactions between organelles.

In summary, these observations on endocytic processes in type II cells cultured in monolayer raise questions about the suitability of these cells for detailed studies of receptor-mediated endocytosis. It will be important to test key experimental findings with studies on intact lungs or less-altered type II cells.

References

1   Mason, R.J.; Williams, M.C.; Widdicombe, J.H.; Sanders, M.J.; Misfeldt, D.S.; Berry, L.C., Jr.: Transepithelial transport by pulmonary alveolar type II cells in primary culture. Proc. natn. Acad. Sci. USA *79:* 6033–6037 (1982).
2   Brown, S.E.S.; Kim, K.J.; Goodman, B.E.; Wells, J.R.; Crandall, E.D.: Sodium-amino acid cotransport by type II alveolar epithelial cells. J. appl. Physiol. *59:* 1616–1622 (1985).
3   Sano, K.; Voelker, D.R.; Mason, R.J.: Tetradecanoylphorbol acetate and terbutaline stimulate surfactant secretion in alveolar type II cells without changing the membrane potential. Biochim. biophys. Acta *902:* 317–326 (1987).

4 Sugahara, J.; Voelker, D.R.; Mason, R.J.: Insulin stimulates amino acid transport by alveolar type II epithelial cells in primary culture. Am. Rev. resp. Dis. *135:* 617–621 (1987).

5 Chauncey, J.B.; Peters-Golden, M.; Simon, R.H.: Arachidonic acid metabolism by rat alveolar epithelial cells. Lab. Invest. *58:* 133–140 (1988).

6 Dobbs, L.G.; Mason, R.J.: Pulmonary alveolar type II cells isolated from rats: release of phosphatidylcholine in response to beta-adrenergic stimulation. J. clin. Invest. *63:* 378–387 (1979).

7 Brown, L.A.S.; Longmore, W.J.: Adrenergic and cholinergic regulation of lung surfactant secretion in the isolated perfused rat lung and in alveolar type II cells in culture. J. biol. Chem. *256:* 66–72 (1981).

8 Mettler, N.R.; Gray, M.E.; Schuffman, S.; Lequire, V.S.: Beta-adrenergic induced synthesis and secretion of phosphatidylcholine by isolated pulmonary alveolar type II cells. Lab. Invest. *45:* 575–586 (1981).

9 Dobbs, L.G.; Wright, J.R.; Hawgood, S.; Gonzalez, R.; Venstrom, K.; Nellenbogen, J.: Pulmonary surfactant and its components inhibit secretion of phosphatidylcholine from cultured rat alveolar type II cells. Proc. natn. Acad. Sci. USA *84:* 1010–1014 (1987).

10 Pian, M.S.; Dobbs, L.G.; Duzgunes, N.: Positive correlation between cytosolic free calcium and surfactant secretion in cultured rat alveolar type II cells. Biochim. biophys. Acta *960:* 43–53 (1988).

11 Sano, K.; Cott, G.R.; Voelker, D.R.; Mason, R.J.: The Na^+/H^+ antiporter in rat alveolar type II cells and its role in stimulated surfactant secretion. Biochim. biophys. Acta *939:* 449–458 (1988).

12 Williams, M.C.: Endocytosis in alveolar type II cells. Effect of charge and size of tracers. Proc. natn. Acad. Sci. USA *81:* 6054–6058 (1984).

13 Williams, M.C.: Uptake of lectins by pulmonary type II cells. Subsequent deposition into lamellar bodies. Proc. natn. Acad. Sci. USA *81:* 6383–6387 (1984).

14 Fisher, A.B.; Chander, A.; Reicherter, J.: Uptake and degradation of natural surfactant by isolated rat granular pneumocytes. Am. J. Physiol. *253:* C792–C796 (1987).

15 Brandt, A.E.: Cell surface saccharides of rat lung alveolar type I and type 2 cells (Abstract). Fed. Proc. *41:* 755 (1982).

16 Dobbs, L.G.; Williams, M.C.; Brandt, A.E.: Changes in biochemical characteristics and pattern of lectin binding of alveolar type II cells with time in culture. Biochim. biophys. Acta *846:* 155–166 (1985).

17 Williams, M.C.: Conversion of lamellar body membranes into tubular myelin in alveoli of fetal rat lungs. J. Cell Biol. *72:* 260–277 (1977).

18 Hubbard, A.L.: Receptor-mediated endocytosis of asialoglycoproteins in the hepatocyte. Ciba Fdn Symp. *92:* 109–112 (1982).

19 Goldstein, J.L.; Brown, M.S.; Anderson, R.G.W.; Russell, D.W.; Schneider, W.J.: Receptor-mediated endocytosis: concepts emerging from the LDL receptor system. A. Rev. Cell Biol. *1:* 1–39 (1985).

20 Dobbs, L.G.; Geppert, E.F.; Williams, M.C.; Greenleaf, R.D.; Mason, R.J.: Metabolic properties and ultrastructure of alveolar type II cells isolated with elastase. Biochim. biophys. Acta *618:* 510–523 (1980).

21   Dobbs, L.G.; Gonzalez, R.; Williams, M.C.: An improved method for isolating type
     II cells in high yield and purity. Am. Rev. resp. Dis. *134:* 141–145 (1986).
22   Stahl, P.; Schwartz, A.L.: Receptor-mediated endocytosis. J. clin. Invest. *77:* 657,
     662 (1986).
23   Lwebuga-Mukasa, J.: Characterization of polyclonal antibodies to *Maclura pomifera*
     agglutinin (MPA) and *Ricinus communis* agglutin I (RCAI) receptors of rat lungs
     (Abstract). Am. Rev. resp. Dis. *137:* 277 (1988).
24   Funkhouser, J.D.; Cheshire, L.B.; Ferrara, F.B.; Peterson, R.D.A.: Monoclonal anti-
     body identification of a type II alveolar epithelial cell antigen and expression of the
     antigen during lung development. Devl Biol. *119:* 190–198 (1987).
25   Edelson, J.D.; Shannon, J.M.; Voelker, D.R.; Mason, R.J.: Differentiation of adult
     human alveolar type II cells is modulated by cell-matrix interactions (Abstract). Am.
     Rev. resp. Dis. *137:* 280 (1988).
26   Dobbs, L.G.; Williams, M.C.; Gonzalez, R.: Monoclonal antibodies specific to api-
     cal surfaces of rat alveolar type I cells bind to surfaces of cultured, but not freshly
     isolated, type II cells. Biochim. biophys. Acta *970:* 146–156 (1988).

Mary C. Williams, PhD, Department of Anatomy, School of Medicine,
University of California, San Francisco, CA 94143 (USA)

Wichert P von, Müller B (eds): Basic Research on Lung Surfactant.
Prog Respir Res. Basel, Karger, 1990, vol 25, pp 91–95

Isolation of a Low Molecular Weight Hydrophobic Surfactant Apoprotein and Its Effects on the Formation of Tubular Myelin

Yasuhiro Suzuki, Kumiko Kogishi, Yoko Fujita

Department of Pathology, Chest Disease Research Institute, Kyoto University, Kyoto, Japan

Tubular myelin (TM) is a unique structure in the lung and is reported to be formed from lamellar bodies [1]. Calcium ions and one of the surfactant apoproteins having a nominal molecular weight (MW) of 35 kdaltons (SP35) reportedly have some relationship for the formation of that structure [2] but no detailed studies have been done on the mechanism of TM formation. Recently, hydrophobic low MW proteins, important for the expression of high surface activity, were isolated from pulmonary surfactant. One of these proteins (SP15) we reported previously has similar properties and localizes in the lamellar bodies [3]. In this report, we investigated the effects of SP15 on the formation of TM by reconstituting it from synthetic lipids and proteins in vitro.

Materials and Methods

Pig surface-active material (SAM) was obtained by sucrose density gradient ultracentrifugation and lyophilized. SAM was dissolved in 5% Triton X-100 and SP15 was isolated by a method described elsewhere [4]. It was further purified by an affinity column using a specific monoclonal antibody [3]. SP35 was extracted with 5 mM borate buffer (pH 10) from the precipitates of 5% Triton X-100 extraction and further purified by a DEAE-cellulose column.

Reconstitution of lipid-protein complexes is as follows: dipalmitoylphosphatidylcholine (DPPC) and phosphatidylglycerol from egg lecithin (EPG) (2:1 by weight) were mixed with SP15 and solvents were evaporated. The residue was solubilized with 100 mM octylglucopyranoside and dialyzed against 10 mM Tris-HCl buffer (pH 7.4) containing

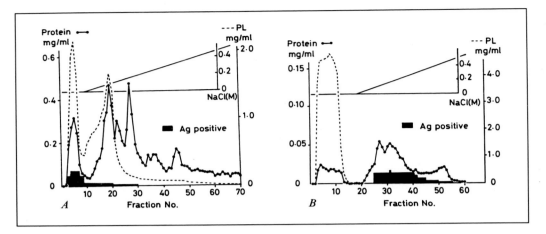

Fig. 1. Elution profile of proteins (solid line) and phospholipids (dashed line) by DEAE-cellulose column *(A)* and by CM-cellulose column *(B)*. Black rectangles indicate the fractions that reacted with a specific monoclonal antibody to pig SP15 [3]. Proteins are eluted with a continuous NaCl gradient.

0.9% NaCl and 1 mM EDTA for 3 days at 4 °C. White amorphous materials formed were precipitated by centrifugation and incubated with various concentrations of SP35 with or without Ca at 37 °C for 5 h. The lipid-protein complex was precipitated and fixed with 2.5% glutaraldehyde in 0.1 M phosphate buffer (pH 7.2) containing 1% tannic acid at 4 °C overnight. It was dehydrated with a graded ethanol series after postfixation with 1% OsO$_4$ for 2 h and embedded in Lubeak 812. Thin sections were cut with a LKB ultramicrotome and examined with a JEOL 100CX electron microscope after staining with uranyl acetate and lead citrate.

Protein and phosphorus assays and SDS-polyacrylamide gel electrophoresis were done by methods described elsewhere [4].

Results and Discussion

SP15 was eluted from a DEAE-cellulose column as a breakthrough fraction (fig. 1A) and eluted with a concentration of 0.1–0.3 M NaCl in a CM-cellulose column (fig. 1B). The eluate from the CM-cellulose column contained a broad band at 15 kdaltons and a minor band at 10 kdaltons, and the latter was removed by using an affinity column with a specific antibody. Amino acid analysis revealed that SP15 corresponded to pig surfactant protein type 1 reported by Curstedt et al. [5].

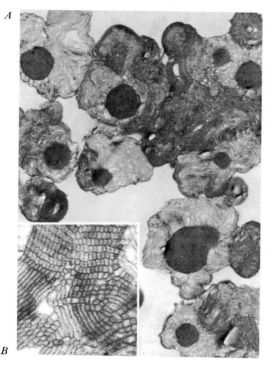

Fig. 2. A lipid-protein complex reconstituted from DPPC, EPG, SP15 and SP35. The DPPC to EPG ratio was 2:1, lipids to SP15 ratio was 5:1 by weight, the concentration of SP35 was 0.22 mg/ml, and calcium concentration was 10 mM. $A \times 9,000$; $B \times 36,000$.

As shown in figure 2, reconstituted material (lipids to SP15, 5:1; concentration of SP35, 0.22 mg/ml; 10 mM CaCl$_2$) showed well-formed lattice structures in several places together with parallel membranes. At higher magnification, these lattices are associated with one to several 8-nm particles, which closely simulate those presented by Williams [1]. Without SP15, no lattice structures were formed but the lipid membranes were coated with similar particles. The results suggest that these particles are derived from SP35. The amount of lattice formed was proportionally related to the concentrations of SP35 and 8.5% of total area showed lattices when concentrations of SP35 were above 0.5 mg/ml. However, at these concentrations, the lattice structures were not rectangular but hexagonal (fig. 3). In the presence of EDTA, no lattices were formed.

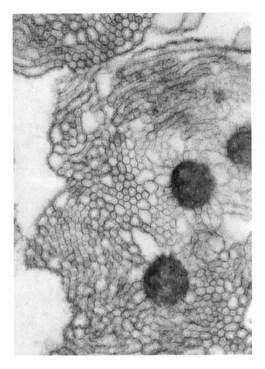

Fig. 3. Structures obtained by incubation of lipid-SP15 complex with 0.9 mg/ml of SP35. Other conditions were the same as in figure 2. Lattice structures are hexagonal. × 36,000.

From these results, it is concluded that SP35 may be a regulatory protein for the formation of TM and SP15 may be directly involved in forming membrane crossings. Membrane fusion may be one of the mechanisms of the membrane crossings, since Schiffer et al. [6] showed that SP5-18 induced fusion of egg PG vesicles in vitro.

References

1   Williams, M.C.: Conversion of lamellar body membranes into tubular myelin in alveoli of fetal rat lungs. J. Cell Biol. *72:* 260–277 (1977).
2   Hook, G.E.R.; Gilmore, L.B.; Talley, F.A.: Dissolution and reassembly of tubular myelin-like multilamellated structures from the lungs of patients with pulmonary alveolar proteinosis. Lab. Invest. *55:* 194–208 (1986).

3   Suzuki, Y.; Kogishi, K.; Fujita, Y.; Kina, T.; Nishikawa, S.: A monoclonal antibody
    to the 15,000 dalton protein associated with porcine pulmonary surfactant. Exp.
    Lung Res. *11*: 61–73 (1986).
4   Kogishi, K.; Kurozumi, M.; Fujita, Y.; Murayama, T.; Kuze, F.; Suzuki, Y.: Isola-
    tion and partial characterization of human low molecular weight protein associated
    with pulmonary surfactant. Am. Rev. resp. Dis. *137*: 1426–1431 (1988).
5   Curstedt, T.; Johansson, J.; Barros-Soderling, J.; Robertson, B.; Nilsson, G.; West-
    berg, M.; Jornvall, H.: Low-molecular-mass surfactant protein type 1. The primary
    structure of a hydrophobic 8-kDa polypeptide with eight half-cystine residues. Eur.
    J. Biochem. *172*: 521–525 (1988).
6   Schiffer, K.; Hawgood, S.; Duzgunes, N.; Goerke, J.: Interaction of the low molecu-
    lar weight group of surfactant-associated proteins (SP5-18) with pulmonary surfac-
    tant lipids. Biochemistry *27*: 2689–2695 (1988).

Dr. Yasuhiro Suzuki, Department of Pathology, Chest Disease Research Institute,
Kyoto University, Sakyo-ku, Kyoto 606 (Japan)

Surfactant Synthesis and Secretion

Wichert P von, Müller B (eds): Basic Research on Lung Surfactant.
Prog Respir Res. Basel, Karger, 1990, vol 25, pp 96–103

Phosphatidylcholine Synthesis in Type II Cells and Regulation of the Fatty Acid Supply

J.J. Batenburg, J.N. den Breejen, M.J.H. Geelen, C. Bijleveld, L.M.G. van Golde

Laboratory of Veterinary Biochemistry, Utrecht University, Utrecht, The Netherlands

Pulmonary surfactant, required to lower the surface tension at the alveolar surface in order to prevent collapse of the alveoli during expiration [7, 8], is synthesized by the alveolar type II epithelial cells [5, 8]. It consists for the greatest part of phospholipids, among which phosphatidylcholine (PC) makes up 70–80% [10]. Of the total PC, approximately 60% is the dipalmitoyl species. This dipalmitoyl-PC (DPPC) is the major surface-active component of pulmonary surfactant [10]. The pathways of surfactant lipid formation and the perinatal development and hormonal regulation of the enzymes required for these pathways have received a lot of attention during the past 15 years. This paper will focus on our recent studies of the pathways involved in DPPC synthesis in type II cells. In addition, data are presented concerning the development and regulation of enzymes involved in the supply of fatty acids for phospholipid formation in the perinatal lung.

Pathways Involved in Dipalmitoyl Phosphatidylcholine Synthesis

In the type II cell the synthesis de novo of PC proceeds via the CDPcholine pathway [2, 18, 22]. The last step in this pathway is the formation of PC from CDPcholine and diacylglycerol by the enzyme cholinephosphotransferase. The diacylglycerol is formed from phosphatidic acid

(PA) by PA phosphatase. PA in turn can be synthesized from glycerol-3-phosphate by stepwise acylation. This process is catalyzed by glycerol-phosphate acyltransferase and 1-acylglycerolphosphate acyltransferase. Many studies have shown that direct synthesis de novo cannot be responsible for the formation of all surfactant DPPC and that part of the DPPC is formed by remodeling of unsaturated PC synthesized de novo [2, 18, 22]. However, the relative importance of the direct synthesis de novo and the route involving remodeling for the production of DPPC in type II cells is unknown. Recent studies on the substrate specificity of glycerol-3-phosphate acylation in our laboratory indicated that type II cell microsomes are capable of synthesizing the dipalmitoyl species of PA, but it could also be concluded that the percentage dipalmitoyl species in the synthesized PA, and thereby that in the PC synthesized de novo, will probably depend on the relative availability of the various acyl-CoA species [3]. Therefore, we decided to determine the acyl-CoA composition of microsomes and cytosol from alveolar type II cells and to study which species of PA were formed by the type II cell microsomes in incubations with exogenous acyl-CoA species mixed in the proportions in which they were found in this cell fraction.

Microsomes and cytosol were prepared from type II cells isolated from adult rat lung. Acyl-CoAs were isolated from the cell fractions, converted to fatty acid methyl esters and analyzed by gas chromatography. In the microsomes we found 49% palmitoyl-CoA, 2% myristoyl-CoA, 21% stearoyl-CoA, 5% palmitoleoyl-CoA, 16% oleoyl-CoA, 5% linoleoyl-CoA and 2% arachidonoyl-CoA. The acyl-CoA composition of the cytosol was practically the same. Subsequently, we incubated type II cell microsomes with acyl-CoA species mixed in these proportions and with [U-^{14}C]glycerol-3-phosphate. The result is shown in table 1. It can be seen that at all acyl-CoA concentrations added, about 40% of the label was found in the disaturated phosphatidic acid species. Subsequent studies in which the palmitoyl-CoA and the stearoyl-CoA in the acyl-CoA mixture were radioactively labeled with ^{3}H and ^{14}C, respectively, showed that upon their utilization in the acylation of glycerol 3-phosphate the palmitoyl and stearoyl moieties were distributed in practically the same way among the various species of PA synthesized. However, palmitoyl residues were used 4 times as fast as stearoyl residues for the formation of both total and disaturated PA. On the basis of its availability myristoyl-CoA can be expected to contribute only to a minor extent to the formation of disaturated PA. Therefore, the chance that one of the acyl groups inserted into disaturated PA is a palmi-

Table 1. Distribution of [^{14}C]glycerol-3-phosphate radioactivity among the species of phosphatidic acid synthesized by microsomes from isolated type II cells

Acyl-CoA concentration	Label in phosphatidic acid species, %				
	Δ0	Δ1	Δ2	Δ>2	
66 μM	35.1±0.05	37.7±7.7	12.2±0.1	14.9±7.8	(2)
33 μM	44.5±0.9	40.1±1.5	9.6±0.9	5.8±0.8	(4)
16.5 μM	42.0±1.9	39.1±1.5	10.8±0.5	8.1±0.7	(6)
8.3 μM	39.5±2.0	33.8±1.1	13.3±0.7	13.1±3.2	(3)

Microsomes from adult rat type II cells were incubated with [U-^{14}C]glycerol-3-phosphate and acyl-CoAs mixed in the proportions in which they were found in type II cell microsomes (see text). Values represent mean ± SE for the number of determinations indicated in parentheses.

toyl group is approximately 4/5 and the chance that both acyl groups are palmitoyl groups is 16/25. Hence, it can be calculated that of the PA molecules formed by type II cell microsomes approximately 16/25 × 40 = 26% are dipalmitoyl-PA. Results from earlier studies with microsomes from whole lung indicated that the PA phosphatase involved in PC formation has very little specificity with regard to the various molecular species of its substrate [20]. From studies with whole lung and isolated type II cells the same conclusion was reached for cholinephosphotransferase [22]. If PA phosphatase and cholinephosphotransferase do not discriminate against certain substrate species, the species composition of PC molecules synthesized de novo will be a reflection of the species composition of the synthesized PA and will, therefore, contain 26% of the dipalmitoyl species. Since about 60% of the PC in surfactant is constituted by the dipalmitoyl species [10], this would mean that about 45% of the surfactant DPPC could be made via synthesis de novo.

Developmental Pattern of Enzymes Involved in Fatty Acid Synthesis

Towards the end of gestation the production of surfactant phospholipids increases dramatically [15, 18, 22]. There are indications that during prenatal development de novo fatty acid synthesis is the primary supplier

of palmitate for the formation of these phospholipids [12]. In slices of
fetal rat lung the synthesis of fatty acids as estimated from the incorporation of ^3H_2O was found to peak at days 21 and 22 of gestation [11, 19]
(term is at day 22). The developmental profile of fatty acid synthase in
whole fetal rat lung was reported to be similar in that it peaked at day 21
of gestation [13]. There is disagreement in the literature on the developmental profile of acetyl-CoA carboxylase in perinatal rat lung tissue.
While one group [11] observed a peak in the activity of this enzyme on
the last day of gestation, a gradual decrease in its activity throughout the
perinatal period was reported by others [13]. Both groups used an assay
in which $^{14}CO_2$ was incorporated into acid-stable product presumed to
be malonyl-CoA. However, in crude extracts other carboxylations may
also take place. Therefore, we decided to reinvestigate the developmental
profile of acetyl-CoA carboxylase in perinatal rat lung using an assay in
which the enzyme converted [1-^{14}C]acetyl-CoA into [^{14}C]malonyl-CoA,
which was converted into labeled fatty acid by an excess of purified fatty
acid synthase added to the incubation medium [6]. Figure 1 shows that
in our hands not only fatty acid synthase, but also acetyl-CoA carboxylase shows a peak in activity in whole fetal rat lung at day 21 of gestation.

Patterson et al. [12] showed that in explants of fetal rat lung tissue
fatty acid synthesis is highly dependent on the activity of ATP-citrate
lyase. To our knowledge, the developmental profile of this lipogenic
enzyme in perinatal rat lung has not yet been reported. From figure 1 it can
be seen that the profile of this enzyme is quite different from that of fatty
acid synthase and acetyl-CoA carboxylase, with a two-fold increase in
activity immediately after birth, followed by a decline.

There is abundant evidence that the rise in fetal plasma glucocorticoids at the end of gestation is involved in the acceleration of surfactant
synthesis by the fetal lung in that period [1, 21]. Glucocorticoids have been
found to increase fatty acid synthesis in fetal rat lung in vivo [19] and to
augment the specific activity [13] and amount [14] of fatty acid synthase in
cultured fetal rat lung. The observation that the peak in fatty acid synthase
in fetal rat lung at day 21 of gestation is accompanied by a similar peak in
the activity of acetyl-CoA carboxylase may indicate that not only the activity of fatty acid synthase but also that of acetyl-CoA carboxylase is regulated by glucocorticoids. The observation that the developmental pattern
of ATP-citrate lyase is completely different suggests that its activity may be
regulated by other hormone(s).

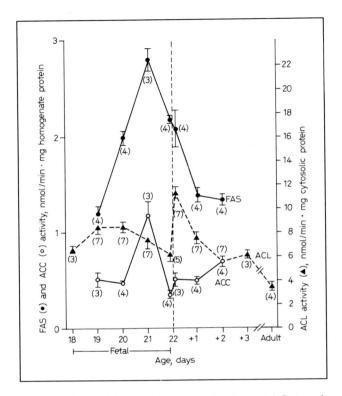

Fig. 1. Activity of fatty acid synthase (FAS), acetyl-CoA carboxylase (ACC) and ATP-citrate lyase (ACL) in perinatal rat lung as a function of development. FAS and ACC were measured in homogenates. ACL was measured in 10^5 g supernatants from homogenates. Data are mean $\pm$ SE (bar) from the number of litters indicated in parentheses.

Regulation of Fatty Acid Synthesis in Fetal Type II Cells

To our knowledge the stimulatory effect of glucocorticoid on fatty acid synthase observed in fetal rat [13, 14] and human [9] lung in explant culture has not yet been shown in isolated fetal type II cells and has not yet been shown to be accompanied by an increase in fatty acid synthesis. Recently, we have described the isolation of alveolar type II cells from fetal rat lung by differential adherence in monolayer culture [4]. We decided to use these cells to study the possible effects of cortisol on fatty acid synthesis.

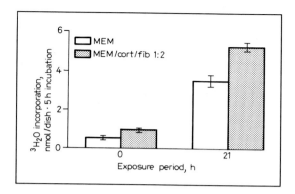

Fig. 2. Effect of fibroblast-conditioned cortisol-containing medium on fatty acid synthesis by type II cells isolated from fetal rat lung. After isolation of the type II cells by differential adherence [4] their rate of ^3H_2O incorporation into fatty acids during a 5-hour period was measured in control medium (MEM) and in $10^{-7}\,M$ cortisol-containing medium which had been conditioned by exposure to fibroblasts for 24 h (MEM/cort/fib). In part of the experiment (shown on the right-hand side of the figure) the cells were preincubated for 21 h in the medium in which they were assayed for their rate of fatty acid synthesis. Data are mean ± SE (bar) from incubations in fourfold.

There is considerable evidence that the stimulation of disaturated PC synthesis in fetal type II cells by corticoid is mediated by a fibroblast-pneumocyte factor (FPF) which is induced in lung fibroblasts when these cells are exposed to corticoid [21, 22]. In agreement with earlier work by others [17], we observed that in type II cells isolated by our method from fetal rat lungs at day 19 of gestation, [Me-^3H]choline incorporation into disaturated PC was not stimulated by inclusion of cortisol ($10^{-7}\,M$) in the culture medium (minimal essential medium; MEM), but was stimulated by cortisol-containing medium which had previously been conditioned by exposure to fetal lung fibroblasts for 24 h. Therefore, we started the studies on the possible effect of cortisol on fatty acid synthesis by conditioning the cortisol-containing medium with fibroblasts. Figure 2 shows results of an experiment in which fatty acid synthesis in fetal type II cells was assayed by measuring the incorporation of ^3H_2O. It is clear that the rate of fatty acid synthesis was considerably increased by culturing type II cells in MEM for 21 h after the 21-hour adherence period. Both when a 21-hour pre-exposure period was employed and without prior exposure of the cells, ^3H_2O incorporation into fatty acids during a 5-hour period was greater in the fibroblast-conditioned cortisol-containing medium than in the control medium. Similar effects were noted when fatty acid synthase activity was measured (not shown). At this moment we do not know whether cortisol and fibroblast conditioning were both necessary for the effect of fibroblast-conditioned cortisol-containing medium on fatty acid synthesis in the fetal type II cells. However, on the basis of the observations described above we speculate that in type II cells glucocorticoids, via the action of FPF, do not only increase the activity of phosphocholine cytidylyltransferase [16] but also the rate of fatty acid synthesis.

Acknowledgements

The authors are much indebted to Mr. R.H. Elfring and Mrs. M.G.J. Schmitz for excellent technical assistance. The investigations described in this paper were supported by the Netherlands Foundation for Chemical Research (SON) with financial aid from the Netherlands Organization for Scientific Research (NWO).

References

1   Ballard, P.L.: Hormonal aspects of fetal lung development; in Farrell, Lung development: biological and clinical perspectives, vol. II, pp. 205–253 (Academic Press, New York 1982).
2   Batenburg, J.J.: Biosynthesis and secretion of pulmonary surfactant; in Robertson, Van Golde, Batenburg, Pulmonary surfactant, pp. 237–270 (Elsevier, Amsterdam 1984).
3   Batenburg, J.J.; Den Breejen, J.N.; Yost, R.W.; Haagsman, H.P.; Van Golde, L.M.G.: Glycerol 3-phosphate acylation in microsomes of type II cells isolated from adult rat lung. Biochim. biophys. Acta *878:* 301–309 (1986).
4   Batenburg, J.J.; Otto-Verberne, C.J.M.; Ten Have-Opbroek, A.A.W.; Klazinga, W.: Isolation of alveolar type II cells from fetal rat lung by differential adherence in monolayer culture. Biochim. biophys. Acta *960:* 441–453 (1988).
5   Batenburg, J.J.; Van Golde, L.M.G.: Formation of pulmonary surfactant in whole lung and in isolated type II alveolar cells; in Scarpelli, Cosmi, Reviews in perinatal medicine, vol. 3, pp. 73–114 (Raven Press, New York 1979).
6   Bijleveld, C.; Geelen, M.J.H.: Measurement of acetyl-CoA carboxylase activity in isolated hepatocytes. Biochim. biophys. Acta *918:* 274–283 (1987).
7   Clements, J.A.: Function of the alveolar lining. Am. Rev. resp. Dis. *115:* suppl., pp. 67–71 (1977).
8   Goerke, J.: Lung surfactant. Biochim. biophys. Acta *344:* 241–261 (1974).
9   Gonzales, L.; Ballard, P.L.; Ertsey, R.; Froh, D.K.: Effect of dexamethasone (dex) and cAMP on fatty acid synthetase in human fetal lung explants. FASEB J. *2:* A492 (1988).

10   King, R.J.; Clements, J.A.: Surface active materials from dog lung. II. Composition and physiological correlations. Am. J. Physiol. *223:* 715–726 (1972).

11   Maniscalco, W.M.; Finkelstein, J.N.; Parkhurst, A.B.: De novo fatty acid synthesis in developing rat lung. Biochim. biophys. Acta *711:* 49–58 (1982).

12   Patterson, C.E.; Davis, K.S.; Rhoades, R.A.: Regulation of fetal lung disaturated phosphatidylcholine synthesis by de novo palmitate supply. Biochim. biophys. Acta *958:* 60–69 (1988).

13   Pope, T.S.; Rooney, S.A.: Effects of glucocorticoid and thyroid hormones on regulatory enzymes of fatty acid synthesis and glycogen metabolism in developing fetal rat lung. Biochim. biophys. Acta *918:* 141–148 (1987).

14   Pope, T.S.; Smart, D.A.; Rooney, S.A.: Hormonal effects on fatty-acid synthase in cultured fetal rat lung; induction by dexamethasone and inhibition of activity by triiodothyronine. Biochim. biophys. Acta *959:* 169–177 (1988).

15   Possmayer, F.: Biochemistry of pulmonary surfactant during fetal development and in the perinatal period; in Robertson, Van Golde, Batenburg, Pulmonary surfactant, pp. 295–355 (Elsevier, Amsterdam 1984).

16   Post, M.; Barsoumian, A.; Smith, B.T.: The cellular mechanism of glucocorticoid acceleration of fetal lung maturation. Fibroblast-pneumonocyte factor stimulates cholinephosphate cytidylyltransferase activity. J. biol. Chem. *261:* 2179–2184 (1986).

17   Post, M.; Torday, J.S.; Smith, B.T.: Alveolar type II cells isolated from fetal rat lung organotypic cultures synthesize and secrete surfactant-associated phospholipids and respond to fibroblast-pneumonocyte factor. Exp. Lung Res. *7:* 53–65 (1984).

18   Rooney, S.A.: The surfactant system and lung phospholipid biochemistry. Am. Rev. resp. Dis. *131:* 439–460 (1985).

19   Rooney, S.A.; Gobran, L.I.; Chu, A.J.: Thyroid hormone opposes some glucocorticoid effects on glycogen content and lipid synthesis in developing fetal rat lung. Pediat. Res. *20:* 545–550 (1986).

20   Rüstow, B.; Kunze, D.; Rabe, H.; Reichmann, G.: The molecular species of phosphatidic acid, diacylglycerol and phosphatidylcholine synthesized from *sn*-glycerol 3-phosphate in rat lung microsomes. Biochim. biophys. Acta *835:* 465–476 (1985).

21   Smith, B.T.: Pulmonary surfactant during fetal development and neonatal adaptation: hormonal control; in Robertson, Van Golde, Batenburg, Pulmonary surfactant, pp. 357–381 (Elsevier, Amsterdam 1984).

22   Van Golde, L.M.G.; Batenburg, J.J.; Robertson, B.: The pulmonary surfactant system: biochemical aspects and functional significance. Physiol. Rev. *68:* 374–455 (1988).

J.J. Batenburg, PhD, Laboratory of Veterinary Biochemistry, Utrecht University, PO Box 80.176, NL–3508 TD Utrecht (The Netherlands)

Wichert P von, Müller B (eds): Basic Research on Lung Surfactant.
Prog Respir Res. Basel, Karger, 1990, vol 25, pp 104–108

Surfactant and Non-Surfactant Phosphatidylcholine Biosynthesis in Cultivated Fetal Rat Alveolar Type II Cells: Effect of Growth Factors

Jacques Bourbon, Edith Doucet, Michel Rieutort, Léa Marin

Centre de Biologie Cellulaire, CNRS, Ivry, France

The incorporation of labeled precursors into phospholipids has been largely used as an index of lung surfactant biosynthesis, in vivo as well as in vitro. This approach, however, is not specific of surfactant, since the same molecular species are present in membranes, even if some phospholipids are predominant in surfactant.

In this report, we present an adaptation of a method previously designed for isolating surfactant from small amounts of lung tissue [1] in order to determine the incorporation of a labeled precursor into surfactant of cultivated fetal alveolar type II cells.

Epidermal growth factor (EGF) has been shown to accelerate fetal type II cell maturation in vivo [2, 3] and in explant culture [4]. The similar increase of saturated and unsaturated phosphatidylcholines (PCs) induced by EGF in cells derived from fetal rat lung [5] suggested, however, that this effect could be unspecific of surfactant. We have used type II cell culture and the method of surfactant extraction to determine if EGF exerted its effects on type II cells themselves, if it specifically stimulated surfactant biosynthesis, and if effects of other growth factors could be evidenced.

Material and Methods

The cell culture system derives from that described by Post et al. [6]. In brief, a cell suspension (25–30·10⁶ cells) was prepared from aseptically dissected lungs of 19.5-day-old rat fetuses (Wistar strain) with collagenase/DNAase and enriched with type II cells by

4-day organotypic culture on collagen matrix pads ('Gelfoam' sponge) in 2% fetal calf serum (FCS)-supplemented Eagle's minimum essential medium (MEM). The type II cells were then recovered by collagenase digestion of the pads and grown to confluence as monolayers in multiwell plates with 10% FCS-supplemented MEM. The medium was then replaced either by unsupplemented MEM (control) or by growth factor-containing MEM (EGF, 100 ng/ml, or fibroblast growth factor = FGF, 10 ng/ml, or bovine brain-derived growth factor = BDGF, 10 units[1]/ml, or platelet-derived growth factor = PDGF, 2 units[2]/ml) and containing either 0.09 μCi of [^{14}C]-thymidine or 1 μCi of [^{3}H]-choline chloride per ml. Incorporation was allowed for 48 h.

For thymidine incorporation, DNA was extracted from lysed cells by 10% TCA, pelleted and counted. For choline incorporation, the cells were scraped in Tris-HCl, EDTA, NaCl buffer pH 7.4, homogenized in the presence of unlabeled lungs from rat fetuses at term used as carrier tissue, and centrifuged twice on discontinuous sucrose gradients as described previously [1]. Lipids were extracted from surfactant (S) and non-surfactant (R=residual) fractions by chloroform:methanol, separated by one-dimensional thin-layer chromatography on silica gel, and PC spots were eluted and counted as described previously [1].

Results and Discussion

The morphological aspect of cells grown on collagen sponges shown in figure 1 was very similar to that of type II pneumocytes in vivo.

The incorporation of [^{3}H]-choline in S-fraction phospholipid runs with 3 different cell pools indicates the reproducibility of the method: 7,355 ± 480 dpm/10^5 cells (n = 7), 5,918 ± 511 (n = 16) and 5,370 ± 1,630 (n = 6), respectively.

The effects of growth factors on precursor incorporations into DNA and PC are summarized in table 1. EGF and BDGF stimulated thymidine incorporation into DNA and choline incorporation into both S and R fractions to a similar extent. FGF somewhat stimulated thymidine incorporation but did not modify choline incorporation. No effect of PDGF was evidenced.

Thus, both EGF and BDGF appear to act directly upon the fetal type II pneumocyte. Surfactant biosynthesis, however, was apparently not spe-

---

[1] A unit is the amount of BDGF which induced a 50% of maximal increase of [^{3}H]-thymidine incorporation in isolated corneal cells in culture.

[2] A unit is the amount of PDGF which induced 50% of maximal increase of [^{3}H]-thymidine incorporation in cultivated fibroblasts from mouse embryo, strain C3H/1OT1/2 (data furnished by the purchaser).

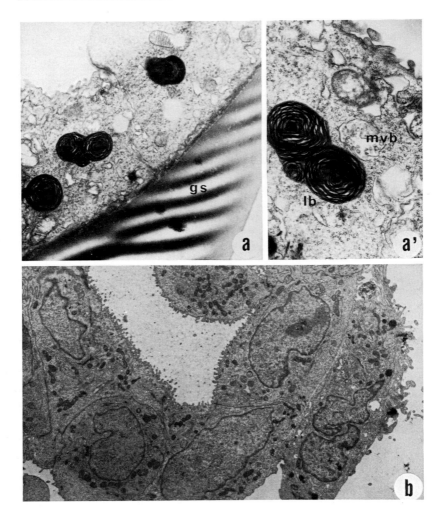

Fig. 1. Morphological aspect in electron microscopy of fetal alveolar type II cells after 4 days of culture as pseudoalveolar structures reconstituted on collagen matrix (Gelfoam sponge). *a* Aspect of the cells adherent to the Gelfoam substrate (gs). The cells appear healthy with numerous lamellar bodies and mitochondria and a developed endoplasmic reticulum. × 8,200. *a'* Higher magnification showing the typical structure of lamellar bodies (lb) and multivesicular bodies (mvb). × 20,500. *b* Pseudoalveolar structure after digestion of the collagen matrix. Epithelial cells appear polarized with typical microvilli facing the lumen of the structure, they are surrounded by a few fibroblastic cells (on the right). × 2,050.

Table 1. Effects of the four growth factors on the incorporation of [^{14}C]-thymidine into DNA and of [^{3}H]-choline into PC of surfactant (S) and residual (R) fractions of cultivated type II pneumocytes in 48 h (dpm incorporated in 10^5 cells): mean ± SEM of 9 different culture experiments

	Control medium	EFG	BDGF	FGF	PDGF
[^{14}C]-thymidine	130 ±22	256 ±28[b]	444 ±70[c]	216 ±30[a]	140 ±35
[^{3}H]-choline					
S fraction	2,650 ±249	4,806 ±1,003[a]	5,802 ±1,199[a]	2,093 ±267	3,046 ±655
R fraction	27,736 ±4,442	48,226 ±3,109[a]	50,542 ±9,005[a]	25,037 ±3,109	26,225 ±5,148

Significant difference as compared with control medium (t test for paired values) for: [a] p < 0.05; [b] p < 0.01; [c] p < 0.005.

cifically stimulated; its increase seems to reflect an overall stimulation of growth and phospholipid biosynthesis of the cells. EGF could, however, have additional indirect effects in vivo, since it presents a corticotropin-releasing activity in the fetus [7].

The absence of effects of FGF and PDGF is surprising since, on the one hand, FGF and BDGF are related molecules, and on the other PDGF has previously been reported to enhance the growth of type II cells [8]. The source of the factors could be critical, as well as the status of the cells in each study since PDGF is a competence factor whereas EGF is a progression factor.

Summary

In order to discriminate between the biosyntheses of surfactant and non-surfactant phospholipids, we determined the incorporation of labeled choline by cultivated alveolar type II cells into surfactant and extra-surfactant fractions separated by ultracentrifugation. Two growth factors, EGF and BDGF increased the incorporation of the precursor in both fractions, which suggests that they exerted an overall stimulation of cell growth rather than a specific enhancement of surfactant biosynthesis.

References

1   Rieutort M, Farrell PM, Engle MJ, et al: Changes in surfactant phospholipids in fetal rat lungs from normal and diabetic pregnancies. Pediatr Res 1986;20:650–654.

2   Catterton WZ, Escobedo MB, Sexson WR, et al: Effect of epidermal growth factor on lung maturation in fetal rabbits. Pediatr Res 1979;13:104–108.

3   Sundell HW, Serenius FS, Escobedo MB, et al: Effects of epidermal growth factor on lung maturation in fetal lamb. Am J Pathol 1980;100:707–726.

4   Gross I, Dynia DW, Rooney SA, et al: Influence of epidermal growth factor on fetal rat lung development in vitro. Pediatr Res 1986;20:473–477.

5   Leheup BP, Gray ME, Stahlman MT, et al: Synergistic effect of epidermal growth and retinoic acid on lung phospholipid synthesis. Pediatr Res 1983;17:381A.

6   Post M, Torday JS, Smith BT: Alveolar type II cells isolated from fetal lung organotypic cultures synthesize and secrete surfactant-associated phospholipids and respond to fibroblast-pneumocyte factor. Exp Lung Res 1984;7:53–65.

7   Polk DH, Ervin MG, Padbury JF, et al: Epidermal growth factor acts as a corticotropin-releasing factor in chronically catheterized fetal lambs. J Clin Invest 1987;79: 984–988.

8   Stiles AD, Smith BT, Post M: Reciprocal autocrine and paracrine regulation of growth of mesenchymal and alveolar epithelial cells from fetal lung. Exp Lung Res 1986;11:165–177.

Jacques Bourbon, MD, Centre de Biologie Cellulaire, CNRS,
67, rue Maurice-Günsbourg, F–94205 Ivry-Seine (France)

Wichert P von, Müller B (eds): Basic Research on Lung Surfactant.
Prog Respir Res. Basel, Karger, 1990, vol 25, pp 109–114

Hormonal Control of Surfactant Synthesis

Barry T. Smith, Martin Post

Division of Neonatology, Department of Paediatrics and Research Institute,
The Hospital for Sick Children, Toronto, Canada

Physiologic preparation of the fetal lung for birth involves hormonal signals which quantitatively increase type II cell surfactant synthesis. In some pathological states, other hormonal signals are down-regulators and delay lung maturation as defined by surfactant synthesis. In this chapter, our consideration of hormonal regulation of surfactant synthesis will only address surfactant lipid (primarily saturated phosphatidylcholine; PC) synthesis. Although, as noted elsewhere in this volume, the role of surfactant-associated proteins is receiving increased attention, relatively little is yet known regarding the physiologic hormonal regulation of these important surfactant constituents in the fetal lung. This remains an area of important endeavour.

Glucocorticoids

Glucocorticoids play a central role in signalling the onset of surfactant synthesis in the fetal lung [1]. Briefly, endogenous fetal glucocorticoids (largely in the 11-oxidized form, e.g. cortisone) increase in the fetal circulation in late gestation [2]. Within the lung (indeed, in the fetal lung fibroblast [3]), they are reduced at the eleventh carbon to their biologically active form [4], and interact with glucocorticoid receptors in the fetal lung [5]. This complex interaction results in enhanced surfactant lipid synthesis [1] in preparation for air breathing after birth. The physiologic relevance of these mechanisms has been demonstrated by experiments in which chemically [6] or surgically [7] induced hypoadrenalism, blockade of activation by 11-reductase activity [8], or inhibition of receptor binding [9] delay fetal lung maturation as defined by surfactant synthesis.

Based on the above observations, it is now known that administration of synthetic glucocorticoids to the mother in premature labour can precociously induce surfactant synthesis and reduce the incidence of neonatal respiratory distress syndrome (RDS) in the neonate [10]. While this benefit is statistically significant and use of such therapy is encouraged, the effect is nevertheless modest in terms of a significant residual pool of affected infants despite apparently adequate therapy [10]. The major reasons for non-response appear to include the relatively slow action of antenatal glucocorticoids, a lesser response in male fetuses, and an abrogated response in infants of diabetic mothers.

Thyroid Hormones

Although the evidence is more limited than with glucocorticoids, it appears that thyroid hormones also play a role in the physiologic timing of fetal lung maturation [1]. Circulating tri-iodothyronine (T3) and thyroxine (T4) levels are lower in premature infants with RDS than in unaffected infants of similar gestational age [11] and experimental hypothyroidism in lambs delays lung maturation [12]. The fetal lung contains specific receptors for T3 [13]. Limited and poorly controlled studies have suggested that antenatal administration of T4 may reduce the incidence of RDS [14]. The route of administration was necessitated by the fact that thyroid hormones cross the placenta poorly. In animals, however, thyrotropin-releasing hormone (TRH) does cross the placenta, resulting in increased thyroid hormone levels and accelerated lung maturation [15]. The effect of thyroid hormones on surfactant synthesis in vitro appears to be additive [16] or synergistic [17]. Based on this, two controlled trials, in the United States and New Zealand, are underway to determine if combined administration of TRH and glucocorticoids will improve the response in terms of reduction of the incidence of RDS after prenatal administration of these agents to mothers in premature labour.

Cellular Basis of Hormone Action

As noted above, endogenous (11-oxidized) fetal glucocorticoids are activated within the fetal lung fibroblast to active glucocorticoid [3]. These mesenchymal cells play a further active role in the fetal lung response to glucocorticoid, for, with respect to surfactant lipid synthesis, these hormones do not act directly upon the fetal alveolar type II cell [18]. Instead, glucocorticoids act upon the fetal lung fibroblast to induce production of *fibroblast-pneumonocyte factor* (FPF), which in turn induces surfactant lipid synthesis by the fetal type II cell [18]. FPF is a low molecular weight polypeptide whose production is restricted to the fetal lung [19], but which is of limited species specificity [20]. The potential physiologic role of FPF is demonstrated by its ability to accelerate lung maturation in vivo [21] and by delay of lung maturation in vivo after administration of monoclonal antibodies which recognize FPF and block its biologic activity [22].

The production of FPF by the fetal lung fibroblast in response to glucocorticoid is abrogated by blockers of protein and DNA synthesis [23]. Floros et al. [23] have isolated a fraction of messenger RNA (mRNA) of approximately 400 bases whose primary translation products possess FPF bioactivity. However, for immune recognition (with the antibodies noted above [22]), some post-translational processing appears to be required, as activity of the translation products was only immunologically inhibited after microinjection of the mRNA into *Xenopus oocytes,* not after cell-free translation [23].

The action of FPF on the fetal type II cell is mediated by stimulation of the rate-regulatory enzyme in surfactant PC synthesis, cholinephosphate cytidylyltransferase [24]. This effect is presumably post-translational, since maximum stimulation of enzyme activity is seen within 60 min [24].

The synergistic action of thyroid hormones and glucocorticoids, noted above, may be explained on the basis of this cellular interaction: thyroid hormone alone appears to have little effect on isolated fetal type II cells (and does not stimulate production of FPF by the fibroblast), but appears to enhance the response of the fetal alveolar type II cell to FPF [17].

Clinical Correlates

The indirect action of glucocorticoid on fetal type II cell surfactant lipid synthesis, as mediated by FPF production may underlie some of the clinical limitations of glucocorticoid therapy for the prevention of RDS, including temporal factors, sex differences in response to glucocorticoids, and limited efficacy in infants of diabetic mothers.

Temporal Factors

A major limitation of glucocorticoid acceleration of fetal lung maturation is time: in clinical studies, at least 24 h appears to be required to demonstrate efficacy [10]. Indeed, in the largest clinical study, over 90% of candidates for maternal glucocorticoid therapy were eliminated from study because delivery was expected within 24 h [10]. As noted above, FPF stimulates surfactant lipid synthesis in the type II cell within 60 min [24], and the major portion of the delayed response to glucocorticoid appears to be the time required for synthesis of FPF by the fetal lung fibroblast [23].

Sex Differences

Male fetuses demonstrate an increased incidence of RDS and respond less well to antenatal glucocorticoid administration [10]. It appears that fetal androgen levels, required for male phenotypic differentiation, delay the pulmonary mechanisms required for lung maturation [25]. Indeed, androgen inhibits FPF production by the fetal lung fibroblast at a pretranslational level [26].

The Infant of the Diabetic Mother

The infant of the diabetic mother has about a 6-fold increased risk of RDS [27], particularly where the maternal disease is poorly controlled and results in fetal hyperglycemia and fetal hyperinsulinemia. Carlson et al. [28] have shown that insulin inhibits FPF production by the fetal lung fibroblast in response to cortisol.

Future Directions

Despite the low abundance of FPF and its limited species specificity, rendering it a poor immunogen, attempts are underway to isolate a complementary DNA probe for the molecule. This would provide more precise reagents for study of its regulation and action for further study of this phenomenon. In the more distant future, clinical application could accelerate our ability to regulate fetal lung maturation and bypass the site of the defect (FPF production) in the response of the male fetus and the infant of the diabetic mother to glucocorticoids.

References

1 Smith, B.T.: Pulmonary surfactant during fetal development and neonatal adaptation: Hormonal control; in Robertson, Van Golde, Batenburg. Pulmonary surfactant, pp. 357–381 (Elsevier, Amsterdam 1984).

2 Kitterman, J.A.; Liggins, G.C.; Campos, G.A.; Clements, J.A.; Forster, C.S.; Lee, C.H.; Creasy, R.K.: Prepartum maturation of the lung in fetal sheep. Relation to cortisol. J. appl. Physiol. *51:* 384–390 (1981).

3 Torday, J.S.; Post, M.; Smith, B.T.: Compartmentalization of 11-oxidoreductase within the fetal lung alveolus. Am. J. Physiol. *249:* C173–C176 (1985).

4 Smith, B.T.; Torday, J.S.; Giroud, C.J.P.: The growth promoting effect of cortisol on human fetal lung cells. Steroids *22:* 515–524 (1974).

5 Ballard, P.L.; Ballard, R.A.: Cytoplasmic receptor for glucocorticoids in lung of the human fetus and neonate. J. clin. Invest. *53:* 477–486 (1974).

6 Vidyasagar, D.; Chernick, V.: Effect of metopirone on the synthesis of lung surfactant in does and fetal rabbits. Biol. Neonate *27:* 1–16 (1975).

7 Blackburn, W.R.; Travers, H.; Potter, M.: The role of the pituitary-adrenal-thyroid axes in lung differentiation. Lab. Invest. *26:* 306–318 (1972).

8 Smith, B.T.: The role of pulmonary corticosteroid 11-reductase activity in lung maturation in the fetal rat. Pediat. Res. *11:* 412–414 (1977).

9 Guettari, M.; Marin, L.; Bourbon, J.; Dufour, M.E.; Rieutort, M.; Tordet, C.: Effects of the antiglucocorticoid RU486 on the maturation of fetal rat lung surfactant. Exp. Lung Res. (in press).

10 Collaborative Group on Antenatal Steroid Therapy: Effect of antenatal dexamethasone administration on the prevention of respiratory distress syndrome. Am. J. Obstet. Gynec. *141:* 276–287 (1981).

11 Cuestas, R.A.; Lindall, A.; Engel, R.R.: Low thyroid hormones and respiratory distress syndrome of the newborn. New Engl. J. Med. *295:* 297–302 (1976).

12 Erenberg, A.; Rhodes, M.L.; Weinstein, M.M.; Kennedy, R.L.: The effect of fetal thyroidectomy on ovine fetal lung maturation. Pediat. Res. *13:* 230–235 (1979).

13 Gonzales, L.; Ballard, P.L.: Nuclear 3,5,3′-triiodothyronine receptors in rabbit lung: characterization and developmental changes. Endocrinology *111:* 542–552 (1982).

14 Mashiach, S.; Barkai, G.; Sack, J.; Stern, E.; Goldman, B.; Bristi, M.; Serr, D.M.: Enhancement of fetal lung maturity by intra-amniotic administration of thyroid hormone. Am. J. Obstet. Gynec. *130:* 284–289 (1978).

15 Rooney, S.A.; Marino, P.A.; Gobran, L.; Gross, I.; Warshaw, J.B.: TRH effects on fetal lung maturation. Pediat. Res. *18:* 623–625 (1979).

16 Gross, I.; Wilson, C.M.: Fetal lung in organ culture. IV. Supra-additive hormone interactions. J. appl. Physiol. *52:* 1420–1425 (1975).

17 Smith, B.T.; Sabry, K.: Glucocorticoid-thyroid synergism in lung maturation. A mechanism involving epithelial-mesenchymal interaction. Proc. natn. Acad. Sci. USA *80:* 1951–1954 (1983).

18 Smith, B.T.: Fibroblast-pneumonocyte factor: Intercellular mediator of glucocorticoid effect on fetal lung; in Stern, Neonatal intensive care, pp. 15–32 (Masson, New York 1978).

19 Smith, B.T.: Optimization of production of fibroblast-pneumonocyte factor by rat
    fetal lung fibroblasts; in Ritzen, The biology of normal human growth (Raven Press,
    New York 1981).

20 Smith, B.T.: Lack of species specificity in production of fibroblast-pneumonocyte
    factor by perinatal lung fibroblasts; in Monset-Couchard, Minkowski, The physio-
    logical and biochemical basis for perinatal medicine, pp. 54–58 (Karger, Basel
    1981).

21 Smith, B.T.: Lung maturation in the fetal rat. Acceleration by the injection of fibro-
    blast-pneumonocyte factor. Science 204: 1094–1095 (1979).

22 Post, M.; Floros, J.; Smith, B.T.: Inhibition of lung maturation by monoclonal anti-
    bodies against fibroblast-pneumonocyte factor. Nature, Lond. 308: 284–286
    (1984).

23 Floros, J.; Post, M.; Smith, B.T.: Glucocorticoids affect the synthesis of pulmonary
    fibroblast-pneumonocyte factor at a pretranslational level. J. biol. Chem. 260: 2265–
    2267 (1985).

24 Post, M.; Barsoumian, A.; Smith, B.T.: The cellular mechanism of glucocorticoid
    acceleration of fetal lung maturation. Fibroblast-pneumonocyte factor stimulates
    cholinephosphate cytidylyltransferase activity. J. biol. Chem. 261: 2179–2185
    (1986).

25 Nielsen, H.C.; Zinman, H.M.; Torday, J.S.: Dihydrotestosterone inhibits fetal rabbit
    pulmonary surfactant production. J. clin. Invest. 69: 611–616 (1982).

26 Floros, J.; Nielsen, H.C.; Torday, J.S.: Dihydrotestosterone blocks fibroblast-pneu-
    monocyte factor at a pretranslational level. J. biol. Chem. 262: 13592–13598
    (1987).

27 Robert, M.F.; Neff, R.K.; Hubbell, J.P.; Taeusch, H.W.; Avery, M.E.: Association
    between maternal diabetes and the respiratory distress syndrome in the newborn.
    New Engl. J. Med. 294: 357–360 (1976).

28 Carlson, K.S.; Smith, B.T.; Post, M.: Insulin acts on the fibroblast to inhibit gluco-
    corticoid stimulation of lung maturation. J. appl. Physiol. 57: 1577–1579 (1984).

Barry T. Smith, MD, Division of Neonatology, Department of Paediatrics,
555 University Avenue, Toronto, Ont. M5G 1X8 (Canada)

Wichert P von, Müller B (eds): Basic Research on Lung Surfactant.
Prog Respir Res. Basel, Karger, 1990, vol 25, pp 115–121

Regulation of Surfactant Secretion in Cultured Type II Cells[1]

L.G. Dobbs[a], *H.R.W. Wirtz*[a], *M.S. Pian*[b]

Departments of [a]Medicine and [b]Pediatrics, Cardiovascular Research Institute, University of California San Francisco, Calif., USA

Background

Various experimental models have been used to study surfactant secretion. Studies with intact animals and perfused lungs have demonstrated that many different stimuli, including deep lung inflations [1–4], adrenergic stimuli [4–8], thyroxine [9, 10], estrogen [11], and prostaglandins [12, 13] increase surfactant secretion. Over the past 10 years, isolated type II cells have been used to study surfactant secretion in vitro. Type II cells cultured for 24 h secrete material that is morphologically similar to lamellar forms seen in fluid-filled alveoli of fetal rats. The secreted material is similar in lipid composition to surfactant obtained by endobronchial lavage; it also contains SP-A. Type II cell secretions are surface active, lowering surface tension to less than 10 mN/m [14].

Conventionally, PC has been used as a marker for surfactant secretion in isolated type II cells. Agonists that stimulate secretion of PC in isolated type II cells are listed in table 1. These agonists fall into various classes. One class of agonists, represented by beta-adrenergic agents [15–17], forskolin [18], methylxanthines [15, 16], and cholera toxin [19], causes an increase in cellular cAMP [16, 19]. Experimental evidence suggests that both protein kinase C [20] and intracellular calcium $(Ca^{2+})_i$ [21, 22] also mediate stimulated secretion in type II cells. Treatment of type II cells with

[1] This work was supported in part by NIH grants HL-24075 and HL-34356.

Table 1. Stimuli of surfactant secretion

Physiologic stimuli	Stimulus in isolated type II cells	Cellular mediator
? Circulating catecholamines	beta-adrenergic agonists, cAMP, forskolin, methylxanthines, cholera toxin	cAMP
?	tetradecanoyl phorbol acetate (TPA)	protein kinase C
?	ionophores (A23187, ionomycin)	Ca^{2+}
?	leukotrienes, arachidonic acid	?
?	adenosine compounds	cAMP, Ca^{2+}
Deep inflation	?	?

TPA (1-0-tetradecanoyl-13-phorbolacetate) increases intracellular protein kinase C, while treatment with ionophores increases $(Ca^{2+})_i$. Arachidonic acid [23] and its metabolites [24] have been found to stimulate surfactant secretion in isolated type II cells; the mechanism by which this stimulation occurs is unknown. Adenosine compounds, including ATP, are also potent stimulators of surfactant secretion in vitro [25, 26]. Although type II cell cAMP content increases as a result of treatment with adenosine and ATP [26], it has been suggested that intracellular second messengers other than cAMP may also be important in mediating secretion caused by these secretagogues.

Although many secretagogues and several putative intracellular 'second messengers' have been identified for surfactant secretion, little is known about the physiologic correlates of these biochemical events. Because there is a paucity of nerve endings at the alveolar level [27], it seems unlikely that the nervous system plays an important role in regulating surfactant secretion by direct action on type II cells. Circulating catecholamines may be at least one physiologic stimulus that causes an increase in intracellular levels of cAMP. Although it has been proposed that the effect of deep inflation is mediated by protein kinase C [20], there is no evidence to support this hypothesis.

It is difficult, from the findings summarized in table 1, to hypothesize how surfactant secretion might be controlled locally. Although it could be argued that local control of surfactant secretion is not necessary because rapid diffusion of surfactant among alveoli occurs, it could also be argued that the effective regulation of surfactant synthesis and secretion might involve local control mechanisms. For example, local deficits of surfactant might stimulate secretion and local surpluses might inhibit secretion. ATP, leukotrienes, or arachidonic acid could all potentially be involved in mediating such responses, but obvious cellular sources or triggers of release of these substances have not been identified.

We have recently investigated the possibility that local factors may be involved in the regulation of surfactant secretion. One possibility is that, in deep inflations of the lung, alveoli may actually be stretched. Mechanical factors have been shown in other cell systems to modulate cellular biochemical events. We have developed a method by which type II cells can be stretched in vitro in order to test the hypothesis that stretch may stimulate secretion [28]. Another possible method by which secretion could be regulated locally is that secreted surfactant may itself play a role in modulating surfactant secretion. Previously, we have shown that isolated surfactant inhibits secretion of phosphatidylcholine by type II cells [29]. Of the surfactant components, SP-A has proven to be the most potent inhibitor of surfactant secretion [29–31]. These findings suggest that a feedback control mechanism may exist for control of surfactant secretion and suggest that local factors may be involved in the regulation of surfactant secretion.

Materials and Methods

Methods

Methods for the studies have been described in detail elsewhere. Briefly, type II cells were isolated from rat lungs by digestion with elastase and differential adherence on plates coated with IgG [32]. Secretion of phosphatidylcholine by type II cells cultured for 22 h was measured as previously described [33]. Intracellar calcium $(Ca^{2+})_i$ was measured in type II cells loaded with quin2 [22]. Type II cells were stretched as described by Wirtz and Dobbs [28]. SP-A was prepared by the method of Hawgood et al. [34].

Materials

Tissue culture medium and fetal bovine serum were obtained from the University of California Cell Culture Facility. We purchased TPA from consolidated Midland (Brewster, N.Y.), terbutaline from Merrill-Dow (Cincinnati, Ohio), and ^3H-choline from Amersham. Ionomycin and quin2/AM were purchased from Calbiochem (La Jolla, Calif.).

Results

Effects of SP-A on Secretion

Previous studies have shown that SP-A inhibits secretion in rat type II cells. Both basal secretion and secretion stimulated by agonists that act via separate pathways are inhibited [29] (table 2). SP-A isolated from a patient with alveolar proteinosis has a very similar inhibitory effect to SP-A isolated from rats (table 2).

Effect of SP-A on Intracellular Calcium

Treatment with ionomycin (10^{-6} M) causes a rise in $(Ca^{2+})_i$ in type II cells [22]. Pretreating type II cells with SP-A (1 µg/ml) reduced the $(Ca^{2+})_i$ increase by 46% [35].

Effect of Stretch on Secretion

Type II cells cultured for 24 h on a stretchable membrane were stretched by applying hydrostatic pressure to the inferior surface of the membrane. A single stretch that increased the calculated surface area of membrane by 21% for 20 s stimulated secretion by 75% over basal levels. There was no observed increase in the activity of LDH released into the medium.

The stimulation of secretion caused by stretching was inhibited by SP-A [see Wirtz and Dobbs, this vol.].

Table 2. Percent of PC secreted in 3 H (mean ± SD; n = 3)

	Concentration, µg protein/ml						
	0	0.01	0.03	0.10	0.30	1	3
Rat SP-A							
Control	1.7±0.2	–	–	–	–	–	–
TPA (10^{-8} M)	12.7±1.9	10.2±0.2	10.3±1.8	6.4±1.1	4.6±2.9	2.4±0.1	4.5±1.5
Proteinosis SP-A							
Control	2.4±0.2	–	–	–	–	–	–
TPA (10^{-8} M)	12.5±2.9	10.9±1.9	8.8±1.7	7.0±1.4	3.1±0.1	1.7±0	1.3±0.2

Summary

Various secretagogues have been shown in vitro to stimulate surfactant secretion. The physiologic correlates of most of these chemical agents are unknown. We have shown that stretching type II cells in culture stimulates the secretion of PC. One of the surfactant proteins, SP-A, potently inhibits basal secretion and secretion stimulated by a variety of agents, including stretch. SP-A partially inhibits the rise in $(Ca^{2+})_i$ that is caused by ionomycin. The observed stimulatory effects of stretch and inhibitory effects of SP-A represent two possible mechanisms by which local factors within the lung may regulate surfactant secretion.

References

1   Faridy, E.E.: Effect of distension on release of surfactant in excised dogs' lungs. Resp. Physiol. 27: 99–114 (1976).

2   Nicholas, T.E.; Barr, H.A.: Control of release of surfactant phospholipids in the isolated perfused rat lung. J. appl. Physiol. 51: 90–98 (1981).

3   Massaro, G.D.; Massaro, D.: Morphologic evidence that large inflations of the lung stimulate secretion of surfactant. Am. Rev. resp. Dis. 127: 235–236 (1983).

4   Oyarzún, M.J.; Clements, J.A.: Control of lung surfactant by ventilation, adrenergic mediators, and prostaglandins in the rabbit. Am. Rev. resp. Dis 117: 879–891 (1978).

5   Lawson, E.E.; Brown, E.B.; Torday, J.S.; Madansky, D.L.; Taeusch, H.W.: Influence of epinephrine on fetal pulmonary fluid production and surfactant release. Physiologist 20: 55a (1977).

6   Kero, P.; Hirvonen, T.; Valimaki, I.: Prenatal and postnatal isoxsuprine and respiratory-distress syndrome. Lancet ii: 198 (1973).

7   Wyszogrodski, I.; Taeusch, H.W.; Avery, M.E.: Isoxsuprine-induced alterations of pulmonary pressure-volume relationships in premature rabbits. Am. J. Obstet. Gynec. 119: 1107–1111 (1974).

8   Corbet, A.J.S.; Flax, P.; Rudolph, A.J.: Role of autonomic nervous system controlling surface tension in fetal rabbit lungs. J. appl. Physiol. 43: 1039–1405 (1977).

9   Olsen, D.B.: Neurohumoral-hormonal secretory stimulation of pulmonary surfactant in the rat. Physiologist 15: 230a (1972).

10 Das, D.K.; Ayromlooi, B.; Bandyopadhyay, S.: Potentiation of surfactant release in fetal lung by thyroid hormone action. J. appl. Physiol. 56: 1621–1626 (1984).

11 Thuresson-Klein, A.; Moawad, A.H.; Hedqvist, P.: Estrogen stimulates formation of lamellar bodies and release of surfactant in the rat fetal lung. Am. J. Obstet. Gynec. 151: 506–514 (1985).

12 Oyarzún, M.J.; Clements, J.A.: Ventilatory and cholinergic control of pulmonary surfactant in the rabbit. J. appl. Physiol. 43: 39–45 (1977).

13 Oyarzún, M.J.; Donoso, P.; Arias, M.; Quijada, D.: Thromboxane mediates the increase in alveolar surfactant pool induced by free fatty acid infusion in the rabbit. Respiration 46: 231–240 (1984).

14 Dobbs, L.G.; Mason, M.C.; Benson, B.J.; Sueishi, K.: Secretion of surfactant by

primary cultures of alveolar type II cells isolated from rats. Biochim. biophys. Acta *713:* 118–127 (1982).

15   Dobbs, L.G.; Mason, R.J.: Pulmonary alveolar type II cells isolated from rats: Release of phosphatidylcholine in response to β-Adrenergic stimulation. J. clin. Invest. *63:* 378–387 (1979).

16   Brown, L.A.S.; Longmore, W.J.: Adrenergic and cholinergic regulation of lung surfactant secretion in the isolated perfused rat lung and in the alveolar type II cell in culture. J. biol. Chem. *256:* 66–72 (1981).

17   Mettler, N.R.; Gray, M.E.; Schuffman, S.; LeQuire, V.S.: Beta-adrenergic induced synthesis and secretion of phosphatidylcholine by isolated pulmonary alveolar type II cells. Lab. Invest. *45:* 575–586 (1981).

18   Rice, W.R.; Hull, W.M.; Dion, C.A.; Hollinger, B.A.; Whitsett, J.A.: Activation of cyclic AMP-dependent protein kinase during surfactant release from type II pneumocytes. Exp. Lung Res. *9:* 135–149 (1985).

19   Mescher, E.J.; Dobbs, L.G.; Mason, R.J.: Cholera toxin stimulates secretion of saturated phosphatidylcholine and increases cellular cyclic AMP in isolated rat alveolar type II cell. Exp. Lung Res. *5:* 173–182 (1983).

20   Sano, K.; Voelker, D.R.; Mason, R.J.: Involvement of protein kinase C in pulmonary surfactant secretion from alveolar type II cells. J. biol. Chem. *260:* 12725–12729 (1985).

21   Sano, K.; Voelker, D.R.; Mason, R.J.: Effect of secretagogues on cytoplasmic free calcium in alveolar type II epithelial cells. Am. phys. Soc. *253:* C679–C686 (1987).

22   Pian, M.; Dobbs, L.G.; Düzgunes, N.: Positive correlation between cytosolic free calcium and surfactant secretion in cultured rat alveolar type II cells. Biochim. biophys. Acta *960:* 43–53 (1988).

23   Gilfillan, A.M.; Rooney, S.A.: Arachidonic acid metabolites stimulate phosphatidylcholine secretion in primary cultures of type II pneumocytes. Biochim. biphys. Acta *833:* 336–341 (1985).

24   Gilfillan, A.M.; Rooney, S.A.: Leukotrienes stimulate phosphatidylcholine secretion in cultured type II pneumocytes. Biochim. biophys. Acta *876:* 22–27 (1986).

25   Rice, W.R.; Singleton, F.M.: P2-purinoceptors regulate surfactant secretion from rat isolated alveolar type II cells. Br. J. Pharmacol. *89:* 485–491 (1986).

26   Gilfillan, A.M.; Rooney, S.A.: Functional evidence for adenosine A_2 receptor regulation of phosphatidylcholine secretion in cultured type II pneumocytes. J. Pharmac. exp. Res. *241:* 907–916 (1987).

27   Meyrick, B.; Reid, L.: Nerves in rat intra-acinar alveoli. An electron microscopic study. Resp. Physiol. *11:* 367–377 (1971).

28   Wirtz, H.; Dobbs, L.G.: Phosphatidylcholine secretion is stimulated by a single mechanical stretch of rat alveolar type II cells cultured on silastic membranes. FASEB J. *2:* A708 (1988).

29   Dobbs, L.G.; Wright, J.R.; Gonzalez, R.; Venstrom, K.; Nellenbogen, J.: Pulmonary surfactant and its components inhibit secretion of phosphatidylcholine from cultured rat alveolar type II cells. Proc. natn. Acad. Sci. USA *84:* 1010–1014 (1987).

30   Kuroki, Y.; Mason, R.J.; Voelker, D.R.: Pulmonary surfactant apoprotein A structure and modulation of surfactant secretion by rat alveolar type II cells. J. biol. Chem. *263:* 3388–3394 (1988).

31 Rice, W.R.; Ross, G.F.; Singleton, F.M.; Dingle, S.; Whitsett, J.A.: Surfactant-associated protein inhibits phospholipid secretion from type II cells. Am. phys. Soc. *63:* 692–698 (1987).

32 Dobbs, L.G.; Gonzalez, R.F.; Williams, M.C.: An improved method for isolating type II cells in high yield and purity. Am. Rev. resp. Dis. *134:* 141–145 (1986).

33 Dobbs, L.G.; Gonzalez, R.F.; Marinari, L.A.; Mescher, E.J.; Hawgood, S.: The role of calcium in the secretion of surfactant by rat alveolar type II cells. Biochim. biophys. Acta *877:* 305–313 (1986).

34 Hawgood, S.; Benson, B.J.; Hamilton, R.L.: Effects of a surfactant-associated protein and calcium ions on the structure and surface activity of lung surfactant lipids. Biochemistry *24:* 184–190 (1985).

35 Pian, M.; Dobbs, L.G.: Surfactant secretion stimulated in cultured rat alveolar type II cells correlates with levels of cytosolic free calcium. FASEB J. *2:* A957 (1988).

L.G. Dobbs, MD, Cardiovascular Research Institute, Department of Medicine, University of California, San Francisco, CA 94143 (USA)

Wichert P von, Müller B (eds): Basic Research on Lung Surfactant.
Prog Respir Res. Basel, Karger, 1990, vol 25, pp 122–126

Effect of Beta-Mimetic Drugs on Phospholipid Synthesis in the Isolated Lung

W. Bernhard, B. Müller, P. von Wichert

Medizinische Poliklinik, Zentrum für Innere Medizin, Philipps-Universität, Marburg, BRD

Pulmonary surfactant (PSA) is composed mainly of phosphatidylcholine with two palmitoyl residues, phosphatidylglycerol (PG), and some lipophilic or hydrophilic apoproteins of 5–18 or 28–36 kdaltons molecular weight, respectively [Wright and Clements, 1987]. To get more insight into the regulation of PSA synthesis, we investigated the influence of terbutaline and ICI 118,551, a competitive β_2-adrenoceptor antagonist [Bilski et al., 1983], on the incorporation of ^{14}C-choline and ^3H-palmitate into total (PC), disaturated (DSPC) and unsaturated (UPC) phosphatidylcholine and of ^3H-Palmitate incorporation into phosphatidylethanolamine (PE). Furthermore, we determined the changes of phospholipid (PL) content in the tissue. To leave the PSA-producing type II pneumocytes in their natural environment, we used the isolated, ventilated, and perfused rat lung (IVPL) model.

Materials and Methods

Male Wistar rats of 250 g body weight were given 60 mg/kg pentobarbital and 1,500 IU/kg heparin intraperitoneally. Isolation and perfusion of the lungs were performed similar to Post et al. [1983]. Terbutaline (10 μM) and ICI 118,551 (1, 10 or 50 μM) alone were added to the perfusate before starting the experiment. If the perfusion was done in the presence of both drugs, terbutaline was added 7 min after perfusion with 10 μM ICI 118,551. 15 min after the start 0.925 MBq methyl-^{14}C-choline (1.85 GBq/mmol) and 9.25 MBq 9,10-^3H-palmitic acid (1.04 TBq/mmol) were added. Total perfusion time was 3.15 h.

Bronchoalveolar lavage (BAL) of the tissue except of the right upper and middle lobe was performed with 6 × 5 ml room-tempered 150 mM saline. Alveolar macrophages were separated by centrifugation at 270 *g* at 4 °C for 15 min. Tissue was extracted using the method of Folch et al. [1957], and BAL fluid by the method of Bligh and Dyer [1959]. PLs were separated by one-dimensional thin-layer chromatography (TLC) [Gilfillan et al., 1983]. DSPC was isolated from tissue by column chromatography [Mason et al., 1976] and cleaned by TLC. Quantification of PL was done according to Bartlett [1959]. ^{14}C-choline and ^{3}H-palmitate label of PL aliquots was determined with an LKB Wallac 1219 rackbeta liquid scintillation counter using the external standard method. Data of UPC were calculated from those of PC and DSPC. Cyclic adenosine monophosphate (cAMP) was determined in the perfusate [Bellamy and Tierney, 1984] with a commercial radioimmunoassay. Wet to dry weight ratio was measured by drying the right upper lobe after the experiment at 90 °C to constant weight. Statistics were done using the U test of Wilcoxon, Mann-Whitney [Sachs, 1984].

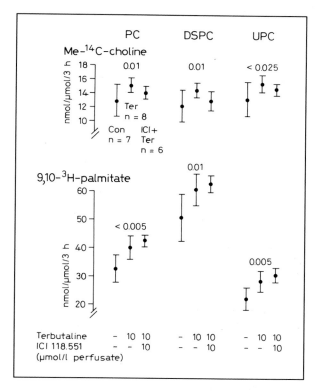

Fig. 1. Bronchoalveolar lavaged lung tissue: Incorporation of ^{14}C-choline and ^{3}H-palmitate into PC, DSPC and UPC. Concentrations of terbutaline and ICI 118,551 are shown at the bottom of the figure. Results are shown in means ± SE. Significance is shown as α-error.

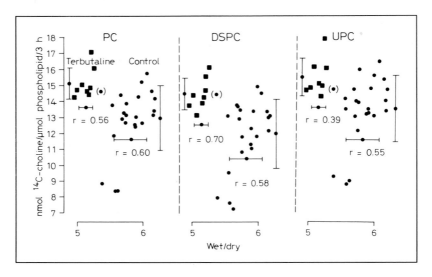

Fig. 2. Lung tissue: Correlations between wet to dry weight ratio of the right upper lobe and the incorporation of ^{14}C-choline into PC, DSPC, and UPC. Each data point within a PL group represents one perfusion experiment. Symbols with bars represent means ± SE. r = Correlation coefficient.

Results

cAMP was increased by terbutaline from 3 ± 4 to 25 ± 10 pmol/ml. Terbutaline enhanced the incorporation of ^{14}C-choline and ^{3}H-palmitate into PC, DSPC and UPC (fig. 1) but decreased PC, DSPC and PG in tissue from 44.2 ± 1.2 to 42.7 ± 1.7% (α = 0.05), 16.2 ± 0.5 to 15.2 ± 0.8% (α = 0.005) and 3.2 ± 0.3 to 2.6 ± 0.3% (α < 0.001), respectively. In PC of BAL fluid ^{14}C-choline was increased from 1.5 ± 0.6 to 2.6 ± 0.5 nmol/μmol (α < 0.001) and ^{3}H-palmitate was increased from 4.6 ± 1.3 to 9.6 ± 1.8 nmol/μmol (α < 0.001) by terbutaline. All effects of terbutaline were antagonized by ICI 118,551 except the increase of ^{3}H-palmitate incorporation into tissue PC, DSPC and UPC. ICI 118,551 selectively and concentration dependently increased the incorporation of ^{3}H-palmitate into PC for 15–22% (α < 0.01) and the ^{3}H-palmitate to ^{14}C-choline incorporation ratio of tissue PC, DSPC and UPC. The incorporation of ^{3}H-palmitate into PE was not altered by either terbutaline or ICI 118,551 + terbutaline

(3.9 ± 0.5, 4.1 ± 0.8 and 3.6 ± 0.3 nmol/µmol PE, respectively). There was a positive correlation between ^{14}C-choline incorporation into PC, DSPC and UPC and the wet to dry weight ratio of the tissue (fig. 2).

Discussion

The terbutaline-induced increase of cAMP in the perfusate implicates that a sufficient β-adrenergic stimulation of the IVPL had occurred [Bellamy and Tierney, 1984]. Furthermore, the decrease of tissue PC and DSPC and the increased appearance of [14]C- and ^{3}H-labeled PC in the BAL fluid by terbutaline is in agreement with the well-known β-adrenergically induced PL secretion. Since the augmentation of ^{14}C-choline and ^{3}H-palmitate in tissue PC, DSPC and UPC exceeded the decreases of their amount in tissue it is suggested that terbutaline increased the incorporation. This effect may be caused by β-adrenoceptors, because ICI 118,551 blocked the terbutaline-induced enhancement of ^{14}C-choline incorporation. The inability of the β-blocker to antagonize the terbutaline effect on ^{3}H-palmitate incorporation may be explained by its own potency to increase ^{3}H-palmitate incorporation into PC.

Since the control lungs had a higher wet to dry weight ratio than the terbutaline-treated lungs, one might argue that the different incorporation rates were caused by edema in the controls. In this case one would expect a negative correlation between the wet to dry weight ratio and the ^{14}C-choline incorporation into PC, DSPC and UPC. But the positive correlation coefficients do not emphasize this hypothesis. The finding that terbutaline caused an enhanced synthesis of UPC as well as DSPC is in agreement with the fact that β-mimetics decrease the activity of phospholipase A_2 and acyl-CoA:lysolecithin-acyltransferase, being responsible for the remodeling of UPC to DSPC [von Wichert et al., 1988]. So we conclude that β-adrenergic stimulation may increase de novo synthesis of DSPC and UPC in the isolated rat lung.

References

Bartlett, G.R.: Phosphorus assay in column chromatography. J. biol. Chem *234*: 466–468 (1959).
Bellamy, P.E.; Tierney, D.F.: Cyclic nucleotide concentrations in tissue and perfusate of isolated rat lung. Exp. Lung Res. *7*: 67–76 (1984).

Bilski, A.J.; Halliday, S.E.; Fitzgerald, J.D.; Wale, J.L.: The pharmacology of a β_2-selective adrenoceptor antagonist (ICI 118,551). J. cardiovasc. Pharm. *5:* 430–437 (1983).

Bligh, E.G.; Dyer, W.J.: A rapid method of total lipid extraction and purification. Can. J. Biochem. Physiol. *37:* 911–917 (1959).

Folch, J.; Lees, M.; Stanley, G.H.S.: A simple method for the isolation and purification of total lipides from animal tissues. J. biol. Chem. *226:* 497–509 (1957).

Gilfillan, A.M.; Chu, A.J.; Smart, D.A.; Rooney, S.A.: Single plate separation of lung phospholipids including disaturated phosphatidylcholine. J. Lipid Res. *24:* 1651–1656 (1983).

Mason, R.J.; Nellenbogen, J.; Clements, J.A.: Isolation of disaturated phosphatidylcholine with osmium tetroxide. J. Lipid Res. *17:* 281–283 (1976).

Post, M.; Batenburg, J.J.; Schuurmans, E.A.J.M.; Oldenborg, V.; Molen, A.J. van der; Golde, L.M.G. van: The perfused rat lung as a model for studies on the formation of surfactant and the effect of ambroxol on this process. Lung *161:* 349–359 (1983).

Wichert, P. von; Müller, B.; Meyer-Ingold, W.: Influence of a β-adrenergic agonist on septic shock-induced alterations of phosphatidylcholine metabolism in rat lung. Lung *166:* 257–267 (1988).

Wright, J.R.; Clements, J.A.: Metabolism and turnover of lung surfactant. Am. Rev. resp. Dis. *135:* 426–444 (1987).

Wolfgang Bernhard, Dr. rer. physiol., Medizinische Poliklinik,
Zentrum für Innere Medizin der Philipps-Universität Marburg, Baldingerstrasse,
D–3550 Marburg (FRG)

Wichert P von, Müller B (eds): Basic Research on Lung Surfactant.
Prog Respir Res. Basel, Karger, 1990, vol 25, pp 127–135

Type II Cell Purinoceptors and Surfactant Secretion[1]

Seamus A. Rooney

Division of Perinatal Medicine, Department of Pediatrics,
Yale University School of Medicine, New Haven, Conn., USA

Adenosine and adenine nucleotides have long been recognized as intracellular entities. Adenosine is a component of nucleic acids and of the intracellular second messenger cAMP while ATP is a cellular energy source critical for many metabolic processes. However, there is also abundant evidence that in several organ systems adenosine and adenine nucleotides have a variety of effects mediated via cell surface receptors [1–4]. Although initially reported some 60 years ago by Drury and Szent-Gyorgyi [5], these phenomena have only recently received widespread attention. The receptors mediating such extracellular effects were initially termed 'purinergic receptors' since the agonists were believed to be released from purinergic nerves. As it has subsequently been recognized that such receptors also exist on cells with no neural connections the term 'purinoceptor' is now utilized [3].

Purinoceptors are divided into two major categories, P_1 and P_2 [2, 6]. The potency order at the P_1 (adenosine) receptor is adenosine $\geq$ AMP $>$ ADP $\geq$ ATP while the reverse is true at the P_2 (ATP) receptor [2]. Both P_1 and P_2 receptors have been further subdivided: P_1 into A_1 and A_2 subtypes [6], although an A_3 subtype has also been suggested [7], and P_2 into at least four subtypes – P_{2x}, P_{2y}, P_{2z} and P_{2t} [3, 8]. This receptor classification has been largely based on the order of potency of various agonists in eliciting a pharmacological response. Radioligand binding studies with agonists and antagonists have also been carried out especially in the case of the A_1 and

[1] This work was carried out in collaboration with Alasdair M. Gilfillan, PhD and Laurice I. Gobran and was supported by a research grant (HL-31175) from the National Heart, Lung and Blood Institute, United States Public Health Service.

A_2 adenosine receptors [9–12]. Isolation and structural characterization of both A_1 [13] and A_2 [14] receptors have been reported.

Action at P_1 receptors is accompanied by changes in cAMP levels and adenylate cyclase activity – decreases in the case of A_1 and increases in the case of A_2 – and is antagonized by methylxanthines [6]. Indeed, many of the pharmacological effects of methylxanthines, such as theophylline and caffeine, appear to be due to antagonism of adenosine receptors rather than to phosphodiesterase inhibition [1, 4]. Signal transduction following occupancy of the P_2 receptor may involve changes in protein kinase C activity [15] or in the levels of intracellular Ca^{++} [16–18] or inositol phosphates [19–21].

The first indication that lung surfactant secretion might be influenced by purinoceptor agonists was the finding by Gilfillan et al. [22] that ATP stimulated phosphatidylcholine secretion in perifused slices of adult rat lung. However, since cAMP was known to stimulate surfactant secretion [23] these workers assumed that the observed effect of ATP was mediated by cAMP following its synthesis from ATP rather than by a purinoceptor. That adenosine may have an effect on lung surfactant production was first suggested by Ekelund and Enhorning [24] who examined the effects of methylxanthines on surfactant secretion. They found that enprofylline stimulated surfactant release in fetal rabbits [24] while aminophylline did not [25]. Since enprofylline was reported to be less potent as an adenosine antagonist than aminophylline [26] they suggested that the latter may antagonize adenosine stimulation of surfactant secretion [24]. Subsequently, the same workers reported that infusion of adenosine into pregnant rabbits increased the amount of phospholipid recovered in lung lavage from the fetuses [27]. Whether this effect of adenosine was directly on the fetal lung and was receptor mediated was not examined, however.

Purinoceptor agonist stimulation of surfactant phospholipid secretion in type II pneumocytes was first reported by Rooney and Gilfillan [28] in 1986. Adenosine, AMP, ADP and ATP all stimulated phosphatidylcholine secretion in a concentration-dependent manner with (table 1) ADP and ATP producing the most stimulation and AMP being the most potent agonist [29]. Similar findings were reported by Rice and Singleton [30]. Although ranking the agonists in order of maximum effect, ADP > ATP >> AMP = adenosine, suggested mediation by the P_2 receptor, this was not supported by the potency order, AMP > ATP $\geq$ adenosine > ADP (table 1). The discrepancy led us to speculate that more than one receptor was involved. However, since adenosine and the adenine nucleotides are

Table 1. Effect of adenosine and adenine nucleotides on phosphatidylcholine secretion in primary cultures of adult rat type II pneumocytes[1]

Agonist[2]	Maximum stimulation % increase over control	EC_{50} concentration[3] μM
Adenosine	141	5.3
AMP	140	0.6
ADP	573	23.0
ATP	414	4.1

[1] These data are adapted from Gilfillan and Rooney [29].
[2] Agonist concentrations were 1 mM which stimulated maximally [29].
[3] The EC_{50} is the concentration which produces 50% of the maximum stimulation.

all metabolically interconvertible [1, 3], it was also possible that the stimulation was effected by a single receptor and that the observed differences were due to differences in the ability to be converted to the active compound. To address these possibilities we examined the effects of nonmetabolizable analogs of adenosine and ATP. The adenosine analogs 5'-N-ethylcarboxyamidoadenosine (NECA) and L-N^6-phenylisopropyladenosine (L-PIA) stimulated phosphatidylcholine secretion to the same extent as adenosine [29] while ATP analogs also stimulated [31]. This suggested that both P_1 and P_2 receptors were involved. Further evidence in support of two receptors came from the finding that adenosine deaminase abolished the stimulatory effect of adenosine on phosphatidylcholine secretion but had little effect on that of ATP [31]. Similarly, the stimulatory effect of NECA was diminished by the methylxanthines theophylline and 8-phenyltheophylline [29, 32] but 8-phenyltheophylline had little effect on ATP stimulation [29, 31]. Finally, the increase in phosphatidylcholine secretion produced by adenosine, NECA and L-PIA was accompanied by an increase in intracellular cAMP [32] the level of which was only minimally increased by ATP [31]. These data showed that the effects of adenosine and ATP are mediated by P_1 and P_2 receptors, respectively.

The fact that adenosine and its analogs increased rather than decreased the level of cAMP showed that its effect is mediated by the A_2 subtype of the P_1 receptor [6]. The potency order of adenosine analogs in stimulating phosphatidylcholine secretion [32] is also consistent with the A_2 receptor [6].

Table 2. Effect of Cibacron Blue on the stimulatory effect of ATP on phosphatidylcholine secretion in adult rat type II cells[1]

ATP, μM	Cibacron Blue[2], μM	% [³H]phosphatidylcholine secreted[3]
0	0	1.40 ± 0.13[3]
10	0	3.27 ± 0.18
10	5	3.61 ± 0.23
10	10	3.72 ± 0.39
10	25	3.90 ± 0.25
10	50	3.84 ± 0.09
10	100	3.27 ± 0.03

[1] Type II cells, isolated by the method of Dobbs et al. [35], were incubated with [methyl-³H]choline overnight, transferred to fresh medium and incubated with the indicated concentration of ATP and Cibacron Blue for 90 min after which [³H]phosphatidylcholine in the cells and medium was measured [32]. Secretion is expressed as the amount of [³H]phosphatidylcholine in the medium as % of that in the cells + medium.
[2] Cibacron Blue 3GA (Sigma Chemical Co., St Louis, Mo.) was previously termed Reactive Blue 2 [33, 34].
[3] Data are means ± SE from 3 type II cell preparations each of which is from 3 rats.

Which subtype of the P_2 receptor mediates the effect of ATP? The stimulatory effect of ATP was antagonized by α,β-methylene ATP [29, 31] which is consistent with the P_{2x} receptor [8]. The potency order [31], however, ATP > β,γ-methylene ATP = methylthio ATP = 2-deoxy ATP ≥ 8-bromo ATP > α,β-methylene ATP, rules it out since the methylene ATPs are more potent than ATP at P_{2x} [8]. The P_{2y} receptor also appears to be ruled out as 2-methylthio ATP was reported to be much more potent than ATP at that site [8]. In addition, Cibacron Blue was reported to be a selective P_{2y} antagonist [33, 34] and this agent had no effect on the stimulatory effect of ATP in type II cells (table 2). It is unlikely to be the P_{2z} receptor since approximately 50 μM ATP is required to obtain a response at this site [3] and the ATP EC_{50} is 4.1 μM in the type II cell (table 1). Since ADP is the agonist at the P_{2t} receptor while ATP is an antagonist [3], it cannot mediate the ATP effect in type II cells. Thus, the P_2 receptor which mediates the stimulatory effect of ATP on phosphatidylcholine secretion in type II cells does not appear to be identical to any of those previously described. Recently, a further P_{2s} subtype was reported in smooth muscle [36]. However, it also differed from the type II cell receptor in that

Table 3. Effects of ADP and UTP, alone and in combination with ATP, on phosphatidylcholine secretion in type II cells[1]

Experiment[2]	Agonist, (1 mM)[3]	% [^{3}H]phosphatidylcholine secreted
A	none	1.19 ± 0.22 (4)[4]
	ATP	3.84 ± 0.43 (4)*
	ADP	3.97 ± 0.60 (4)*
	ATP + ADP	3.73 ± 0.32 (4)
B	none	0.98 ± 0.12 (5)
	ATP	4.53 ± 0.29 (5)**
	UTP	2.28 ± 0.23 (5)**
	ATP + UTP	4.72 ± 0.29 (3)

* $p < 0.05$, ** $p < 0.01$, significantly different from the controls.
[1] Experimental details are as described in table 2.
[2] A and B are two different series of experiments.
[3] The ATP concentration was 0.01 mM in one of the A experiments.
[4] The data are from the number of experiments indicated in parentheses and were analyzed statistically with Student's paired t test [29] corrected for multiple comparisons.

α,β-methylene ATP was not an antagonist [36]. Whether other subtypes of the P_2 receptor also exist remains to be established. Clarification of P_2 receptor subtypes must await radioligand binding studies with specific agonists and antagonists.

Although ADP and ATP differed with respect to the maximum stimulatory effect and EC_{50} values (table 1), the effects of these two nucleotides were not additive (table 3) and are thus mediated by the same mechanism. Recently, a UTP receptor different from the ATP receptor has been reported in rat liver [37] and rabbit ear artery [38]. In the type II cell, however, although UTP stimulated phosphatidylcholine secretion to a lesser extent than did ATP, the effects of the two agonists were not additive (table 3) suggesting mediation by the same receptor.

Do purinoceptors play a physiological role in regulation of surfactant secretion? As shown in table 4, the EC_{50} values of purinoceptor agonists are comparable to those of other physiological agents in the type II cell. In support of a physiological role for purinoceptor agonists in regulation of surfactant secretion was the finding that 8-phenyltheophylline decreased the amount of surfactant phospholipid in newborn rabbit lung lavage [41]. This

Table 4. EC_{50} values for purinoceptor agonists and other agents which stimulate phosphatidylcholine secretion in isolated type II cells

Secretagogue	EC_{50}, M	Reference
Adenosine	5.3×10^{-6}	29
NECA	8.9×10^{-8}	32
ATP	4.1×10^{-6}	29
	1.0×10^{-6}	16
	1.0×10^{-7}	30
Terbutaline	1.6×10^{-7}	39
	8.0×10^{-7}	23
Isoproterenol	3.2×10^{-9}	39
Leukotriene E_4	3.6×10^{-12}	40
Leukotriene D_4	4.5×10^{-11}	40

suggested that the ventilation-induced increase in surfactant secretion [42] is mediated, at least in part, by adenosine. In this context it is worth noting that Nicholas and Barr [43] were unable to block the ventilation-induced increase in surfactant secretion in the perfused adult rat lung with propranolol, atropine or indomethacin, suggesting that agents other than β-agonists, cholinergic agonists and prostaglandins play a role in this process.

The above data provide evidence for the existence of an A_2 adenosine receptor and an ATP receptor in the type II pneumocyte. Recently, we have found that inclusion of adenosine deaminase in the incubation medium increased the stimulatory effects of terbutaline, NECA and forskolin on phosphatidylcholine secretion in type II cells [44]. One explanation for this finding is that adenosine deaminase destroys adenosine acting at an inhibitory A_1 receptor. Further studies are needed to clarify the existence and functional significance of such a receptor.

Summary

Adenosine and adenine nucleotides stimulate phosphatidylcholine secretion in adult rat type II pneumocytes. The stimulatory effect of adenosine is mediated by the A_2 subtype of the P_1 purinoceptor and is associated with an increase in intracellular cAMP. The effect of ATP is mediated by a P_2 purinoceptor. 8-Phenyltheophylline, an adenosine receptor antagonist, diminished the ventilation induced increase in the amount of surfactant in newborn rabbit lung lavage. These data from tissue culture and in vivo studies suggest that purinoceptor agonists play a role in the regulation of lung surfactant secretion.

References

1 Arch, J.R.S.; Newsholme, E.A.: The control of the metabolism and the hormonal role of adenosine. Essays Biochem. *14:* 82–123 (1978).

2 Satchell, D.: Purine receptors: classification and properties. Trends Pharmacol. Sci. *5:* 340–343 (1984).

3 Gordon, J.L.: Extracellular ATP: effects, sources and fate. Biochem. J. *233:* 309–319 (1986).

4 Stiles, G.L.: Adenosine receptors: structure, function and regulation. Trends Pharmacol. Sci. *7:* 486–490 (1986).

5 Drury, A.N.; Szent-Gyorgyi, A.: The physiological activity of adenine compounds with especial reference to their action upon the mammalian heart. J. Physiol. *68:* 213–237 (1929).

6 Burnstock, G.; Buckley, N.J.: The classification of receptors for adenosine and adenine nucleotides. Method. Pharmacol. *6:* 193–212 (1985).

7 Ribeiro, J.A.; Sebastiao, A.M.: Adenosine receptors and calcium. Basis for proposing a third (A_3) adenosine receptor. Prog. Neurobiol. *26:* 179–209 (1986).

8 Burnstock, G.; Kennedy, C.: Is there a basis for distinguishing two types of P_2-purinoceptor? Gen. Pharmacol. *16:* 433–440 (1985).

9 Daly, J.W.; Padgett, W.; Thompson, R.D.; Kusachi, S.; Bugni, W.J.; Olsson, R.A.: Structure-activity relationships for N^6-substituted adenosines at a brain A_1-adenosine receptor with a comparison to an A_2-adenosine receptor regulating coronary blood flow. Biochem. Pharmac. *35:* 2467–2481 (1986).

10 Bruns, R.F.; Lu, G.H.; Pugsley, T.A.: Characterization of the A_2 adenosine receptor labeled by [^{3}H]NECA in rat striatal membranes. Molec. Pharmacol. *29:* 331–346 (1986).

11 Bruns, R.F.; Fergus, J.H.; Badger, E.W.; Bristol, J.A.; Santay, L.A.; Hartman, J.D.; Hays, S.J.; Huang, C.C.: Binding of the A_1-selective adenosine antagonist 8-cyclopentyl-1,3-dipropylxanthine to rat brain membranes. Arch. Pharmacol. *335:* 59–63 (1987).

12 Lohse, M.J.; Klotz, K.N.; Lindenborn-Fortino, J.; Reddington, M.; Schwabe, U.; Olsson, R.A.: 8-Cyclopentyl-1,3-dipropylxanthine (DPCPX). A selective high affinity antagonist radioligand for A_1 adenosine receptors. Arch. Pharmacol. *336:* 204–210 (1987).

13 Yeung, S.M.H.; Perez-Reyes, E.; Cooper, D.M.F.: Hydrodynamic properties of adenosine R_i receptors solubilized from rat cerebral-cortical membranes. Biochem. J. *248:* 635–642 (1987).

14 Lohse, M.J.; Elger, B.; Lindenborn-Fortino, J.; Klotz, K.N.; Schwabe, U.: Separation of solubilized A_2 adenosine receptors of human platelets from non-receptor [^{3}H]NECA binding sites by gel filtration. Arch. Pharmacol. *337:* 64–68 (1988).

15 Rice, W.R.: P_2-Purinoceptors regulate surfactant secretion via mobilization of cytosolic calcium and activation of C-kinase. Am. Rev. resp. Dis. *135:* A357 (1987).

16 Rice, W.R.; Singleton, F.M.: P_{2y}-purinoceptor regulation of surfactant secretion from rat isolated alveolar type II cells is associated with mobilization of intracellular calcium. Br. J. Pharmacol. *91:* 833–838 (1987).

17 McMillan, M.K.; Soltoff, S.P.; Cantley, L.C.; Talamo, B.R.: Extracellular ATP elevates intracellular free calcium in rat parotid acinar cells. Biochem. biophys. Res. Commun. *149:* 523–530 (1987).

18   El-Moatassim, C.; Dornand, J.; Mani, J.C.: Extracellular ATP increases cytosolic
     free calcium in thymocytes and initiates the blastogenesis of the phorbol 12-myris-
     tate 13-acetate-treated medullary population. Biochim. biophys. Acta *927:* 437–444
     (1987).
19   Daniel, J.L.; Dangelmaier, C.A.; Selak, M.; Smith, J.B.: ADP stimulates IP$_3$ forma-
     tion in human platelets. FEBS Lett. *206:* 299–303 (1986).
20   Blachier, F.; Malaisse, W.J.: Effect of exogenous ATP upon inositol phosphate pro-
     duction, cationic fluxes and insulin release in pancreatic islet cells. Biochim. bio-
     phys. Acta *970:* 222–229 (1988).
21   Haggblad, J.; Heilbronn, E.: P$_2$-purinoceptor-stimulated phosphoinositide turnover
     in chick myotubes. Calcium mobilization and the role of guanyl nucleotide-binding
     proteins. FEBS Lett. *235:* 133–136 (1988).
22   Gilfillan, A.M.; Hollingsworth, M.; Jones, A.W.: The pharmacological modulation
     of [^{3}H]-disaturated phosphatidylcholine overflow from perifused lung slices of adult
     rats. A new method for the study of lung surfactant secretion. Br. J. Pharmacol. *79:*
     363–371 (1983).
23   Dobbs, L.G.; Mason, R.J.: Pulmonary alveolar type II cells isolated from rats.
     Release of phosphatidylcholine in response to β-adrenergic stimulation. J. clin.
     Invest. *63:* 378–387 (1979).
24   Ekelund, L.; Enhorning, G.: Pulmonary surfactant release in fetal rabbits as affected
     by enprofylline. Pediat. Res. *19:* 1000–1003 (1985).
25   Ekelund, L.; Burgoyne, R.; Brymer, D.; Enhorning, G.: Pulmonary surfactant release
     in fetal rabbits as affected by terbutaline and aminophyllin. Scand. J. clin. Lab.
     Invest. *41:* 237–245 (1981).
26   Persson, C.G.A.; Andersson, K.E.; Kjellin, G.: Effects of enprofylline and theophyl-
     line may show the role of adenosine. Life Sci. *38:* 1057–1072 (1986).
27   Ekelund, L.; Andersson, K.E.; Enhorning, G.: Release of fetal lung surfactant: effects
     of xanthines and adenosine; in Andersson, Persson, Anti-asthma xanthines and
     adenosine, pp. 202–208 (Excerpta Medica, Amsterdam 1985).
28   Rooney, S.A.; Gilfillan, A.M.: Do purinergic receptors play a role in the regulation of
     lung surfactant production? Fed. Proc. *45:* 1665 (1986).
29   Gilfillan, A.M.; Rooney, S.A.: Purinoceptor agonists stimulate phosphatidylcholine
     secretion in primary cultures of adult rat type II pneumocytes. Biochim. biophys.
     Acta *917:* 18–23 (1987).
30   Rice, W.R.; Singleton, F.M.: P$_2$-purinoceptors regulate surfactant secretion from rat
     isolated alveolar type II cells. Br. J. Pharmacol. *89:* 485–491 (1986).
31   Gilfillan, A.M.; Rooney, S.A.: Functional evidence for involvement of P$_2$ purinocep-
     tors in the ATP stimulation of phosphatidylcholine secretion in type II alveolar
     epithelial cells. Biochim. biophys. Acta *959:* 31–37 (1988).
32   Gilfillan, A.M.; Rooney, S.A.: Functional evidence for adenosine A$_2$ receptor regu-
     lation of phosphatidylcholine secretion in cultured type II pneumocytes. J. Pharmac.
     Exp. Ther. *241:* 907–914 (1987).
33   Burnstock, G.; Warland, J.J.I.: P$_2$-purinoceptors of two subtypes in the rabbit mes-
     enteric artery: reactive blue 2 selectively inhibits responses mediated via the P$_{2y}$ –
     but not the P$_{2x}$ – purinoceptor. Br. J. Pharmacol. *90:* 383–391 (1987).
34   Hopwood, A.M.; Burnstock, G.: ATP mediates coronary vasoconstriction via P$_{2x}$-
     purinoceptors and coronary vasodilation via P$_{2y}$-purinoceptors in the isolated per-
     fused rat heart. Eur. J. Pharmacol. *136:* 49–54 (1987).

35   Dobbs, L.G.; Gonzalez, R.; Williams, M.C.: An improved method for isolating type II cells in high yield and purity. Am. Rev. resp. Dis. *134:* 141–145 (1986).

36   Wiklund, N.P.; Gustafsson, L.E.: Indications for P_2-purinoceptor subtypes in guinea pig smooth muscle. Eur. J. Pharmacol. *148:* 361–370 (1988).

37   Haussinger, D.; Stehle, T.; Gerok, W.: Actions of extracellular UTP and ATP in perfused rat liver. A comparative study. Eur. J. Biochem. *167:* 65–71 (1987).

38   von Kugelgen, I.; Haussinger, D.; Starke, K.: Evidence for a vasoconstriction-mediating receptor for UTP, distinct from the P_2 purinoceptor, in rabbit ear artery. Arch. Pharmacol. *336:* 556–560 (1987).

39   Gilfillan, A.M.; Lewis, A.J.; Rooney, S.A.: Effects of thiazinamium chloride and other antihistamines on phosphatidylcholine secretion in rat type II pneumocyte cultures. Biochem. Pharmac. *36:* 277–281 (1987).

40   Gilfillan, A.M.; Rooney, S.A.: Leukotrienes stimulate phosphatidylcholine secretion in cultured type II pneumocytes. Biochim. biophys. Acta *876:* 22–27 (1986).

41   Rooney, S.A.; Gobran, L.I.: Adenosine and leukotrienes have a regulatory role in lung surfactant secretion in the newborn rabbit. Biochim. biophys. Acta *960:* 98–106 (1988).

42   Lawson, E.E.; Birdwell, R.L.; Huang, P.S.; Taeusch, H.W.: Augmentation of pulmonary surfactant by lung expansion at birth. Pediat. Res. *13:* 611–614 (1979).

43   Nicholas, T.E.; Barr, H.A.: Control of release of surfactant phospholipids in the isolated perfused rat lung. J. appl. Physiol. *51:* 90–98 (1981).

44   Rooney, S.A.; Gobran, L.I.: Adenosine can inhibit as well as stimulate phosphatidylcholine secretion in rat type II pneumocytes. J. Cell Biol. *109:* 864a (1988).

Seamus A. Rooney, PhD, Yale Department of Pediatrics, PO Box 3333, 333 Cedar Street, New Haven, CT 06510 (USA)

Intraalveolar Surfactant Metabolism and Recycling

Wichert P von, Müller B (eds): Basic Research on Lung Surfactant.
Prog Respir Res. Basel, Karger, 1990, vol 25, pp 136–143

Intraalveolar Metabolism of Lung Surfactant

Role of SP-A in Regulating Surfactant Pool Size

Jo Rae Wright[a, b], *Stephen L. Young*[a], *Paul A. Stevens*[a], *John A. Clements*[c]

[a] Cardiovascular Research Institute, Departments of [b] Physiology and [c] Pediatrics, University of California, San Francisco, Calif., USA

After surfactant is secreted by the type II cell into the alveolar airspace it undergoes a complex series of transformations that alter both its morphological and biochemical properties. The secreted lamellar body contents can form the complex lattice-like structure called tubular myelin which is thought by many to be a precursor to the surface film. The sequence of events that occurs after or during surface film formation is not well understood. It does appear, however, that some component or components of surfactant are recycled or internalized by the type II cell [1].

It is perhaps noteworthy that in spite of its complex life cycle, the pool size of surfactant in adult animals that are in a 'steady state' (i.e. not exercising) is remarkably constant. The yield of phospholipid that can be recovered by lavage varies little from day to day or laboratory to laboratory for a given species. It seems that an interesting challenge faces the type II cell. How is secretion and clearance balanced so that pool size is maintained constant in a steady state?

One of the major hypotheses that has been guiding our recent research efforts is that the surfactant protein SP-A (M_r 26,000–36,000) plays an important role in regulating the alveolar pool size of surfactant. We have previously reported that subfractions of surfactant that contain SP-A are taken up by type II cells to a greater extent than are subfractions that do not contain SP-A [2]. SP-A enhances the uptake of lipids by isolated alveolar type II cells [3]. Several groups have now demonstrated that SP-A can inhibit the secretion of phosphatidylcholine by isolated type II cells in primary culture [4–6]. All of these observations when considered together are consistent with the hypothesis that SP-A may regulate pool size by a push-pull mechanism. That is, when there is an excess of SP-A in the alveoli, further secretion is inhibited until the surfactant is taken up by the type II cell. When there is a deficiency of SP-A, secretion commences until sufficient SP-A is available to inhibit further secretion and enhance clearance.

A major emphasis of our recent research has been to study factors that regulate surfactant clearance, turnover, and pool size. The purpose of this article is to review our most recent results on this subject. The three specific questions we have investigated are: (1) Is SP-A, as well as surfactant lipids, taken up by the type II cell and reincorporated into lamellar bodies in adult rats? (2) Do isolated type II cells have a receptor for SP-A? (3) What are the kinetics of secretion and clearance in the immediate newborn period, that is, how are secretion and clearance adjusted so that pool size increases immediately after birth?

Methods

We have described the methods that were used in these studies in detail elsewhere [7–9], and they will be summarized only briefly in this article.

SP-A was isolated by sequential butanol/detergent extraction of rat surfactant or human surfactant obtained from the lavage of a patient with alveolar proteinosis [3]. The SP-A was radioiodinated with Enzymobeads (Bio-Rad Laboratories, Richmond, Calif.).

In order to study the uptake of SP-A in adult rats, the rats were anesthetized with methohexital, and a mixture of liposomes and radioiodinated SP-A (10:1 by weight) were instilled into the tracheal lumen through a 25-gauge needle. Each rat received 80 µl of the mixture which contained 80 µg of phospholipid, a dose which was approximately 10% of the endogenous phospholipid pool. The rats were sacrificed, and lavage and lamellar body fractions were isolated as previously described [10] and analyzed for radioactivity.

In order to characterize the binding of SP-A to type II cells, cells were isolated from adult rat lungs according to the methods of Dobbs et al. [11]. Freshly isolated cells were incubated at 4 °C with ^{125}I-SP-A. Cells and medium were separated by centrifugation and the cells were washed twice. The cells were resuspended in a small amount of medium, and the cell number and radioactivity were determined. In some experiments, potential inhibitors of binding of SP-A were included in the incubation medium.

The kinetics of secretion and clearance were studied in newborn rabbits. Pregnant rabbits (30 days gestation) received ^{3}H-choline intravenously at 8 h prior to delivery. The fetuses were delivered by Caesarean section. Some newborn rabbits received radioactively

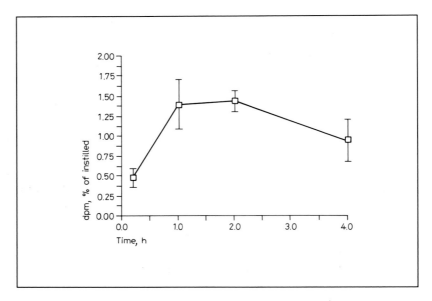

Fig. 1. Uptake of ^{125}I-SP-A into isolated lamellar bodies. Values shown are mean ±
SE for 4 experiments.

labelled natural surfactant intratracheally at the time of delivery. Lavage and lamellar
body pool sizes and specific activities of [3]H (newly synthesized phospholipids) and [14]C
('recycled' phospholipids) were determined at later times. The rate of surfactant secretion
and clearance were determined using a tracer analysis for non-steady state conditions that
has been described in detail in Stevens et al. [9].

Results and Discussion

Is SP-A Taken Up by the Type II Cell and Reincorporated into
Lamellar Bodies in Adult Rats?

We observed a rapid association of intratracheally instilled human
SP-A with the lung tissue and a rapid removal from the lavage fluid. However, after this initial rapid incorporation, there is little change with time
in the tissue-associated radioactivity. There is a time-dependent increase
of radioactivity up to 2 h postinstillation in the lamellar body fraction
(fig. 1). Although the maximal incorporation that we measured was only
approximately 1.5% of the instilled dose of radioactivity, we have esti-

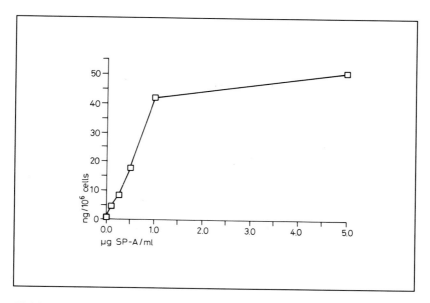

Fig. 2. Specific binding of ^{125}I rat SP-A as a function of SP-A concentration. Values shown are from a representative experiment.

mated that our recovery of lamellar bodies is approximately 16% [2]. Therefore, the actual total amount taken up could be as high as 9% per hour of the instilled dose. Analysis of lamellar body associated proteins by polyacrylamide gel electrophoresis showed that the radioactive labeled proteins co-migrated with intact SP-A. These results are consistent with the hypothesis that SP-A is taken up from the alveoli and incorporated into the lamellar bodies, and eventually resecreted.

Do Type II Cells Have a Receptor for SP-A?

Studies were done to characterize the binding of SP-A to isolated type II cells and to determine if the binding properties were consistent with those of a receptor-mediated process. The specific binding of SP-A to freshly isolated type II cells as a function of SP-A concentration is shown in figure 2. Specific binding was calculated by subtracting nonspecific binding, which was determined by curve stripping of the total binding curve, from total binding. In these experiments, SP-A was incubated with the cells at 4 °C in order to inhibit internalization of the SP-A. The amount of SP-A

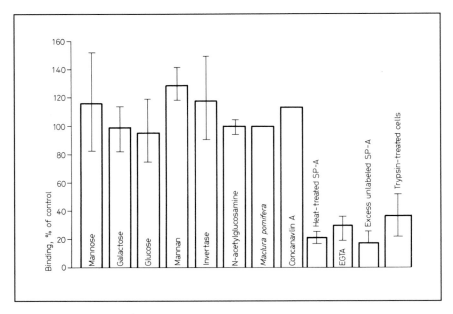

Fig. 3. Potential inhibitors of binding of SP-A to freshly isolated type II cells. Cells were incubated with 1 μg/ml of ^{125}I-rat SP-A alone (control) or in the presence of mannose, galactose, glucose, or N-acetylglucosamine (60 mM) or mannan or invertase (0.5 mg/ml), or the lectins *Maclura pomifera* or concanavalin A (100 μg/ml). Heat-treated SP-A was warmed to 90 °C for 3 min prior to the addition of SP-A. EGTA (0.2 mM) was added just prior to the addition of SP-A. Excess SP-A was added in a 10-fold excess compared to unlabelled SP-A. Type II cells were treated with 0.25% trypsin (from bovine pancreas, type III, Sigma Chemical Co.) for 10 min at room temperature after which time trypsin inhibitor (from soybean, type II-S, Sigma) was added. Cells and medium were separated by centrifugation and cells were resuspended in 1 ml of medium containing radiolabelled protein. All values are expressed at a percentage of the binding obtained in the absence of treatment or potential inhibitors. Values shown are mean ± SE for 3–6 experiments, except the values with the lectins which are the averages of two experiments.

bound to the cells increased with increasing concentrations and reached a plateau at approximately 1 μg/ml. Kuroki et al [12] recently reported that binding of SP-A to type II cells that have been maintained in primary culture for 16–18 h is also saturable.

The effects of various potential inhibitors of binding of SP-A are shown in figure 3. Because SP-A is a carbohydrate-binding protein [15], the effects of various sugars, glycoproteins and lectins on binding were tested. None of the substances tested inhibited binding. Although this observation raises the possibility that SP-A binds to cells via mechanisms other than binding to cell-surface carbohydrates, it does not exclude the possibility that we have not tested the appropriate sugar or that the affinity of SP-A for the cell-surface binding site is much greater than that of its affinity for the sugars. Heat treating the SP-A at 90 °C for 3 min reduced binding by approximately 80%. The inclusion of the calcium chelator, EGTA, or excess unlabelled SP-A also significantly reduced binding. In addition, trypsin treatment of the type II cell surface reduced binding by approximately 60%.

These results demonstrate that binding of SP-A to isolated type II cells displays properties of a receptor-mediated process: binding is saturable, can be reduced by excess unlabelled SP-A, and is reduced by treatment of the cell surface with a proteolytic enzyme. We speculate that the interaction of SP-A with the type II cell is a regulated process that plays an important role in regulating the pool size of alveolar surfactant.

What Are the Kinetics of Secretion and Clearance in the Immediate Newborn Period?

We as well as many other groups observed that there is a large increase in the pool size of alveolar surfactant immediately following birth. We used a tracer analysis for non-steady state conditions to calculate surfactant secretion and clearance rates in the 12 h after delivery. Before birth both secretion and clearance rates were approximately 1.8 µg PL (phospholipid)/g body weight (BW)/h. Shortly after birth the secretion rate increased almost 4-fold to 37.7 µg PL/g BW/h. Between 1.5 and 2 h after birth, secretion fell to prebirth levels and then increased slowly to 6.0 µg PL/g BW/h between 11.5 and 12 h. After birth the clearance rate increased but to a lesser extent than the secretion rate. The maximum clearance rate we calculated was 24.7 µg PL/g BW/h between 0.5 and 1 h after birth. The clearance rate then followed the same general pattern as the secretion rate but the two rates did not become equal in the first day of life.

Conclusions

Our recent studies suggest that SP-A is internalized by type II cells, incorporated into lamellar bodies, and that it is not substantially degraded during the uptake process. Our results also suggest that type II cells contain a receptor for SP-A. The factors that regulate the changes in pool size immediately after birth are not yet understood. Our results suggest that there may be a phasic relationship between secretion and clearance in the immediate newborn period. The role of SP-A, if any, in this process remains to be elucidated.

Acknowledgments

This work was supported in part by Grants HL-30923, HL-24075, and HL-32188 from the National Heart, Lung, and Blood Institute, the Veterans Administration, the American Lung Association-California, and American Heart Association. J.A. Clements is a Career Investigator of the American Heart Association and P.A. Stevens was a Fellow of the American Lung Association, California.

References

1   Wright, J.; Clements, J.; Metabolism and turnover of lung surfactant. Am. Rev. resp. Dis. *135:* 426–444 (1987).

2   Wright, J.; Wager, R.; Hamilton, R.; Huang, M.; Clements, J.: Uptake of lung surfactant subfractions into lamellar bodies of adult rabbit lungs. J. appl. Physiol. *60:* 817–825 (1986).

3   Wright, J.; Wager, R.; Hawgood, S.; Dobbs, L.; Clements, J.: Surfactant Apoprotein M_r = 26,000-36,000 enhances uptake of liposomes by type II cells. J. biol. Chem. *262:* 2888–2894 (1987).

4   Dobbs, L.; Wright, J.; Hawgood, S.; Gonzalez, R.; Venstrom, K.; Nellenbogen, J.: Pulmonary surfactant and its components inhibit secretion of phosphatidylcholine from cultured rat alveolar type II cells. Proc. natn. Acad. Sci. USA *84:* 1010–1014 (1987).

5   Rice, W.; Ross, G.; Singleton, F.; Dingle, S.; Whitsett, J.: Surfactant-associated protein inhibits phospholipid secretion from type II cells. J. appl. Physiol. *63:* 692–698 (1987).

6   Kuroki, Y.; Mason, R.; Voelker, D.: Pulmonary surfactant apoprotein A structure and modulation of surfactant secretion by rat alveolar type II cells. Proc. natn. Acad. Sci. USA *85:* 5566–5570 (1988).

7   Young, S.; Wright, J.; Clements, J.: Cellular uptake and processing of surfactant lipids and apoprotein SP-A by rat lung J. appl. Physiol. 66: 1336–1342 (1989).

8   Wright, J.; Borchelt, J.; Hawgood, S.: Lung surfactant apoprotein SP-A (M_r 26–36 kDa) binds with high affinity to isolated alveolar type II cells. Proc. natn. Acad. Sci. USA (in press, 1989).

9   Stevens, P.; Wright, J.; Clements, J.: Secretion and clearance of surfactant in the newborn (submitted, 1989).

10   Baritussio, A.; Magoon, M.; Goerke, J.; Clements, J.: Precursor-product relationship
      between rabbit type II cell lamellar bodies and alveolar surface-active material. Bio-
      chim. biophys. Acta *666:* 382–393 (1981).
11   Dobbs, L.; Gonzalez, R.; Williams, M.: An improved method for isolating type II
      cells in high yield and purity. Am. Rev. resp. Dis. *134:* 141–145 (1986).
12   Kuroki, Y.; Mason, R.; Voelker, D.: Rat alveolar type II cells express a high affinity
      receptor for surfactant protein A (SP-A). Proc. natn. Acad. Sci. USA *85:* 5566–5570
      (1988).
13   Haagsman, H.; Hawgood, S.; Sargeant, T.; Buckley, D.; White, R.; Drickamer, K.;
      Benson B.: The major lung surfactant protein, SP 28-36, is a calcium-dependent,
      carbohydrate-binding protein. J. biol. Chem. *262:* 13877–13880 (1987).

Jo Rae Wright, PhD, Cardiovascular Research Institute, University of California,
Box 0130, San Francisco, CA 94143 (USA)

Wichert P von, Müller B (eds): Basic Research on Lung Surfactant.
Prog Respir Res. Basel, Karger, 1990, vol 25, pp 144–148

Surfactant Catabolism and Recycling in the Perinatal Period

Alan Jobe

Department of Pediatrics, Harbor-UCLA Medical Center, Torrance, Calif., USA

Surfactant is being used for the treatment of infants with respiratory distress syndrome and may be used in adult patients with a variety of lung diseases. The fate of exogenous surfactant has not been well studied and any studies will be complicated by the complex metabolism of surfactant components within the type II cell and within the alveolus [1]. Surfactant metabolism is particularly intriguing in the perinatal period because of differences in surfactant metabolism between developing and adult animals and because of the transitional events that occur with the onset of air breathing at birth [2]. Intravascularly injected radiolabeled precursors of surfactant phosphatidylcholine and intratracheal injections of trace doses of labeled surfactant were used to measure the turnover time and efficiency of recycling of surfactant phosphatidylcholine in 3-day-old rabbits [2, 3]. In contrast to adult rabbits, the turnover time of surfactant phosphatidylcholine was somewhat longer at about 11 h versus estimates of 3–11 h for adult animals, the efficiency of recycling was over 90%, and very little catabolism of surfactant phosphatidylcholine could be demonstrated (fig. 1). Radiolabeled surfactant phosphatidylcholine given to term lambs at delivery also was lost from the lung with a very long half-life of about 6 days [4]. About 50% of the labeled phosphatidylcholine remaining in the lungs of the lambs was recovered by alveolar wash at all times. The pattern of labeling of the lamellar body pool was consistent with recycling in these term lambs; however, estimates of recycling efficiencies were not made.

Surfactant metabolism at birth is characterized by increased secretory activity resulting in increased alveolar surfactant pool sizes. Alveolar surfactant is thought to have a 'life cycle' characterized by changes in aggregate size, structure, and surfactant-specific protein content of the alveolar

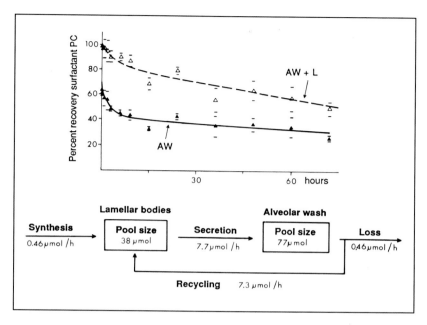

Fig. 1. Clearance of labeled phosphatidylcholine (PC) from a trace dose of surfactant given into the airways of 3-day-old rabbits and estimates of pool sizes and fluxes of surfactant PC per kilogram body weight for these rabbits. The 3-day-old rabbits weigh about 56 g. There was a slow loss of labeled PC from the alveolar wash (AW) and the total lung calculated as the sum of recoveries in the lung tissue after alveolar wash and the alveolar wash (AW+L). The fluxes were calculated using a turnover time of 10 h based on data from Jacobs et al. [2, 3].

surfactant [1]. Recent experiments by Stevens et al. [5] and Bruni et al. [6] showed that the surfactant secreted at birth was in the form of essentially intact lamellar bodies and that the smaller aggregate forms appeared subsequently. The surfactant subfractions came to distribute similarly to those found in adult rabbits within a few hours of birth. These transformations presumably represent the initiation of the alveolar 'life cycle' of surfactant. However, surfactant reuptake also was detected in the immediate newborn period by Stevens et al. [7] using radiolabeled tracers and a nonsteady-state analysis to compensate for changes in pool sizes that occur with birth. Thus, even during increased secretory activity, recycling of surfactant components occurred. These experiments point out the very dynamic nature of surfactant metabolism.

Recycling activity with little demonstrable catabolic activity also occurred in the preterm lung. Modestly preterm lambs that were delivered by cesarean section at 138 days' gestational age required about 50% oxygen and 20 cm H_2O peak inspiratory ventilator pressures to achieve pO_2 values of about 100 mm Hg and pCO_2 values of 35 mm Hg [8]. These lambs received a trace amount of radiolabeled natural surfactant that was mixed with the fetal lung fluid at birth. Soon after birth and ventilation only about 55% of the labeled surfactant could be recovered by alveolar wash. There was decreased recovery with time until only about 20% was in the alveolar wash 24 h after delivery (fig. 2). However, the labeled surfactant became 'lung associated' in that it was not recovered by alveolar wash. There also was an increased specific activity in lamellar body fractions. The sum of the recoveries in the alveolar wash and lung tissue indicated no net loss of the label from the lungs.

These same lambs also received an intravascular injection of [^{14}C]choline at 1 h of age to label the de novo synthetic and secretory pools. The appearance of radiolabeled phosphatidylcholine in the air spaces was detected, but the total alveolar saturated phosphatidylcholine pool size did not change significantly from 11 min to 24 h after birth. The entrance of de novo synthesized phosphatidylcholine into the air spaces seemed to contribute very little to the pool size of surfactant in the air spaces. When the various measurements from this experiment were combined, the constant alveolar pool size despite the loss of labeled surfactant from that pool and the secretion of de novo synthesized surfactant phosphatidylcholine were consistent with active recycling of surfactant. Presumably alveolar surfactant pool size was being maintained by phosphatidylcholine from intracellular stores and not primarily by de novo synthesis [9]. It should also be noted that lung function and in vitro surface properties of the surfactant did not deteriorate over the 24-hour study period. Surfactant subfractions did not demonstrate the predominance of large aggregate forms described for the term rabbit [5, 6], perhaps because of alterations in surfactant metabolism caused by prematurity or mechanical ventilation [8].

These results in term newborn and preterm animals were reassuring in terms of surfactant treatment for respiratory distress syndrome. There was very little catabolic activity in the prenatal lung. Although much of the trace doses of surfactant became 'lung associated' and thus was not in the alveolar pool, the behavior of surfactant was consistent with recycling in the perinatal lung. In support of the idea that surfactant used for treatments could be acting as substrate to prime the metabolic pathways, pre-

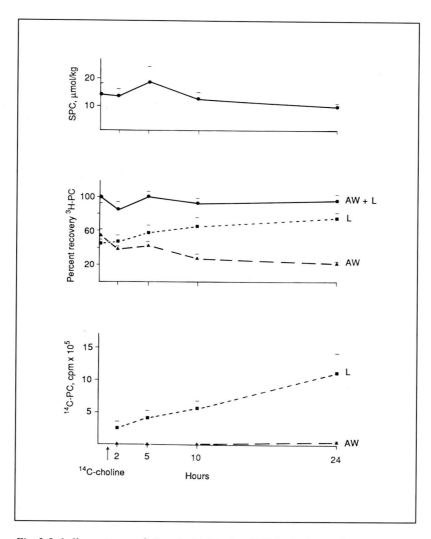

Fig. 2. Labeling patterns of phosphatidylcholine (PC) in the lungs of preterm ventilated lambs. The lambs were studied at 138 days' gestational age. The figure is labeled to
indicate the alveolar saturated PC (SPC) pool size per kilogram, the percent recovery of a
trace dose of labeled PC from labeled surfactant that had been mixed with the fetal lung
prior to birth, and the labeling of endogenously synthesized and secreted PC following an
intravascular injection of radiolabeled choline at 1 h of age. The total lung recoveries
(AW+L) were the sum of the alveolar wash (AW) plus lung tissue (L) recoveries. The
labeling patterns were consistent with recycling of surfactant PC by the lungs. (Figure
redrawn from Jobe et al. [8] with permission.)

term lambs treated with either natural sheep surfactant or Surfactant-TA had stable respiratory function and maintained surfactant aggregate size distributions comparable to those found in healthy adult animals [10]. The movements of radiolabeled phosphatidylcholine from de novo synthetic pools and from the air spaces to the lung were consistent with recycling activities. There were no remarkable differences between the natural surfactant or the lipid solvent extracted Surfactant-TA, and neither surfactant was lost from the lungs, again indicating that there was very little catabolic activity present in the preterm lung. The clinical effectiveness of surfactant treatments may be explained in part by the ability of the preterm lung to metabolize the replacement surfactant to maintain surfactant function.

References

1   Wright, J.; Clements, J.: Metabolism and turnover of lung surfactant. Am. Rev. respir. Dis. *136:* 426–444 (1987).

2   Jacobs, H.; Jobe, A.; Ikegami, M.; Jones, S.: Surfactant phosphatidylcholine source, fluxes and turnover times in 3-day-old, and adult rabbits. J. biol. Chem. *257:* 1805–1810 (1982).

3   Jacobs, H.; Jobe, A.; Ikegami, M.; Conaway, D.: The significance of reutilization of surfactant phosphatidylcholine. J. biol. Chem. *258:* 4159–4165 (1983).

4   Glatz, T.; Ikegami, M.; Jobe, A.: Metabolism of exogenously administered natural surfactant in newborn lamb. Pediat. Res. *16:* 711–715 (1982).

5   Stevens, P.; Wright, J.; Clements, J.: Changes in quantity, composition and surface activity of alveolar surfactant at birth. J. appl. Physiol. *63:* 1049–1057 (1987).

6   Bruni, R.; Baritussio, A.; Quaglino, D.; Gabelli, C.; Benevento, M.; Ronchetti, J.: Postnatal transformations of alveolar surfactant in the rabbit. Changes in pool size, pool morphology and isoforms of the 32–38 kd apoprotein. Biochim. biophys. Acta *958:* 255–267 (1988).

7   Stevens, P.; Wright, J.; Clements, J.: Phasic lung surfactant secretion and clearance in the newborn rabbit FASEB J. *2:* A1274 (1988).

8   Jobe, A.; Ikegami, M.; Seidner, S.; Pettenazzo, A.; Ruffini, L.: Surfactant phosphatidylcholine metabolism and surfactant function in preterm ventilated lambs. Am. Rev. respir. Dis. *139:* 352–359 (1989).

9   Jobe, A.: The role of surfactant in neonatal adaption. Semin. Perinatol. *12:* 113–123 (1988).

10 Ikegami, M.; Jobe, A.; Yamada, T.; Seidner, S.; Priestly, A.; Ruffini, L.: Surfactant metabolism in surfactant treated ventilated lambs. Pediat. Res. *23:* 510A (1988).

Alan Jobe, MD, PhD, Department of Pediatrics, Bldg. A-17 Annex, Harbor-UCLA Medical Center, Torrance, CA 90509 (USA)

Wichert P von, Müller B (eds): Basic Research on Lung Surfactant.
Prog Respir Res. Basel, Karger, 1990, vol 25, pp 149–157

Modulation of Lung Surfactant Clearance

Aron B. Fisher, Avinash Chander, Chandra Dodia

Institute For Environmental Medicine, University of Pennsylvania,
Philadelphia, Pa., USA

Recent work has elucidated the intracellular routes for synthesis and secretion of the components of lung surfactant. Much less information is available concerning the pathways for surfactant clearance. Potential pathways for removal of surfactant components from the alveolar space include: (1) degradation in alveolar space (extracellular); (2) phagocytosis and degradation by alveolar macrophages; (3) removal by tracheobronchial mucociliary clearance, and (4) uptake by epithelium.

The failure to detect metabolic products or degradative enzymes in lung lavage fluid indicates that surfactant components are not degraded in the extracellular space [5]. Alveolar macrophages phagocytize and degrade surfactant, but analysis of cells lavaged from the lungs suggests that this is a relatively minor pathway for surfactant clearance [2]. Likewise, the removal of surfactant via the tracheobronchial mucociliary clearance mechanism appears to contribute little to overall surfactant clearance [8]. The most convincing recent evidence indicates that the major pathway for removal of surfactant from the alveolar space is by uptake through the alveolar epithelium [3]. Isolated granular pneumocytes (type II epithelium) internalize surfactant components suggesting that these cells participate in surfactant reuptake [4]. A possible role for membranous (type I) epithelium has not yet been adequately investigated. Theoretically, uptake of surfactant could be nonspecific or could occur through a specific receptor-mediated process. These possibilities are under investigation and different uptake mechanisms could be operative for different lipid and protein components of surfactant.

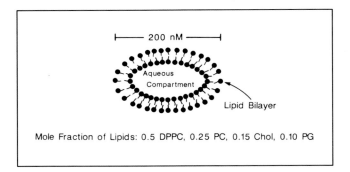

Fig. 1. Schematic representation of unilamellar liposomes prepared by sonication followed by centrifugation to remove large aggregates [5]. The lipid composition was chosen to approximate that of surfactant. [³H]-DPPC was included as a tracer. In some experiments [¹⁴C]-sucrose was encapsulated in the aqueous compartment.

In this study, we evaluated some factors that might control the rate of surfactant clearance. Our basic hypothesis is that surfactant secretion and reuptake are linked in the normal lung through (undefined) cellular mechanisms in order to regulate the extracellular concentration of surfactant in the alveolar space. To test this hypothesis, we evaluated known secretagogues for their effect on phospholipid uptake by lungs. The model chosen for these studies was the isolated perfused rat lung which permitted study of surfactant secretion and clearance without the confounding effects of systemic variables.

Methods

Clearance of lipid was evaluated by using unilamellar liposomes with a composition that approximated that of natural surfactant with [³H]-dipalmitoyl-phosphatidylcholine (DPPC) as a radioactive tracer (fig. 1). Liposomes suspended in a small volume of saline (0.1 ml) were instilled in the anesthetized rat through an endotracheal cannula at the level of the carina. The rat was permitted to breathe spontaneously for several minutes and then the lungs were removed for isolated lung perfusion. The liposomes appear to be distributed during this initial period throughout the tracheobronchial tree [5]. Perfusion was begun approximately 5 min after instillation of liposomes and this is regarded as 'zero' time since it is when the lungs begin exposure to perfusate mediators. After varying duration of perfusion, the lungs were lavaged with saline. The radioactivity remaining in the lavaged lung was measured and was considered lipid uptake.

Table 1. Radioactivity remaining in lung after 5 lavages at 5 min ('zero time') after endo-tracheal instillation of liposomes

Tracer	Condition	n	% of instilled dpm
[^{3}H]-DPPC	in liposomes	4	10.3 ± 0.37
[^{14}C]-Dextran	extraliposomal	4	0.68 ± 0.11
[^{14}C]-Sucrose	extraliposomal	4	2.2 ± 0.13
[^{14}C]-Sucrose	liposomal (aqueous)	3	11.4 ± 0.84
[^{3}H]-PC	natural surfactant	3	32.2 ± 1.7

In some experiments, radiolabeled natural surfactant was instilled into lungs in order to compare its uptake with that of liposomes. Radiolabeled natural surfactant was prepared by perfusing isolated rat lungs for approximately 6 h with [^{3}H-methyl]-choline. Surfactant was isolated from the lung lavage fluid by density gradient centrifugation, lyophilized, and resuspended in saline by sonication.

In order to measure degradation of internalized lipids, tissue was extracted with chloroform:methanol and the radioactivity was measured in organic and aqueous fractions and further analyzed by thin-layer chromatography. Disaturated PC (DSPC, representing DPPC plus other species of PC with 2 saturated fatty acids) was measured by osmication. Microsomal and lamellar body fractions were isolated from lung homogenates by density gradient centrifugation. The methods used for these studies have been described previously [5, 6].

Results and Discussion

Initial Rate of Lipid Uptake

Lungs studied at 5 min after instillation of radiolabelled liposomes showed a significant fraction of lipid that could not be removed by lung lavage (table 1). By contrast, the extraliposomal aqueous marker was recovered almost totally. When the aqueous marker was encapsulated in the liposome, recovery of aqueous and lipid labels were similar. Therefore, some fraction of instilled liposomal lipids rapidly became inaccessible to lung lavage suggesting rapid cell association by internalization or membrane fusion. The percent of instilled lipid rapidly internalized was significantly greater for biosynthesized (natural) surfactant compared with synthetic liposomes (table 1).

Table 2. Effect of agonists on uptake of liposomes by isolated perfused rat lung

Additions to perfusate	[³H]-DPPC retention at 2 h, % of instilled	
	total uptake[1]	corrected for 'zero' time[2]
No additions	18.6 ± 0.60	8.3
Isoproterenol, 500 μM	26.8 ± 1.45*	16.5
+ Propranolol, 500 μM	20.4 ± 0.51	10.1
cAMP (8-Br), 100 μM	32.9 ± 0.65*	22.6
cGMP (8-Br), 100 μM	21.8 ± 0.48	11.5

* $p < 0.05$ vs. no addition by ANOVA.
[1] Mean ± SE for n = 4 for each condition.
[2] 'Zero' time was obtained 5 min after endotracheal instillation as in table 1.

Effect of Secretagogues on Lipid Uptake

We evaluated the effect of known secretagogues for lung surfactant on the uptake of liposomes by the isolated lung during 2 h of perfusion. Both isoproterenol and a cAMP analogue resulted in significant increase in retention of liposomes in the lung after lavage; a cGMP analogue had no significant effect (table 2). The effect of isoproterenol was blocked by the beta-receptor antagonist, propranolol. In subsequent experiments, addition of terbutaline (0.1 mM), adenosine triphosphate (ATP; 1 mM), or phorbol ester (TPA, 30 ng/ml) also significantly stimulated uptake of liposomes to approximately the same extent as 8-Br-cAMP. Uptake of natural surfactant at 2 h was 46.6 ± 0.8% of instilled in control lungs and 59.8 ± 2.0% with 8-Br-cAMP.

Beta-adrenergic agonists (terbutaline, cAMP), ATP, and TPA all significantly stimulated surfactant secretion by the isolated perfused rat lung in the present study (data not shown) and in previous reports [1, 7, 10–12]. The present results indicate that secretagogues for lung surfactant increased the uptake of phospholipids by lung epithelium and support a possible linkage between secretion and uptake of phospholipids.

Fate of Internalized Lipid

We next considered the fate of the phospolipid accumulated by the perfused lung. The possibilities include: (1) transport to interstitium and then into lung perfusate; (2) degradation of phospholipids and reutilization

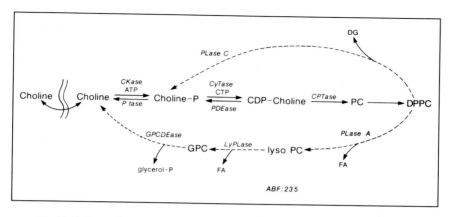

Fig. 2. Pathway for choline incorporation into DPPC and for DPPC degradation. Synthetic pathways are indicated by the solid arrows and degradative pathways by the dashed lines. The intermediates in heavy type indicate choline-containing compounds. The curvy lines at the left represent a cell membrane. FA = Fatty acid; DG = diglyceride.

of components for surfactant synthesis; (3) recycling of intact phospholipids into lung surfactant.

The possibility of transport to the perfusate was evaluated by analysis of perfusate radioactivity during lung perfusion. With control lungs, perfusate radioactivity increased to approximately 3% of the instilled liposomal radioactivity at 1 h and then remained essentially unchanged during the next 2 h. Similar levels of perfusate radioactivity were found in the presence of each of the secretagogues. Perfusate activity could acount for less than 15% of the total radioactivity taken up by the lung during the 2-hour perfusion. Perfusate analysis indicated that all of the radioactivity was in the form of free choline indicating that intact lipid was not transported across the interstitium into the perfusate.

Since free radiolabeled choline appeared in the perfusate, we further evaluated the metabolic fat of the internalized radiolabelled lipid by analysis of the lung homogenate. Potential pathways for degradation of DPPC are shown in figure 2. In the initial liposomal suspension, all of the radiolabel was in DPPC. At 2 h of perfusion, significant label was also present in lysoPC and in aqueous soluble metabolites indicating DPPC degradation (table 3). The appearance of label in aqueous metabolites was approximately doubled in the presence of 8-Br-cAMP (table 3) and with ATP, TPA or terbutaline (data not shown). We also noted appearance of radio-

Table 3. Degradation of instilled [^{3}H]-DPPC in isolated rat lung at 2 h of perfusion

Fraction of lung extract	% of homogenate dpm	
	control	cAMP (8-Br)
Organic soluble		
PC	88.5 ± 0.09	73.4 ± 0.6
DSPC	75.0 ± 0.8	56.7 ± 0.7
LysoPC	1.5 ± 0.1	7.3 ± 0.6
Aqueous soluble		
Choline	6.7 ± 0.1	6.0 ± 0.5
Choline-P	1.6 ± 0.08	4.4 ± 0.4
CDP-Choline	1.1 ± 0.04	4.6 ± 0.4
GPC	0.6 ± 0.04	4.0 ± 0.3

Results are mean $\pm$ SE for 8 experiments under each condition. Values are expressed as % of dpm recovered in lung homogenate. GPC = Glycerophosphoryl choline.

label in PC with at least one unsaturated fatty acid (indicated in table 3 as total PC minus DSPC). The appearance of radiolabel in CDP-choline and unsaturated PC indicates reutilization of components for phospholipid resynthesis. Thus, degradation with reutilization appears to be a major fate of internalized phospholipids.

The reutilization of the phospholipid components led us to evaluate the appearance of radioactivity in the microsomal and lamellar body fractions from lung homogenates (table 4) representing the sites of lung surfactant phospholipid synthesis and storage. Control studies in which known amounts of labelled liposomes were added to lung homogenates indicated that the 'zero' time values could be accounted for by contamination of fractions with extracellular lipid. There was a significant increase of radioactivity in each of these fractions during lung perfusion and appearance in both fractions was increased by 8-Br-cAMP. Recovery of dpm at 2 h as a % of total lung homogenate was $9.4 \pm 0.7\%$ for lamellar body and $10.9 \pm 1.01\%$ for microsomal fractions (mean $\pm$ SE; n = 4), although this probably underestimates the actual transfer to these compartments because of incomplete recoveries of organelles.

The appearance of radioactivity in lamellar bodies and microsomes can be accounted for in part by resynthesis from radiolabelled degradation

Table 4. Appearance of radioactivity in isolated lamellar body and microsomal fractions of perfused lung

	Lamellar bodies		Microsomes	
	dpm/g protein	unsat. PC % of PC	dpm/g protein	unsat. PC % of PC
'Zero' time[1]	273 ± 8	–	12.0 ± 0.4	–
2-hour control	1,200 ± 85	9.5	30.8 ± 8.4	20.7
2-hour 8-Br-cAMP	1,900 ± 221	29.7	41.6 ± 3.3	45.8

Results are mean ± SE for n = 4 for each condition.
Unsaturated PC = Total PC-DSPC. These are mean values (n = 4) and are corrected for contamination with extracellular lipid (100% of label in DSPC) by subtracting the 'zero' time value.
[1] Measured 5 min after endotracheal instillation of liposomes.

products from DPPC. At 2 h of perfusion, approximately 20% of total PC in microsomes and 10% in lamellar bodies was present in an unsaturated form and this percentage was significantly increased in the presence of 8-Br-cAMP (table 4). At 2 h of perfusion in the presence of 8-Br-cAMP, approximately 30% of lamellar body radiolabelled PC is unsaturated and presumably synthesized de novo. The lamellar body fraction contains 2:1 DSPC:unsaturated PC, and, assuming that these lipids are packaged at the same rate, de novo synthesized DSPC at 2 h with 8-Br-cAMP is estimated at 60% of radiolabelled PC.

Consequently, total de novo synthesized radiolabelled PC (saturated plus unsaturated) is estimated at 90% of total PC, suggesting that for this condition, very little of the lamellar body DPPC has been directly recycled without degradation and resynthesis. Using similar assumptions, de novo synthesized radiolabelled DPPC would account for only 30% of lamellar body radiolabelled PC under control conditions at 2 h of perfusion. These findings suggest that the surfactant lipid in lamellar bodies is derived from both directly recycled DPPC without degradation plus de novo synthesized PC (and DSPC) utilizing [3H]-choline derived from DPPC breakdown. Recycling of intact surfactant phospholipid by the lung has been demonstrated previously [9].

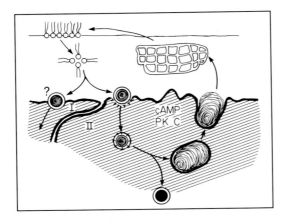

Fig. 3. Schematic representation of the putative events in the surfactant cycle. I and II indicate types 1 and 2 lung epithelial cells, respectively. Surfactant is stored in lamellar bodies, is secreted by exocytosis, and undergoes transformation to tubular myelin and then a monolayer at the air-water interface. Subsequent events include formation of vesicles, internalization of lipid through coated pits with formation of an endosome, and then degradation in a lysosomal compartment or transfer to lamellar bodies. Cyclic AMP and activators of protein kinase C increase secretion as well as internalization of surfactant components. The endocytic pathway remains speculative and the possible involvement of type 1 cells in uptake has not been adequately evaluated.

Conclusions

These results with isolated perfused rat lung indicate that exogenous phospholipids in the alveolar space are internalized by lung epithelium and in part degraded and reutilized for lung surfactant phospholipid synthesis. The rate of uptake, degradation, and reutilization is increased in the presence of secretagogues for lung surfactant. These results suggest that the surfactant cycle is regulated in order to provide optimal concentrations of extracellular material. A model incorporating these control mechanisms is shown in figure 3. The precise mechanism for regulation and integration of the surfactant cycle has not yet been determined.

This study raises several questions which have not yet been answered for our system. What is the mechanism for the initial rapid rate of lipid uptake and for the greater initial rate of uptake of natural surfactant compared with synthetic lipid? Are the membranous pneumocytes (type 1 epithelium) involved in phospholipid uptake? Are surfactant proteins internalized, either separately or as a complex with phospholipid, and what is their metabolic fate? Is the uptake of lipid ± protein receptor mediated? We expect that answers to these questions will be available by the time of the next International Surfactant Symposium, providing important new insights into the cell biology of this fascinating system.

References

1   Brown, L.; Longmore, W.: Adrenergic and cholinergic regulation of lung surfactant secretion in isolated perfused rat lung and in the alveolar type II cell in culture. J. biol. Chem. *256:* 66–72 (1981).
2   Desai, R.; Tetley, T.; Curtis, C.; Powell, G.; Richards, R.: Studies on fate on pulmonary surfactant in the lung. Biochem. J. *176:* 455–462 (1978).
3   Fisher, A.; Chander, A.: Intracellular processing of surfactant lipids in the lung. A. Rev. Physiol. *47:* 789–802 (1985).
4   Fisher, A.; Chander, A.; Reicherter, J.: Uptake and degradation of natural surfactant by isolated rat granular pneumocytes. Am. J. Physiol. *253:* C792–C796 (1987).
5   Fisher, A.; Dodia, C.; Chander, A.: B-Adrenergic mediators increase pulmonary retention of instilled phospholipids. J. appl. Physiol. *59:* 743–748 (1985).
6   Fisher, A.; Dodia, C.; Chander, A.: Degradation and reutilization of alveolar phosphatidylcholine by rat lungs. J. appl. Physiol. *62:* 2295–2299 (1987).
7   Gilfillan, A.; Rooney, S.: Purinoceptor agonists stimulate phosphatidylcholine secretion in primary cultures of adult rat type II pneumocytes. Biochim. biophys. Acta *917:* 18–23 (1987).
8   Hallman, M.; Epstein, B.; Gluck, L.: Analysis of labeling and clearance of lung surfactant phospholipid in rabbits. Evidence of a bidirectional surfactant flux between lamellar bodies and alveolar lavage. J. clin. Invest. *68:* 742–751 (1981).
9   Jacobs, H.; Ikegami, M.; Jobe, A.; Berry, D.; Jones, S.: Reutilization of surfactant phosphatidyl choline in adult rabbits. Biochim. biophys. Acta *837:* 77–84 (1983).
10 Massaro, S.; Clerch, L.; Massaro, G.: Surfactant secretion. Evidence that cholinergic stimulation of secretion is indirect. Am. J. Physiol. *243:* C39–C45 (1982).
11 Rice, W.; Singleton, F.: P2-purinoceptors regulate surfactant secretion from rat isolated alveolar type II cells. Br. J. Pharmacol. *89:* 485–491 (1986).
12 Sano, K.; Voelker, D.; Mason, R.: Involvement of protein kinase C in pulmonary surfactant secretion from alveolar type II cells. J. biol. Chem. *260:* 12725–12729 (1985).

A.B. Fisher, MD, Director, Institute for Environmental Medicine,
University of Pennsylvania, 1 John Morgan Building, 36th and Hamilton Walk,
Philadelphia, PA 19104-6068 (USA)

Wichert P von, Müller B (eds): Basic Research on Lung Surfactant.
Prog Respir Res. Basel, Karger, 1990, vol 25, pp 158–167

Surfactant and Alveolar Micromechanics[1]

H. Bachofen, S. Schürch, M. Urbinelli

Departments of Anatomy and Medicine, University of Berne, Switzerland;
Department of Medical Physiology and Medicine, University of Calgary, Alta.,
Canada

Classical experiments have illustrated the substantial contributions of
alveolar surface forces to lung recoil [6, 9, 11]: elimination of surface tension by filling the lung with liquid or modifications of the alveolar surface
film (by rinsing the lung with a detergent, by covering the alveoli with test
liquid films [1, 14], etc.) change the pressure-volume (P-V) behavior of
lungs in a distinct manner (fig. 1). However, the global elastic properties of
lungs as reflected by P-V curves do not reveal the micromechanics of lung
parenchyma, and, in particular, do not give information about the
influence of surface forces on alveolar surface area, configuration of
peripheral airspaces, and alveolar stability. In order to better define the
structure-function relations at the alveolar level new techniques had to be
developed, i.e. methods to prepare and fix lungs without disturbing the
alveolar surface [3, 17], and to determine alveolar surface tensions in situ
[12, 13]. The ensuing studies demonstrated that alveolar surface forces
exert a moulding effect on the delicate alveolar tissue elements, and regulate by a complex interplay with tissue forces the alveolar surface area and
the dimensions of peripheral air spaces.

Methods

The procedures and methods to prepare and fix excised rabbit lungs under well-
controlled physiological conditions have been given in detail elsewhere [2–4, 8]. In short,
after opening the thorax the lungs were connected to a perfusion system by cannulating

[1] This work was supported by the Swiss National Science Foundation (grants No.
3.128.81 and 3.036.84), and by the Medical Research Council of Canada (grant MA
6435). The authors gratefully acknowledge the superb assistance given by Karl Babl and
Christoph Lehmann.

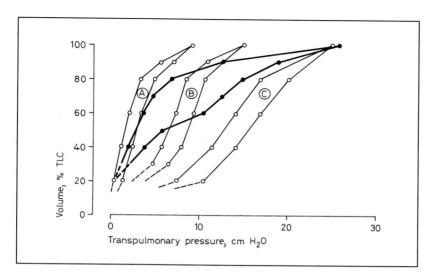

Fig. 1. Pressure-volume curves of a normal lung (heavy line), and of lungs with altered alveolar surface tension. A = Saline-filled lung ($\gamma = 0$ mN/m); B = lung rinsed with a fluorocarbon ($\gamma \sim 15$ mN/m); C = lung rinsed with a detergent ($\gamma \sim 20$–30 mN/m).

the pulmonary artery and the left atrium. Following careful excision of the lungs and heart, studies of lung mechanics were carried out (repeated registrations of P-V cycles in normal air-filled lungs, in saline-filled lungs [8], in air-filled lungs after alveolar lavage with a detergent [2], or with a fluorocarbon test fluid). Thereafter, the lungs were fixed by vascular perfusion with osmium tetroxide and uranyl acetate under controlled physiological conditions with regard to lung volume, lung weight, and perfusion pressures [3]. In some experiments alveolar surface tensions were measured in situ immediately before fixation using an improved microdroplet method [13].

For morphologic and morphometric evaluation lung tissue samples were taken from six different areas according to the principles of stratified random sampling. On the same day, tissue blocks were processed for light and transmission electron microscopy, and embedded in Epon following standard procedures. For morphometric evaluation stereologic methods were employed as described in recent publications [15, 16].

Results and Discussion

Qualitative Findings

Light micrographs and scanning electron micrographs of air- and saline-filled lungs look strikingly different (fig. 2). In saline-filled lungs fixed at medium and lower lung volumes the alveolar septa are undulated,

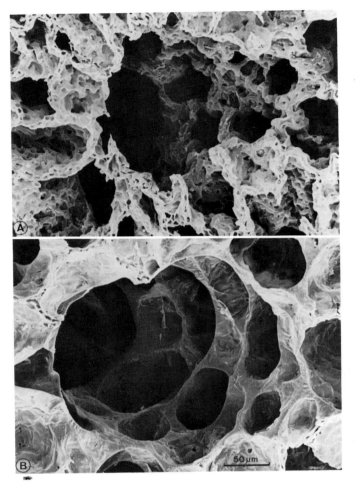

Fig. 2. Saline-filled (A) and air-filled (B) rabbit lungs, fixed at the same volume of about 70% TLC. Note the different alveolar texture and the differences in width of alveolar ducts.

suggesting that part of alveolar septal fibers are not under tension; only at high lung volumes does the entire tissue meshwork appear stretched [8]. In air-filled lungs, in contrast, the alveolar septa appear 'flat' at any degree of inflation. The fiber containing septal midlines form serpentines between the closed-up capillaries, and particularly in alveolar corners folds and pleats may occur: the excess fiber length is accommodated beneath a

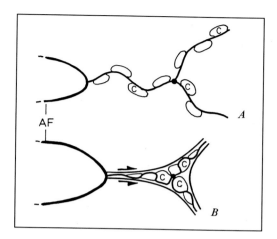

Fig. 3. Septal arrangements in fluid-filled lungs (*A*), and in air-filled lungs (*B*). In the latter, the axial fibers (AF) are stretched by surface forces, the alveolar duct widens, and the alveolar surface area decreases. c = Capillaries which are moulded into a flat surface by surface forces in air-filled lungs.

smooth surface (fig. 3). In saline-filled lungs the capillaries are evenly expanded and bulge into the alveolar lumen, whereas in air-filled lungs the capillaries are fitted to the flat shape of alveolar septa (fig. 3); it is evident that this is due to the moulding effect of surface tension.

Scanning electron micrographs give further evidence of the shaping influence of surface tension on the complex of alveoli and alveolar ducts. In saline-filled lungs alveolar ducts are relatively narrow, and the free edges appear crinkly. In normal air-filled lungs the ducts are more expanded, and a tense line element with fairly uniform radius of curvature can be recognized at the alveolar openings. This transformation is most conspicuous if surface tension is abnormally increased by rinsing the lung with a detergent [2, 18].

Figure 4 illustrates the moulding effect of surface forces in a more gradual way. With increasing surface tension the airspaces become wider, and more septal pleats and folds can be recognized. When subjected to very high surface tensions as in detergent-rinsed lungs the usual alveolar architecture disappears in that masses of piled-up septa separate widened alveolar ducts. Evidently, this transformation is accompanied by a substantial decrease in alveolar surface area [2].

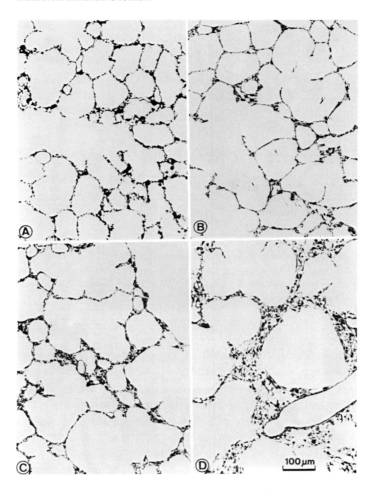

Fig. 4. Light micrographs of air-filled rabbit lungs fixed at 40% TLC. Normal lung on deflation (A) and on inflation (B) with surface tensions of about 1 mN/m and of about 6 mN/m, respectively, as determined by the microdroplet method of Schürch et al. [13]. C Lung rinsed with a fluorocarbon (γ about 15 mN/m). D Detergent-rinsed lung (γ about 20–30 mN/m).

All these findings clearly show that the surface forces markedly influence the peripheral airspaces. Alveoli are not functional units, and hence lung parenchyma cannot be pictured by a collection of balloons. Indeed, larger mechanical structures have to be considered for the design of models which are consistent with anatomy.

A Model of Pulmonary Mechanics

An important part of the functional structure of lung parenchyma is the fiber skeleton of the lung, which shows a particular tripartite organisation [10]: (1) strong *peripheral* tissue elements extend from the pleura into the intersegmental and interlobular septa and eventually deep into the lung parenchyma; (2) *axial* connective tissue consisting of tissue sheaths accompanying the airways penetrate deep into the acini where they find their continuation as the strong fibrous network surrounding the alveolar ducts, i.e. what appears as the lattice of alveolar entrance rings, and (3) a delicate network of *septal* fibers, which is anchored at the elements of both the axial and peripheral connective tissue, interlaces with the capillaries.

All these observations can be incorporated into the model shown in figure 5 [18]. The functional unit is the alveolar duct (or a set of alveolar ducts forming an acinus) embraced by peripheral connective tissue. Fibers instead of tissue membranes are the force-bearing elements, and the dimensions of the peripheral air spaces are governed by an interplay between surface and tissue tensions.

The main force-bearing line elements belong to the peripheral and the axial connective tissue systems. The peripheral fiber network is thought to behave as an interdependent system connected to the pleura; it is largely insensitive to changes in surface tension. In contrast, the force-bearing line elements of the axial connective tissue system that forms the lattice of alveolar entrance rings is extended by surface tension on the alveolar wall. The two-dimensional alveolar walls which are arranged between both strong fiber systems are negligible mechanical components which serve as platforms for surface tension at the air-liquid interface.

Low surface tensions allow a large alveolar surface area between slightly stretched axial fibers. However, when surface tension is abnormally high the axial fibers become more stretched resulting in a disproportionately enlarged duct, in a flattening of alveoli, and in a decrease of the alveolar surface area (fig. 3–5). The major supporting elements of the fiber scaffold are located at the periphery and in the axial parts of the gas exchange units; the gas-exchanging surface is thus relieved from carrying major loads so that the air-blood barrier can be very thin. This advantage has to be paid for by a relative 'instability'.

The Problem of Stability

In this model, alveoli are not individual structures but subunits of the duct system or of an acinus. Even under normal conditions parts of alveo-

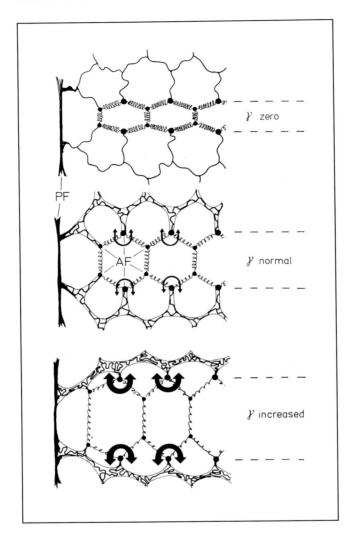

Fig. 5. Model for the mechanical structure of the alveolar duct. Geometry of peripheral air spaces changes as a function of surface tension (γ). At constant volume an increase in surface tension causes diameter of duct to increase, and depth and surface area of alveoli to decrease. PF = Fibers of the peripheral connective tissue system; AF = elements of the axial fiber system that form alveolar entrance rings.

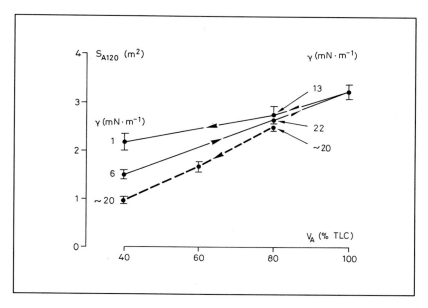

Fig. 6. Relationship between alveolar surface area (S_{A120}; normalized to a TLC of 120 ml) and air volumes of rabbit lungs. Note that the effect of surface tension on surface area is considerable at low lung volumes, but negligible at high volumes (results obtained in normal lungs on inflation and deflation are depicted by full lines; broken line shows relation in detergent-rinsed lung).

lar walls may be folded, especially in alveolar corners (fig. 3). With increasing surface tensions at constant lung volume, the alveolar ducts widen, more septal segments are folded up between adjacent ducts, and the alveolar surface area decreases. This situation is characteristic for lungs with abnormally high surface tensions (fig. 4C, D), and can also be seen in lungs of patients with respiratory distress syndrome [5]. Accordingly, it appears inappropriate to characterize instability by collapse and opening of alveoli, i.e. as a flip-flop mechanism. It is rather a gradual process reflecting the equilibrium configuration where the sum of tissue energy and surface energy is minimum.

Of equal importance is the volume dependence of the shaping influence of surface forces. Indeed, at lung volumes within about the normal range of breathing the peripheral airspace dimensions appear to be governed by the interplay between surface forces and tissue forces, in accordance with the model. At high lung volumes, however, the tissue forces transmitted by the fiber scaffold of the lung become the predominant and even the decisive factor of alveolar micromechanics: at lung volumes of 80% TLC or more, the alveolar surface area-volume relation is largely independent of surface tension (fig. 6). This mechanism has to be borne in mind when tests are done to evaluate surfactant replacement therapy for respiratory distress syndrome: the effect may be apparent only during spontaneous respiration, but can be masked when the lung volume is increased by mechanical ventilation [7].

References

1 Bachofen, H.; Hildebrandt, J.; Bachofen, M.: Pressure-volume curves of air- and liquid-filled excised lungs. Surface tension in situ. J. appl. Physiol. *29:* 422–431 (1970).
2 Bachofen, H.; Gehr, P.; Weibel, E.R.: Alterations of mechanical properties and morphology in excised rabbit lungs rinsed with a detergent. J. appl. Physiol. *47:* 1002–1010 (1979).
3 Bachofen, H.; Ammann, A.; Wangensteen, D.; Weibel, E.R.: Perfusion fixation of lungs for structure-function analysis: credits and limitations. J. appl. Physiol. *53:* 528–533 (1982).
4 Bachofen, H.; Schürch, S.; Urbinelli, M.; Weibel, E.R.: Relations among alveolar surface tension, surface area, volume, and recoil pressure. J. appl. Physiol. *62:* 1878–1887 (1987).
5 Bachofen, M.; Weibel, E.R.: Structural alterations of lung parenchyma in the adult respiratory distress syndrome. Clin. Chest. Med. *3:* 35–56 (1982).
6 Clements, J.A.; Hustead, R.F.; Johnson, R.P.; Gribetz, I.: Pulmonary surface tension and alveolar stability. J. appl. Physiol. *16:* 444–450 (1961).
7 Davis, J.M.; Veness-Meehan, K.; Notter, R.H.; Bhutani, V.K.; Kendig, J.W.; Shapiro, D.L.: Changes in pulmonary mechanics after the administration of surfactant to infants with the respiratory distress syndrome. New Engl. J. Med. *319:* 476–479 (1988).
8 Gil, J.; Bachofen, H.; Gehr, P.; Weibel, E.R.: Alveolar volume-surface area relationship in air- and saline-filled lungs fixed by vascular perfusion. J. appl. Physiol. *47:* 990–1001 (1979).
9 Neergaard, K. von: Neue Auffassungen über einen Grundbegriff der Atemmechanik. Die Retraktionskraft der Lunge, abhängig von der Oberflächenspannung in den Alveolen. Z. Ges. exp. Med. *66:* 373–394 (1928).
10 Orsós, F.: Die Gerüstsysteme der Lunge und deren physiologische und pathologische Bedeutung. 1. Normal-anatomische Verhältnisse. Beitr. klin. Tuberk. *87:* 568–609 (1936).
11 Radford, J.P., Jr.: Mechanical stability of the lung. Archs envir. Hlth *6:* 128–133 (1963).

12   Schürch, S.; Goerke, J.; Clements, J.A.: Direct determination of volume- and time-dependence of alveolar surface tension in excised lungs. Proc. natn. Acad. Sci. USA 75: 3417–3421 (1978).
13   Schürch, S.; Bachofen, H.; Weibel, E.R.: Alveolar surface tensions in excised rabbit lungs: effect of temperature. Resp. Physiol. 62: 31–45 (1985).
14   Smith, J.C.; Stamenovich, D.: Surface forces in lungs. Alveolar surface tension-lung volume relationships. J. appl. Physiol. 60: 1341–1350 (1986).
15   Weibel, E.R.: Stereological methods, vol. 1 (Academic Press, New York 1979).
16   Weibel, E.R.: Morphometric and stereological methods in respiratory physiology including fixation techniques; in Otis, Techniques in respiratory physiology, part I, vol. P4/1, pp. 1–35 (Elsevier, Ireland 1984).
17   Weibel, E.R.; Gil, J.: Electron microscopic demonstration of an extracellular duplex lining layer of alveoli. Resp. Physiol. 4: 42–57 (1968).
18   Wilson, T.A.; Bachofen, H.: A model for mechanical structure of the alveolar duct. J. appl. Physiol. 52: 1064–1070 (1982).

H. Bachofen, MD, Department of Medicine, University of Berne, Inselspital, CH–3010 Berne (Switzerland)

Wichert P von, Müller B (eds): Basic Research on Lung Surfactant.
Prog Respir Res. Basel, Karger, 1990, vol 25, pp 168–175

Induced Structures Obtained from Surface Film Compression of Natural Lung Surfactant and Synthetic Phospholipids

Pierre Tchoreloff, Benoît Denizot, Jacques-Emile Proust, Francis Puisieux[1]

Physicochimie des Surfaces et Innovation en Pharmacotechnie. UA CNRS 1218, Faculté de Pharmacie, Châtenay-Malabry, France

It is well known that lung stability requires the presence of an air/water layer capable of supporting very high surface pressure on dynamic compression (ability to give a very low surface tension) [Clements et al., 1961].

Physiologically, the natural lung surfactant is capable of maintaining a surface tension of about 40 mN/m after full expansion (total lung capacity); during expiration, the surface film is compressed and alveolar surface tension drops to approximately 0 mN/m [Schürch et al., 1982]. Type II pneumocytes release aggregates of surfactant into the aqueous alveolar subphase as lamellar bodies, which then undergo transformation into other structures such as tubular myelin [Keough, 1985]. This tubular myelin conformation which ensures a good continuity between the subphase and the interface is probably responsible for the very fast adsorption rate to the air/water interface of the surfactant.

On the other hand, dipalmitoylphosphatidylcholine (DPPC), which is the principal phospholipidic component of the natural lung surfactant, is known to ensure very low surface tension on compression; it is also known that the post-collapse respreading of this compound is very bad. Knowl-

[1] We thank G. Albrecht for valuable assistance for sample preparation indications.

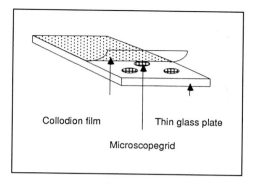

Collodion film | Thin glass plate

Microscopegrid

Fig. 1. Blodgett sampling.

edge of both DPPC and natural lung surfactant (NLS) behavior on surface film, and especially of the induced stuctures obtained by film compression is very important to understand how lung surfactant itself works.

We have used a modified interface sampling methodology of the classical Blodgett technique [Ries, 1961; Ries et al., 1985].

Material and Methods

DPPC was obtained from Sigma Chemical Co., St. Louis, Mo.

NLS was obtained by washing adult ox lung with saline. We used the following procedure: after removal of cells by centrifugation at 500 *g* for 5 min at 4 °C, the supernatant was recentrifuged at 3,000 *g* for 2 h at 4 °C in saline. The pellet was redispersed in tridistilled water, and the obtained phospholipid suspension was dialysed against tridistilled water. The phospholipidic concentration of the final preparation is about 6 mg/ml.

The sample preparation is described in figure 1: thin glass plates are used as a Blodgett sampling support, on which electron-microscope grids are placed and covered by a collodion film.

These collodion-coated microscope grids are then immersed in the subphase of a Langmuir trough (fig. 2). For all experiments, the subphase is tridistilled water. DPPC (in organic solution) or NLS (in aqueous suspension) are spread at the air/water interface, and one or several compression/expansion cycles are monitored. The grids are then slowly withdrawn at a constant controlled surface pressure. The surface layer of phospholipids overlaps the grids, and the dry sample is very gently gold palladium shadowed under a definite angle of 45°. It is now possible to investigate the surface layer structure by transmission electron microscopy (TEM).

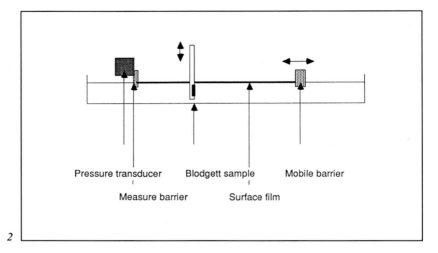

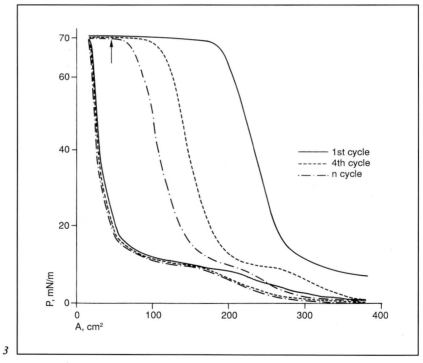

Fig. 2. Blodgett technique on Langmuir trough.
Fig. 3. Surface pressure/area isotherms of DPPC monolayer at 37 °C.

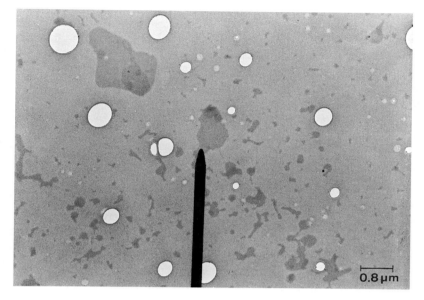

Fig. 4. DPPC, 37 °C, 5th collapse. × 11,000.

Results and Discussion

The high dynamic collapse pressure (72 mN/m) of the DPPC mono-layer isotherm (fig. 3) results from the presence of solid state phospholip-ids. On decompression of the monolayer, a drastic lowering in pressure occurs. On repeated compression (cycle 2), one can observe an important loss of phospholipids evaluated on the graph at 40%. We performed a Blodgett sampling and electron microscope observation of the DPPC layer during the 5th cycle over an area of 50 cm² corresponding to a collapse pressure of 72 mN/m (arrow in fig. 3). The mean superficial density is about 0.4 μg/cm².

The micrograph (fig. 4) of the DPPC layer in this state shows a con-tinuous background which corresponds to a monolayer, and some darker, isolated spots ranging in size from 150 nm to 2 μm. We also note a grada-tion of the electronic density in these spots that would indicate the build up of overlapping layers of phospholipids. (The white circles are artefacts cor-responding to microbubbles in the collodion film).

The results of the same experiments performed with bovine natural lung surfactant are shown in figures 5 and 6.

The maximum pressure obtained at the end of the first compression is 58 mN/m. It seems to be in disagreement with results described using the bubble surfactometer technique [Enhörning, 1977] because the maximum pressure obtained with this technique and with the NLS is 70 mN/m; but this difference could be explained by the differences in the dynamic conditions and in the subphase composition.

The decrease of surface pressure at the collapse levels for the repeated cycles may be explained by the squeezing out of solid state lipids on compression at this temperature. In contrast, with DPPC isotherms, we can also report better respreading of the NLS film on decompression. When about 15 compression/expansion cycles are performed, we obtain a stable isotherm as shown in figure 5 (n cycles). The collapse pressure is then 49 mN/m, the respreading is very efficacious, the hysteresis area is very small, and there is no more loss of material.

Using the Blodgett technique, we obtained a micrograph (fig. 6) showing the layer structure during the first compression, at a pressure of 55 mN/m (1st arrow on the surface pressure isotherm of NLS). It shows the presence of dark areas with some very black spots. Those dark areas are most certainly aggregations and built up of lipid layers, which are 1–2 μm in size. The main difference in comparison with the DPPC compressed layer is that these aggregates are very dense, and that they are interconnected. The increase of superficial accumulated material may be explained by the fact that the sampling was made during the first compression, and also because less phospholipids pass from the surface to the liquid subphase on compression (the mean superficial density is about 1.3 $\mu g/cm^2$).

After stabilisation of the compression/expansion cycle (2nd black arrow in fig. 5), the micrographs (fig. 7, 8) show that some different and new structures are induced on the surface film.

High-density rounded aggregates, 3–6 μm in size appear, and there are much less interconnections. Speckling of the system disappears, and the background becomes more homogeneous at the same time. Sharp separation lines between the round aggregates and the background can be seen.

Fig. 5. Surface pressure/area isotherms of natural bovine surfactant (SN) monolayers at 37 °C.

Fig. 6. SN, 1st collapse, 37 °C. × 11,000.

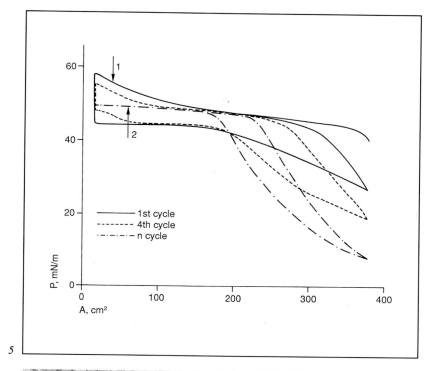

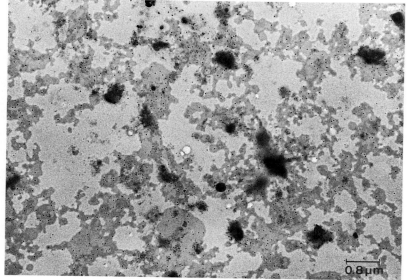

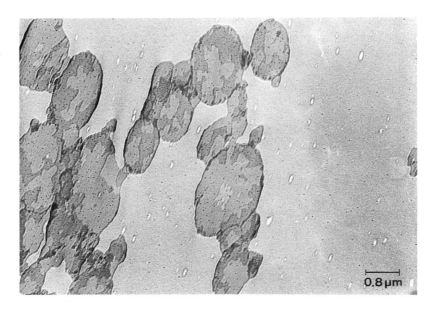

Fig. 7. SN, 37 °C, at collapse. × 11,000.

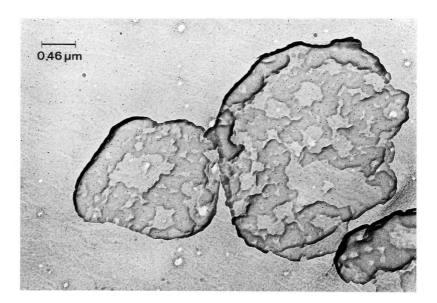

Fig. 8. SN, 37 °C, at collapse. × 19,000.

We can also clearly see the shadowing orientation, showing a rapid variation of the thickness of these aggregates. These structures seem to be characterised by a discontinuous multilayer conformation.

Conclusion

It is possible to investigate the liquid/air interface monolayer or layer structures by this modified Blodgett technique. We demonstrate a remarkable difference in the structure of compressed layers of NLS in relation to DPPC layers. It seems to be very interesting to show a relation between the NLS surface/pressure isotherm evolution and the creation of particular structures which are most certainly responsible for the stabilisation of the NLS isotherm after a number of compression/decompression cycles and of the good respreading of NLS at this level.

At this stage, we cannot find an explanation why these particular structures are created, but we believe that one of the reasons may be the presence of proteins in the composition of NLS. Other experiments which are in progress may bring additional information.

References

Clements, J.A.; Hustead, R.E.; Johnson, R.P.; Gribetz, I.: Pulmonary surface tension and alveolar stability. J. appl. Physiol. *16:* 444–450 (1961).

Enhörning, G.: Pulsating bubble technique for evaluating pulmonary surfactant. J. appl. Physiol. *43:* 198–202 (1977).

Keough, K.M.W.: Physicochemical properties of surfactant lipids. Biochem. Soc. Trans. *13:* 1080–1084 (1985).

Ries, H.E.: Monomolecular films. Scient. Am. *3:* 2–11 (1961).

Ries, H.E.; Albrecht, G.; Ter-Minassian-Saraga, L.: Collapsed monolayers of egg lecithin. Langmuir *1:* 135–137 (1985).

Schürch, S.; Goerke, J.; Clements, J.A.: Direct determination of the surface tension in the lung. Proc. natn. Acad. Sci. USA *73:* 4698–4702 (1982).

Pierre Tchoreloff, MD,
Physicochimie des Surfaces et Innovation en Pharmacotechnie,
UA CNRS 1218, Faculté de Pharmacie, F–92296 Châtenay-Malabry (France)

Wichert P von, Müller B (eds): Basic Research on Lung Surfactant.
Prog Respir Res. Basel, Karger, 1990, vol 25, pp 176–180

Molecular Structure of the Extracellular Surface Layer of the Human Lung Surfactant

An Electrochemical and Immunoelectron Microscopic Study on Langmuir-Blodgett Type Films

Erna Ladanyi[a], *I.R. Miller*[b], *Ronit Popovits-Biro*[c], *Y. Marikovsky*[b], *P. von Wichert*[d], *B. Müller*[d], *K. Stalder*[a]

[a]Department of Occupational Health, University of Göttingen, FRG; Departments of [b]Membrane Research and [c]Structure Chemistry, Weizmann Institute of Science, Israel, and [d]Medical Policlinic, Philipps University, Marburg, FRG

The extracellular lung surfactant surface layer (ELSSL) is the active part of the lung surfactant system. The little available data concerning ELSSL structure refer either to its cross-sectioned transmission electron microscopic (TEM) image or to results on lung lavages containing both ELSSL and hypophase components.

We considered that the isolation of ELSSL from the hypophase is a prerequisite of ELSSL studies and developed recently an appropriate technique [1]. TEM en face visualization of human ELSSL isolated in vitro with this technique revealed a continuous sheet of round-shaped structural elements of two different sizes. The majority of them were identified as DPPC molecules since they occupied an area of 50 Å^2 equal to the area size found with DPPC molecules spread over saline, isolated and visualized with the same technique. The data we report here concern the presence and the role of the surfactant-specific 26–36 kd protein SP-A in the ELSSL.

Materials and Methods

Presence and role of SP-A were studied with electrochemical, Langmuir-Blodgett and TEM techniques on in vitro hypophase-separated [1] natural ELSSL and on ELSSL reassembled at the air/saline interface.

As starting materials served crude human lung lavages (LL), their lipid extracts (LLL), their proteins redissoluted after the lipid extraction (RP), the lipid extract of human amniotic fluid (AL), purified SP-A, and polyclonal anti-human SP-A antibody from rabbit. The electrochemical measurements were carried out on a polarograph (PAR, Model 170; Princeton, N.J., USA) using the A.C. technique on a hanging mercury drop electrode; isolation/reassemblage of the ELSSL occurred with a Fromherz-type Langmuir-Blodgett trough (Mayer-Feintechnik, Göttingen, FRG) equipped with a Wilhelmy balance; TEM images of the specimens lifted on formvar-coated grids, immuno-gold labeled (Janssen, Life Sciences Products, Beerse, Belgium) and uranyl acetate stained were taken on a JEM-100 B electron microscope (JEOL, Tokyo, Japan).

The electrochemical measurements consisted in recording the electrical double-layer capacitance-potential curves for lipid layers with and without previous contact with SP-A or lung lavage LL proteins. The measurements were carried out for layers situated both at the air/saline and at the hanging mercury electrode/saline interfaces thus at the surface and in the bulk. The lipid/protein contact for reassembly of the ELSSL interface occurred alternatively by adsorption of proteins injected beneath the spread lipid layer, by spreading the lipids over the protein solution and by shifting a spread lipid layer over the saline solution of proteins. The obtained values were compared with those for monolayers of individual lipids and for surface layers of LL lipids.

Results and Discussion

Double layer capacitance values C of condensed phospholipid monolayers are well known. They can be regarded as a measure of the compactness/permeability of the layers. The more tight and ordered the molecules are arranged in a sheet or the more numerous the sheets lying one upon the others are, the less is the C value. So, by measuring C for surface layers obtained in different ways we could get information about the architecture of the molecules in the latter. C of the surface layers obtained by spreading of LLL or AL over saline (fig. 1, curve 1) was about half (C = 1.06 μF cm^{-2}; fig. 1, curve 2) of that known for individual phospholipids (C = 1.7–1.9 μF cm^{-2}). Therefore, we concluded that the surfactant lipids alone (of that peculiar composition as they were present in LL and in the AL) were unable to form a stable monolayer but build a multilayer, probably a trilayer. This multilayer was stable for several days. When SP-A was *injected beneath* an AL-formed multilayer, a C value corresponding to a monolayer was obtained for the air/water interface (fig. 1, curve 3). When SP-A or RP were *injected beneath* an LLL-formed multilayer, C augmented more than three times (AL/SP-A: 3.40 μF cm^{-2}, LLL/RP: 3.90 μF cm^{-2}). When the same amount of lipids was *spread over* the mentioned protein solutions, the capacitance achieved in equilibrium the values obtained for equilib-

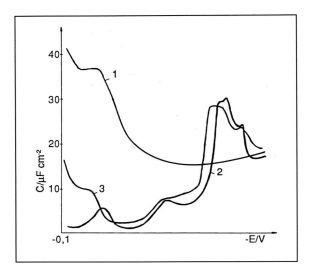

Fig. 1. Electrical double-layer capacitance C vs. potential curves of the reassembled ELSSL at the air/saline interface: (1) saline; (2) lung lavage-extracted lipids spread over saline, surface concentration 2 µg cm^{-2}; (3) surfactant-specific protein SP-A 1.25 µg ml^{-1} injected beneath the lipid layer.

rium adsorption of the LL layers (C = 6.2 µF cm^{-2}). Thus, in all cases a strong lipid-protein interaction could be observed which resulted in a perturbance of the lipid multilayer and in its switch into a more or less compact monolayer. This could occur in two steps: (1) The proteins could attach to/penetrate into the lipid multilayer (when injected beneath) or could be directly embedded in it (when the lipids were spread over a protein solution), or (2) the lipid-bound proteins could remove the excess lipids from the surface into the bulk.

Protein solutions with lipids spread over their surface showed a time-dependent decrease of C values at the mercury electrode/saline interface in the bulk and their final identity with the values found for the surface layer. These results suggest that SP-A has a transport role, thus it could regulate the surface concentration of the lipids at the air/aqueous interface in vivo.

As seen in the electron micrographs of figure 2, SP-A was temporarily present in the ELSSL. After the reassembly of the ELSSL from extracted lipids (fig. 2a) and saline dissoluted SP-A/LL protein (fig. 2b), numerous

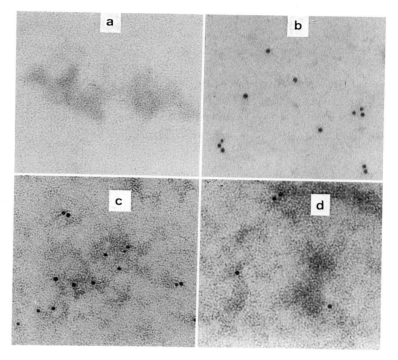

Fig. 2. Transmission electron micrographs of immuno-gold labeled reassembled ELSSL: (a) lung lavage-extracted lipids spread over saline; (b) saline solution of SP-A; (c) lipid layer transported over lung lavage proteins; (d) compressed ELSSL. Uranyl acetate-stained specimens. ×252,000.

randomly distributed gold-labeled SP-A molecules were found in the hypophase-separated and rinsed-up ELSSL (fig. 2c). The penetration of the proteins into the lipid multilayer led to an augmentation of the surface pressure from 22 to 33 mN m^{-1}. When the surface layer was compressed to a surface pressure of 48 mN m^{-1} and to 25% of its initial surface, a part of the label disappeared (fig. 2d), thus the proteins were partially squeezed out from the surface layer. Based on the data obtained with the parallelly used three techniques we can speculate about a change of the molecular architecture of ELSSL during the breathing process. It seems possible that during deflation, when the alveoli are compressed and the alveolar surface is diminished, the SP-A molecules are (partially) squeezed out from the surface layer. They take with them the excess lipid molecules from the surface layer into the hypophase. During inflation with enlarged alveolar surface the SP-A molecules enter again the surface layer bringing with them the supplementary amount of lipids necessary for the augmented monolayer surface.

Our results are consistent with the data indicating that SP-A interacts with surfactant lipids, that it facilitates lipid adsorption as well as with the fact that it reincorporates to lamellar bodies, to type II cells and regulates surfactant pool size [2].

References

1   Ladanyi, E.; Möbius, D.; Stalder, K.; Von Wichert, P.: Structure of isolated lung surfactant monolayer; in Symposium on Membrane Lipids, Sintra 1987. Actas Inst. Bioqim. (in press).
2   Wright, J.; Young, S.; Stevens, P.; Clements, J.: Intraalveolar metabolism of lung surfactant: role of SP-A regulating surfactant pool size. Prog. resp. Res., vol. 25, pp. 136–143 (Karger, Basel 1989).

Dr. Erna Ladanyi, Department of Occupational Health, University of Göttingen, Windausweg 2, D–3400 Göttingen (FRG)

Wichert P von, Müller B (eds): Basic Research on Lung Surfactant.
Prog Respir Res. Basel, Karger, 1990, vol 25, pp 181–190

Alveolar Surface Tension: What Kind of Biophysical Measurements Should Be Done?[1]

Samuel Schürch[a], *Hans Bachofen*[b], *Jon Goerke*[c]

[a] Departments of Medical Physiology and Medicine, University of Calgary,
Health Sciences Centre, Calgary, Alta., Canada;
[b] Departments of Anatomy and Medicine, University of Berne, Switzerland;
[c] Cardiovascular Research Institute, University of California, San Francisco,
Calif., USA

The pulmonary surfactant film at the alveolar air-liquid interface reduces the surface tension to values close to zero on lung deflation. In addition to a low and stable surface tension, interdependence provided by the fibrous continuum enables the lung to maintain a large alveolar surface area, necessary for an efficient gas exchange. A thorough understanding of the mechanical effects of surfactant in the lung has been hindered by several problems: the behavior of films of lung surfactant in vitro, e.g. in the Langmuir-Wilhelmy balance or in the pulsating bubble surfactometer [1, 2], is different from that of the alveolar lining layer in situ [2]. Although minimum surface tensions below 5 $mN \cdot m^{-1}$ in the balance or below 1 $mN \cdot m^{-1}$ in the bubble surfactometer, can be obtained if the compression rate for surfactant films is fast enough, these low surface tensions have much less stability than the alveolar film in situ. The alveolar film itself has, as calculated from pressure-volume studies on lungs, of intact animals or on excised lungs, a far greater stability [3, 4]. In the lungs held at 40% total lung capacity (TLC), the surface tension increased only to 1–2 $mN \cdot m^{-1}$ in 20 min. These results are in line with the surface tension stability data obtained by direct measurement of alveolar surface tension with the microdroplet spreading technique [5–7]. Since film material is spread

[1] This study was supported by the Medical Research Council of Canada, MT-6435, The Alberta Heritage Foundation for Medical Research, and the Swiss National Foundation for Scientific Research.

over a large surface area of more than 100 m² in the human lung, and this
material has a very small escape route up the major airways, the lung
provides an almost leak-proof system for the alveolar surface film.

We will briefly review the alveolar micropuncture technique to measure surface tension in situ. We will then demonstrate that surface tensions
below 1 mN·m⁻¹ that are stable with time can be achieved by monolayers
of air-liquid interfaces of a leak-proof captive bubble. The surface area of
this bubble is changed by varying the pressure in the liquid surrounding the
bubble. There is no capillary to regulate the bubble volume and thus no
pathway for the surfactant to escape from the monolayer. Further, we will
demonstrate that the surface tension values obtained in vitro, with the new
captive bubble method, are in excellent agreement with the values obtained in situ by alveolar micropuncture and drop spreading.

Measurement of Alveolar Surface Tension in situ

Methods

The direct determination of the surface tension in individual alveoli has been
described previously [5, 6]. This method has been improved over the years, especially the
calibration methods that relate the diameters of microdroplets placed onto surfactant
films, to alveolar surface tension [7].

Alveolar surface tension was determined in 24 isolated, continuously perfused lungs
from adult rabbits. Lung preparation, perfusion and physiological monitoring of these
lungs has been described previously [7, 8]. In short, three P-V loops were performed to
establish TLC and to obtain quasi-static pressure readings. The lungs were inflated again
to the volumes selected for fixation; in groups of four the lungs were fixed at 40 and 80%
TLC on the inflation limb, at full inflation, and at 80 and 40% TLC on the deflation limb
of the full P-V curve, four lungs were fixed after being reinflated from 40% on deflation to
80% TLC. In 11 lungs the alveolar surface tension was measured during the fourth P-V
cycle. In each case alveolar micropuncture and the placement of test fluid droplets on
alveolar surfaces were done at 40% TLC on inflation [7].

The details of lung fixation have been described previously [9]. In a recent article,
sampling processing of lung tissue and the morphometric evaluation have been described [8].

Results

Alveolar Surface Tension-Area Relation. Figure 1 represents the trans-
pleural microscopic view of the same alveolus with a microdroplet showing
a larger diameter on inflation than on deflation. This demonstrates that the
surface tension on inflation is higher than on deflation for a given volume,
here at 40 and 80% TLC. The difference in alveolar surface texture is

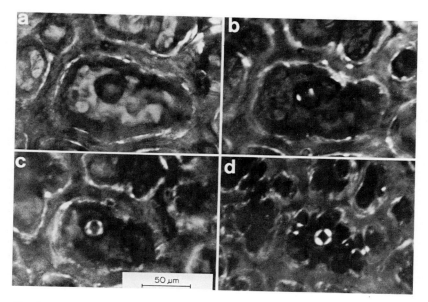

Fig. 1. An FC43 test fluid droplet had been placed onto a subpleural alveolar surface of a continuously perfused rabbit lung. The droplet was placed on inflation from zero pressure at 40% TLC, where the droplet diameter indicates a surface tension of approximately 7 mN·m⁻¹ (*c*). At 80% TLC on inflation, the surface tension was about 23 mN·m⁻¹ (*a*). At 80% TLC on deflation the droplet diameter indicates a surface tension of approximately 13 mN·m⁻¹ (*b*), and at 40% TLC on deflation, the surface tension was less than 2 mN·m⁻¹ (*d*). Difference in droplet size is not perfectly visible because of light reflections by almost hemispherical shape of droplet, typical for close to zero surface tensions.

associated with differences in alveolar surface tension. As mentioned above, the lungs were inflated from zero pressure to 40% TLC. At this point test fluid droplets were placed into subpleural alveoli. A P-V cycle was then performed with a very low rate (0.5 ml/s) to keep the droplet in the field of view.

Figure 2 represents the average surface tension-area relation for a full P-V cycle, starting at zero pressure and for reinflation from 40 to 80% TLC. Note the considerable hysteresis for the cycle starting at zero pressure while for reinflation there is very little hysteresis. Each point was calculated from data obtained from 4 lungs, e.g. the surface tension at 80% TLC on inflation was measured prior to fixation in four lungs.

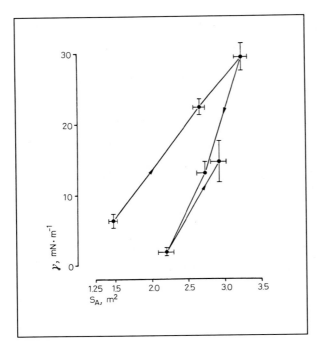

Fig. 2. Surface tension-surface area (SA) relation in air-filled lungs. See text for explanation. Note: very little hysteresis upon reinflation of the lungs from 40% TLC.

In situ Stability of Alveolar Surface Tension. The measurement of alveolar surface tension was measured in lungs that were inflated with very low flow rates (0.5 ml/s). After inflation of the lungs with volume steps of 10% TLC, the flow was interrupted for measuring the drop diameter, which usually required 20–30 s. Therefore, the surface tension measurement followed quasi-static inflation and a waiting period of 20–30 s after each TLC volume step. We were not able to detect any changes in droplet diameter at these points, either at 20 °C or at 37 °C. This indicates that the surface tension had reached a stable value when we measured the droplet diameter. In at least five lungs, at both 22 and 37 °C, we observed the droplet diameters at 40 and 80% TLC on inflation and at 80 and 40% TLC on deflation for more than 10 min. We did not detect any change in droplet diameter. This indicates stable or unchanged surface tensions for at least 10 min at these points.

Lung Surfactant Monolayers Studied with a Captive Bubble

Apparatus

A recently developed captive bubble method to measure surface tensions of lung surface active material or of components related to lung surfactant, such as 1,2-dipalmitoyl-sn-3-glycerophosphoryl choline (DPPC) [10] has been modified to eliminate possible film leakage. In the original method, the bubble size was controlled by air pushed into or withdrawn from the captive bubble through a needle connected to a graduated syringe. By necessity, the needle pierced the bubble air-water interface, thereby creating a potential pathway for film leakage. In the present design, bubble size and thus the surface tension of any insoluble film at the bubble surface is altered by changing the pressure within the closed sample chamber by the addition or withdrawal of aqueous media.

Our sample chamber, figure 3, consists of a plastic (acrylic) cuvette of a cross-section of approximately 1 cm². An outer chamber is provided for the temperature control by an air flow to keep the temperature of the liquid sample in the inner chamber at 37 °C. A conduit connected to a gastight syringe (Hamilton) whose plunger has a threaded shaft, passes through the top of both chambers as well as through a layer of a 1% agar gel. The bubble may be formed directly by sucking air into the liquid-filled chamber through the bottom conduit by simply lowering the pressure of the medium. After squeezing out the undesired air in the bottom conduit, the conduit is closed by the bottom valve.

Operation and Surface Tension Measurements

The sample chamber is filled with an aqueous medium, either a salt solution or a surfactant suspension. After placing an air bubble of atmospheric air, 2–3 mm in diameter, at the center of the agar ceiling, the chamber pressure is reduced to approximately 0.13 atm, causing the bubble to double its original diameter to 4–6 mm. This represents the starting point for adsorption measurements and the beginning of continuous video recording (WVD 5000 camera and AG 6060 recorder Panasonic, Matsushita Canada, Mississauga, Ont). Video images are printed by a video copier 9P60U, Mitsubishi Canada, Toronto, Ont. During the adsorption period the chamber contents are stirred by a magnetic stirrer inside the chamber. After adsorption, the pressure in the sample chamber is increased stepwise or continuously by turning the shaft of the syringe plunger. As the pressure increases, the bubble volume and surface area decrease, compressing the adsorbed surfactant monolayer at the air-water interface. The bubble progressively flattens, indicating a lower surface tension.

The surfactant film may also be formed by spreading the surfactant at the interface with a syringe, e.g. spreading a solution of phospholipid in ethanol. From the video picture of the bubble, the bubble surface tension and area were determined by using the fourth degree polynomial approximation of Malcolm and Elliott [11], for sessile or captive bubbles having a 180% contact angle. To determine the bubble surface areas, we digitized the photographs and calculated the area with computer programs based on the formulas of Rotenberg et al. [12]. These formulas can also be used with bubble contour data to calculate the surface tension, surface area, volume and contact angle of bubbles having higher surface tensions and other contact angles. Application of the Rotenberg formulas to our data resulted in surface tension identical to those obtained with the Malcolm and Elliott [11] method.

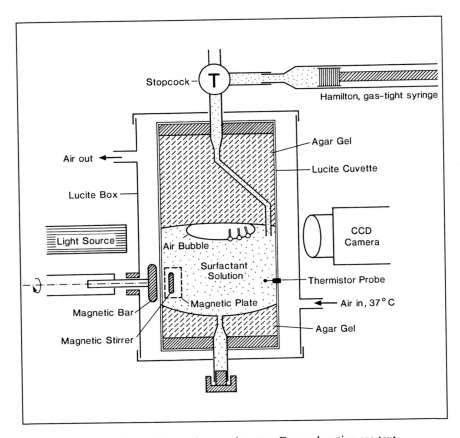

Fig. 3. The captive bubble surface tensiometer. For explanation see text.

Methods

Purified pulmonary surfactant, designated P3 + P4, was prepared from perfused rabbit lungs according to Stevens et al. [13]. Aliquots, each containing 50 µg phosphorus were resuspended in 3 mM Ca^{++} containing phosphate-buffered saline, and filled into the bubble chamber for the surface tension measurements. Before these measurements were performed, the surfactant was stirred at 37 °C for at least 30 min (incubation). A bubble of atmospheric air, 3–4 mm in diameter, was then formed below the agar ceiling, and the chamber was closed pressure tight. The pressure in the chamber was then reduced to expand the bubble to double its original size, and the magnetic stirrer was switched on. This was the starting point for the surfactant adsorption. After an adsorption time of 5 min, we applied pressure stepwise, as described above, to compress the adsorbed surfactant monolayer.

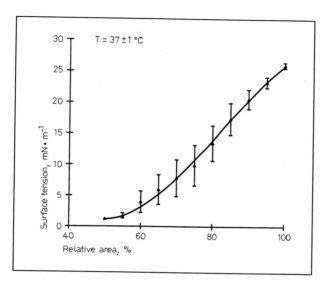

Fig. 4. First compression (after adsorption) isotherm of purified lung surfactant ($P_3 + P_4$) (see text).

Results

Minimum Surface Tension: First Compression. Figure 4 shows a first compression isotherm obtained from 6 independent experiments, each experiment was performed with a separate surfactant filling. The surface tension after 5 min adsorption was between 25 and 26 mN·m⁻¹. The area had to be compressed approximately by 50% in order to achieve a minimum surface tension between 1 and 2 mN·m⁻¹. Figure 5 shows a bubble at three different surface tensions.

Film Stability at Minimum Surface Tension. We investigated the stability of P3+P4 films after these films reached minimum surface tensions between 1 and 2 mN·m⁻¹. Surface tension-time relations were obtained from 4 independent first compression (quasi-static, 3 min) isotherms and from 4 independent isotherms (quasi-static, 3 min) after 5 pulsations. Each pulsation comprised a compression-expansion cycle performed over a 5-second period, between a high surface tension of approximately 30 mN·m⁻¹ and a low surface tension of about 5 mN·m⁻¹.

The stability curves (surface tension-time relations) demonstrate that repeated compression increased stability only slightly. The minimum sur-

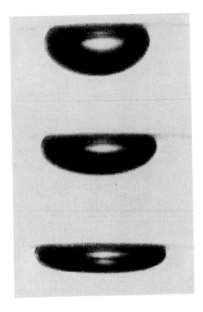

Fig. 5. Bubble shape at 3 different surface tensions, from top to bottom: 20, 12 and 2 mN · m⁻¹. The surfactant was purified lung surfactant ($P_3 + P_4$) (see text).

face tension increased for first compression curves from about 1.5 to 2.8 $mN \cdot m^{-1}$, while for the pulsated bubble, the minimum surface tension increased to 2.3 $mN \cdot m^{-1}$ in a 10-min period (fig. 6).

Discussion

Technical Aspects

By using alveolar micropuncture and drop spreading in continuously perfused rabbit lungs we were able to show that alveolar surface tension reached values below 1–2 $mN \cdot m^{-1}$ at 37 °C, on deflation. Remarkably, a minimum compression rate is not required for alveolar surface tension to fall to near zero values, i.e. minimum surface tensions can be achieved under quasi-static conditions. In contrast, lung surfactant films in the Langmuir-Wilhelmy balance, very slow (quasi-static) compression rates do not yield surface tensions below about 10 $mN \cdot m^{-1}$, presumably because surface film leakage up and around the Teflon barrier is not compensated by a sufficiently rapid compression of the film [1]. On the

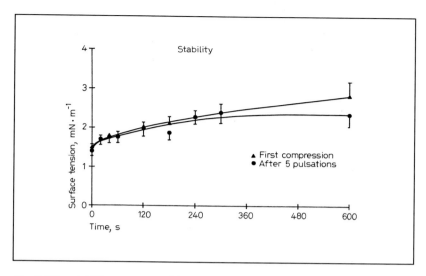

Fig. 6. Film stability as expressed in surface tension-time relations, of purified surfactant films (P_3+P_4) (see text).

other hand, in the pulsating bubble surfactometer [2, 14] much higher cycling rates are used (about 20 cycles/min), and surface film leakage up through the air capillary is usually compensated by rapid film compression.

The proposed captive bubble technique combines the advantages of Pattle's [15] bubble method with those of Langmuir-Wilhelmy balances. Therefore, near zero, stable surface tensions are achieved under quasi-static or dynamic conditions, and surface tension-time or surface tension-area relations can be produced (fig. 4, 6). In particular, the apparatus fulfills two essential conditions [1], i.e. the captive bubble has an absolutely leak-free surface, and the surface film can repeatedly be compressed and expanded. A perfect seal of the bubble implies not only the absence of a communication between the outside and the bubble interior, but also a complete and fully hydrophilic surrounding. The bubble floats against a hydrophilic surface formed by a 1% agar gel.

In summary, figures 4 and 6 show that with the captive bubble method an in vitro film of lung surfactant extract can be formed which closely resembles the steady state alveolar monolayers in situ.

References

1   Goerke, J.: Clements, J.A.: Alveolar surface tension and lung surfactant; in Fishman, Macklem, Mead, Geiger, Handbook of physiology, sect. 3: The respiratory system, vol. III, part 1, pp. 247–261 (American Physiological Society, Bethesda 1986).

2   Enhorning, G.: A pulsating bubble technique for evaluating pulmonary surfactant. J. appl. Physiol. *43:* 198–201 (1977).

3   Mead, J.; Collier, C.: Relation of volume history of lungs to respiratory mechanics in anaesthetized dogs. J. appl. Physiol. *14:* 699–678 (1959).

4   Horie, T.; Hildebrandt, J.: Dynamic compliance, limit cycles, and static equilibria of excised cat lungs. J. appl. Physiol. *31:* 423–430 (1971).

5   Schürch, S.; Goerke, J.; Clements, J.A.: Direct determination of surface tension in the lung. Proc. natn. Acad. Sci. USA *73:* 4698–4701 (1976).

6   Schürch, S.; Goerke, J.; Clements, J.A.: Direct determination of volume- and time-dependence of alveolar surface tension in excised lungs. Proc. natn. Acad. Sci. USA *75:* 3417–3421 (1978).

7   Schürch, S.; Bachofen, H.; Weibel, E.R.: Alveolar surface tension in excised rabbit lungs. Effects of temperature. Resp. Physiol. *62:* 31–45 (1985).

8   Bachofen, H.: Schürch, S.; Urbinelli, M.; Weibel, E.R.; Relations among alveolar surface tension, surface area, volume, and recoil pressure. J. appl. Physiol. *62:* 1878–1887 (1987).

9   Bachofen, H.; Ammann, A.; Wangenstein, D.; Weibel, E.R.: Perfusion fixation of lungs for structure-function analysis: credits and limitations. J. appl. Physiol. *53:* 528–533 (1982).

10 Schürch, S.: Surface tension at low lung volumes. Dependence on time and alveolar size. Resp. Physiol. *48:* 339–355 (1982).

11 Malcolm, J.D.; Elliott, C.D.: Interfacial tension from height and diameter of a single profile drop or captive bubble. Can. J. chem. Eng. *58:* 151–153 (1980).

12 Rotenberg, Y.; Boruvka, L.; Neumann, A.W.: Determination of surface tension and contact angle from the shapes of asymmetric fluid interfaces. J. Colloid Interface Sci. *93:* 169–183 (1983).

13 Stevens, P.A.; Wright, J.R.; Clements, J.A. Changes in quantity, composition, and surface activity of alveolar surfactant at birth. J. appl. Physiol. *63:* 1049–1057 (1987).

14 Possmayer, F.; Yu, S.H.; Weber, J.M.; Harding, P.G.R.: Pulmonary surfactant. Can. J. Biochem. Cell Biol. *62:* 1121–1133 (1984).

15 Pattle, E.R.: The cause of stability of bubbles derived from the lung. Phys. med. Biol. *5:* 11–26 (1960).

Samuel Schürch, PhD, Departments of Medical Pysiology and Medicine, University of Calgary, Health Sciences Centre, 3330 Hospital Drive N.W., Calgary, AB T2N 4N1 (Canada)

Wichert P von, Müller B (eds): Basic Research on Lung Surfactant.
Prog Respir Res. Basel, Karger, 1990, vol 25, pp 191–195

Squeeze-Out and Transition Temperature of Binary Mixtures of Dipalmitoylphosphatidylcholine and Unsaturated Phospholipids

J. Egberts, H. Sloot, A. Mazure[1]

Department of Obstetrics, Leiden University Medical Center,
Leiden, The Netherlands

Dipalmitoylphosphatidylcholine (DPPC), the main component of pulmonary surfactant, will not spread spontaneously on water below 41 °C (gel-to-liquid crystal transition temperature) [1]. Mixing it with unsaturated lipids is required to lower the transition temperature. The mixtures will only reduce surface tension or increase surface pressure during compression if (part of) the unsaturated components are squeezed out of the monolayer [1, 2]. This will result in enrichment of DPPC in the air-liquid interface, and therefore in a higher melting temperature of the monolayer than the original transition temperature. Fluorescence polarization measurements and surface tension experiments were performed to determine the transition or melting temperature of (1) fully hydrated mixtures of phospholipids in suspension, and (2) phospholipids in repeatedly compressed and expanded monolayers. The difference between these two temperatures indicates whether preferential squeeze-out is effective.

Material and Methods

The phospholipids DPPC, phosphatidylcholine (PC) from egg yolk, phosphatidylglycerol (PG) from egg PC, and phosphatidylinositol (PI) from soybean were purchased from Sigma Chemical Co. (St. Louis, Mo.). Vesicles of DPPC and of the unsaturated phospholipids alone or of 7:3 and 9:1 binary mixtures of DPPC and the unsaturated

[1] We thank Dr. D.O.E. Gebhardt for his critical comments.

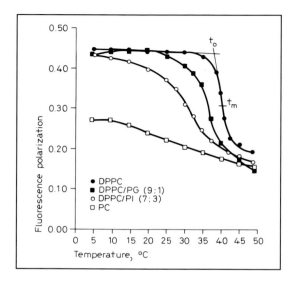

Fig. 1. Temperature-dependent changes in fluorescence polarization of DPPC, PC, and the mixtures DPPC/PG (9:1) and DPPC/PI (7:3). The main melting temperature was read at the deflection point of the curves and the onset of the transition was estimated from the tangents as is shown for DPPC.

phospholipids were prepared ultrasonically in double-distilled water and were incubated at 37 °C for 30 min with DPH (1,6-diphenyl-1,3,5-hexatriene). The fluorescence polarization [3] of these vesicles was measured at different temperatures in a T-shaped fluorometer (excitation 365 nm, emission 460 nm), equipped with polarizers and a temperature-controlled cuvette. The temperature T_o, at which the main   gel-to-liquid crystal transition starts and the main melting temperature (T_m) of the phospholipid (mixture) was read from the fluorescence-temperature (FP-T) curves (fig. 1) [4].

A water-and-air thermostated surfactant balance [5] was used for measuring changes in surface tension of phospholipid monolayers during compression and expansion (area 100 cm^2, cycle period 20 s). The surface temperature was measured by a thermocouple and changes in temperature (approximately 1 °C/min) and surface tension were recorded continuously. The phospholipids were spread in excess at 'room' temperature (20–24 °C) on a surface of deionized and double-distilled water.

Results and Discussion

In figure 1 examples are shown of the relationships between temperature and fluorescence polarization of DPPC, PC, a DPPC/PI (7:3) and a DPPC/PG (9:1) mixture. The FP value of DPPC decreased sharply within

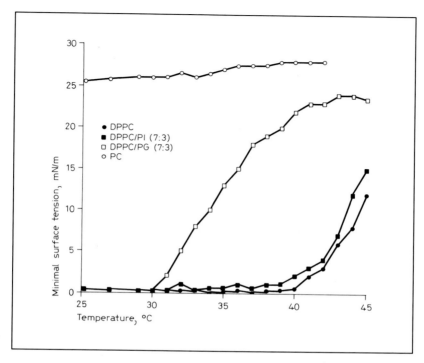

Fig. 2. Temperature-dependent changes in the minimal surface tension of mono-layers of DPPC, PC and the mixtures DPPC/PG (7:3) and DPPC/PI (7:3). Minimal surface tension increased significantly above the melting point of the monolayer. The unsaturated phospholipid PC is already in a fluid state at 20 °C.

the temperature range of 39–42 °C and its main transition temperature of 40.8 °C is in accordance with that found by others [3, 5]. Melting of the DPPC/PI and DPPC/PG mixtures started at lower temperatures ($T_o[FP]$) and occurred over larger ranges of temperature. The 7:3 binary mixtures of DPPC and PI, PG, or PC had onset temperatures of 26–32 °C and the 9:1 mixtures started to melt at 35–37 °C. The main transition temperatures of the 7:3 and 9:1 mixtures lay, respectively, between 32 and 36 °C and 37 and 39 °C.

Figure 2 shows that when the temperature is raised above 40 °C, the minimal surface tension (σmin) of the DPPC monolayer begins to increase. The monolayer cannot withstand high surface pressures any longer, because it melts ($T_o[ST]$). This monolayer melting temperature of DPPC is

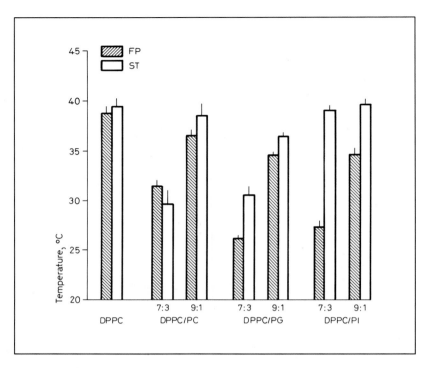

Fig. 3. The onset temperature of the gel to liquid crystal transition, determined by fluorescence polarization (FP) and the melting temperature of monolayers (ST) of DPPC and binary mixtures of DPPC and PC, PI or PG ($\pm$ SD).

in accordance with the temperature $T_o[FP]$ at which the liquid to gel transition starts (fig. 3). The 7:3 mixtures of DPPC/PG and DPPC/PI have significantly different temperatures at which the σmin values start to increase. The DPPC/PG (7:3) monolayer started to melt at approximately 30 °C, whereas the DPPC/PI (7:3) monolayer is rigid till 39 °C. In figure 3 it is shown that the monolayer melting temperatures $T_o[ST]$ of most of the mixtures were higher than their $T_o[FP]$. A difference between $T_o[ST]$ and $T_o[FP]$ is consistent with preferential squeeze-out of the unsaturated component and enrichment of the monolayer with DPPC. The squeeze-out phenomenon was hardly effective in monolayers of the 7:3 mixtures of DPPC/PC or of DPPC/PG which started to melt at 30–32 °C. The 9:1 DPPC/PG mixture became fluid at approximately 37 °C. Preferential squeeze-out was very effective in both the 7:3 and 9:1 DPPC/PI mixtures and also in the 9:1 DPPC/PC mixture. These monolayers started to melt at 39–40 °C, which is even above their main transition temperatures and which approximated the melting temperature of DPPC.

Preferential squeeze-out of the unsaturated component promotes the function of an artificial surfactant composed of DPPC and unsaturated lipids [2]. The above-described techniques can be used to determine the quality of the surfactant in as far as it is determined by the squeeze-out phenomenon.

Conclusion

We found that of the tested mixtures, only the 7:3 DPPC/PI mixture spreads below 37 °C and withstands high surface pressures at higher temperatures. This mixture might therefore be a good basis for a synthetic lung surfactant.

References

1   Watkins, J.C.: The surface properties of pure phospholipids in relation to those of lung extracts. Biochim. biophys. Acta *152:* 293–306 (1968).
2   Bangham, A.D.; Morley, C.J.; Phillips, M.C.: The physical properties of an effective lung surfactant. Biochim. biophys. Acta *573:* 552–556 (1979).
3   Shinitzky, M.; Barenholz, Y.: Fluidity parameters of lipid regions determined by fluorescence polarization J. biol. Chem. *249:* 2652–2657 (1974).
4   Lentz, B.R.; Alford, R.D.; Hoechli, M.; Dombrose, F.A.: Phase behaviour of mixed phosphatidylglycerol/phosphatidylcholine multilamellar and unilamellar vesicles. Biochemistry *21:* 4212–4219 (1982).
5   Schoedel, W.; Slama, H.; Hansen, E.: Zeitabhängiges Verhalten von Filmen von oberflächenaktivem Material aus Lungenalveolen. Pflügers Arch. *322:* 336–346 (1971).

J. Egberts, PhD, Department of Obstetrics, Leiden University Medical Center, Rijnsburgerweg 10, NL–2333 AA Leiden (The Netherlands)

Wichert P von, Müller B (eds): Basic Research on Lung Surfactant.
Prog Respir Res. Basel, Karger, 1990, vol 25, pp 196–199

Individual Molecular Species of Phosphatidylcholine from Fetal and Neonatal Lung[1]

A.N. Hunt[a], M.A. Hall[a], F.J. Kelly[b], I.C.S. Normand[a], A.D. Postle[a]

[a] Child Health and [b] Human Nutrition, Faculty of Medicine, Southampton
General Hospital, Southampton, UK

Measurement of the complete spectrum of individual molecular species of phosphatidylcholine (PC) can potentially provide greater information about the regulation of surfactant metabolism than the measurement of total disaturated PC. It facilitates the measurement of the rates of synthesis and catabolism of different PC species in subcellular fractions of the lung [1], and can allow a more detailed analysis of surfactant PC remodelling in the type II cell and recycling within the alveolus. In this study we have defined the developmental changes in PC species composition that occur in guinea pig and human lung. These methods have then been extended to the analysis of PC species in tracheal aspirate samples from preterm human infants.

Methods

Lungs were removed from fetal guinea pig pups, delivered by Caesarian section between days 50 and 86 (term) of gestation, and from neonatal guinea pigs up to 10 days of age. Samples of human fetal lung were obtained at post-mortem. Immature human amniotic fluid was obtained at 16–18 weeks of gestation at routine amniocentesis, while mature amniotic fluid at term was obtained during artificial rupture of membranes. Nasopharyngeal aspirate samples were taken at delivery from term infants, while tracheal aspirate samples were taken from preterm infants within 4 h of delivery. Samples were

[1] This work was supported by grants from the British Lung Foundation and the Wessex Medical School Trust.

Table 1. PC molecular species in fetal and neonatal guinea pig lung

Gestational age, days		n	PC species, % total (mean ± SEM)				
			14:0/16:0	16:0/16:1	16:0/18:2	16:0/16:0	16:0/18:1
	50	6	10.4±1.6	10.4±0.7	9.4±0.7	24.5±0.6	32.9±2.5
	55	8	10.3±0.8	11.7±0.7	13.2±0.7	26.9±0.5	26.0±2.4
	60	5	9.4±0.5	12.3±0.6	12.0±0.8	32.3±0.3	24.7±0.9
	63	6	13.6±0.8	14.0±0.9	9.1±1.1	35.7±3.5	18.3±0.9
	65	5	15.2±1.2	15.8±0.9	10.5±1.6	33.6±1.9	15.7±2.0
	67	3	16.3±0.9	15.8±0.9	8.4±0.8	38.1±1.7	15.6±1.6
term	68	6	17.7±1.3	12.3±0.6	7.4±1.4	42.9±1.2	13.6±1.2
	+2	5	14.8±1.7	10.4±1.4	9.2±1.5	43.2±3.3	14.6±1.7
	+5	5	7.4±0.4	8.0±1.0	15.3±0.5	35.3±1.4	21.6±0.7
	+10	7	9.9±0.3	9.1±0.9	10.2±0.6	40.9±1.9	18.9±1.2

stored at $-20\,°C$ until analysis, after centrifugation at $1,000\,g$ for 10 min. The incidence of RDS in preterm infants was assessed independently using clinical criteria.

Lipids were extracted with chloroform:methanol [2] using PC14:0/14:0 as internal standard, dried under N_2 and dissolved in chloroform. The PC fraction was obtained by selective elution from 100 mg Bondelut NH_2 columns with chloroform:methanol (3:2, v:v) [3]. PC molecular species were resolved by isocratic reverse-phase high-performance liquid chromatography (HPLC) on a 25-cm Apex II ODS column using a mobile phase of $40\,mM$ choline chloride in 92.5% methanol, 7.5% water at $1\,ml\cdot min^{-1}$. Eluted peaks were quantified by a post-column fluorescence derivatisation procedure [4]. The methanol HPLC stream was mixed with an aqueous stream of 1,6-diphenyl-1,3,5-hexatriene at $3\,ml\cdot min^{-1}$, and the resultant fluorescent peaks were detected by excitation at 340 nm and emission at 460 nm. The lower limit of detection for each molecular species was 200 pmol, while the total analysis time including sample preparation was under 2 h.

Results and Discussion

The same five major molecular species of PC were evident in all samples of guinea pig lung, but their relative proportions changed with gestational age. The fractional proportions of PC16:0/16:0 and PC14:0/16:0 increased towards term (table 1) and these were accompanied by a corresponding decrease in PC16:0/18:1. Total PC concentration increased some 8-fold from 0.73 ± 0.06 to 6.10 ± 1.69 over this same period. This analysis suggested that a comparison of PC16:0/16:0 to PC16:0/18:1 (P/O ratio) might provide a sensitive index for the evaluation of lung surfactant

Table 2. PC species in human lung secretions

Sample	n	Gestational age, weeks	PC 16:0/16:0 %	P/O ratio
Amniotic fluid	7	16–20	15.1 ± 2.0	0.6 ± 0.1
Amniotic fluid	10	37–40	60.1 ± 2.3	7.8 ± 0.8
Pharyngeal aspirate	13	40	59.4 ± 2.5	12.8 ± 1.3
Tracheal aspirate				
No RDS	15	33.5 ± 0.7	44.5 ± 3.1	6.0 ± 0.6
+ RDS	22	30.8 ± 0.7	39.4 ± 2.6	3.1 ± 0.4

status, and hence could be useful as a predictive marker for the development of respiratory distress syndrome (RDS).

The analysis of amniotic fluid and neonatal tracheal aspirate samples provided some support for such a concept (table 2). The most significant aspect was the difference in PC species analysis of initial tracheal aspirate samples between those preterm infants who developed RDS and those who did not. While there was no difference in the fractional content of PC16:0/16:0 between these groups, the P/O ratio was considerably lower in the RDS group of infants.

These results should be interpreted with caution. The composition of the tracheal and bronchial fluid PC might not accurately reflect that of the alveolar surfactant pool. There was an inevitable age bias between the two groups, as the infants with RDS tended to be of lower gestational age than the control group. Nevertheless, the difference in the P/O ratio remained significant even when the results were corrected for the gestational age differential. Sensitivity did not appear to be a significant problem as there was sufficient material for accurate analysis in some 85% of aspirate samples. Work is now being carried out to establish further the usefulness of PC molecular species analysis in the diagnosis and management of RDS in preterm infants, by comparison with alternative techniques such as the L/S ratio [5], surface tension parameters [6] and disaturated PC [7].

References

1   Schlame, M.; Casals, C.; Rustow, B.; Kunze, D.: Molecular species of phosphatidylcholine and phosphatidylglycerol in rat lung surfactant and different pools of pneumocytes type II. Biochem. J. *253:* 209–215 (1988).

2   Bligh, E.G.; Dyer, W.S.: A rapid method of total lipid extraction and purification. Can. J. Biochem. *37:* 911–923 (1959).

3   Caesar, P.A.; Wilson, S.J.; Normand, I.C.S.; Postle, A.D.: A comparison of the specificity of phosphatidylcholine synthesis by human fetal lung maintained in either organ or organotypic culture. Biochem. J. *253:* 451–457 (1988).

4   Postle, A.D.: Method for the sensitive analysis of individual molecular species of phosphatidylcholine by high performance liquid chromatography using post-column fluorescence detection. J. Chromatogr. *415:* 241–251 (1987).

5   Hallman, M.; Merritt, T.A.; Pohjavuori, M.; Gluck, L.: Effect of surfactant substitution on lung effluent phospholipids in respiratory distress syndrome. Evaluation of surfactant phospholipid turnover, pool size and the relationship to severity of respiratory failure. Pediat. Res. *20:* 1228–1235 (1986).

6   Gluck, L.; Kulovich, M.V.; Borer, R.C.; Brenner, P.H.; Andeson, G.C.; Spellacy, W.N.: Diagnosis of respiratory distress syndrome by amniocentesis. Am. J. Obstet. Gynec. *109:* 440–445 (1971).

7   Mason, J.A.; Nellenbogen, J.; Clements, J.A.: Isolation of disaturated phosphatidylcholine with osmium tetroxide. J. Lipid Res. *15:* 525–527 (1976).

A.D. Postle, MD, Child Health, Level G, Centre Block, Southampton General Hospital, Tremona Road, Southampton S09 4XY (UK)

Wichert P von, Müller B (eds): Basic Research on Lung Surfactant.
Prog Respir Res. Basel, Karger, 1990, vol 25, pp 200–203

Comparison of Different Types of Exogenous Surfactant in an Experimental Aspiration Trauma

W. Strohmaier, G. Schlag

Ludwig Boltzmann Institute for Experimental and Clinical Traumatology,
Vienna, Austria

The pulmonary surfactant system plays a crucial role for both onset and progression of the respiratory distress syndrome in the adult (ARDS). Based on the development of various basically different ARDS models [1] some of the mechanisms involved in surfactant deterioration have been elucidated.

Without claiming clinical relevance most of these models have responded excellently to the application of exogenous surfactant.

Aim of the Study

This study wants to contribute the present discussion employing a recently developed aspiration trauma model in rabbits [2]. Furthermore, the therapeutic potential of three basically different surfactant preparations was investigated.

Material and Methods

Animal Procedure

New Zealand white rabbits (2,600–3,200 g) were orally intubated and received 2 ml/kg body weight (BW) of an acid cocktail (group A), which is prepared by dissolving 1 tablet Oroacid (Rösch & Handel, Vienna, Austria) in 40 ml Ringer's solution (RS) and filtered. Sham controls (group SC) received an equal volume of RS. Healthy animals served as controls (group C). After 24 h the animals were intubated again and treated with either one of the three surfactant preparations (groups NS, AS, SAS) or RS as placebo (group P). During the following 24-hour observation period the rabbits received no additional treatment.

Surfactant Preparations

Natural Surfactant (NS). Freshly excised porcine lungs were lavaged extensively with RS. Cells and debris were separated (500 g for 10 min), the surface-active fraction was spun down by a single run (10,000 g for 60 min) and resuspended in RS (7.4 mg phospholipid (PHL) per ml). The dose administered was 12 mg PHL/kg BW.

Artificial Surfactant (AS). The rabbits received 12 mg/kg BW of a suspension of a lyophilisate consisting of dipalmitoylphosphatidylcholine (DPPC):egg-phosphatidylglycerol (PG) 4:1.

Semiartificial Surfactant (SAS). 120 mg AS-PHL were suspended in 7.5 ml RS containing 4.6 mg rabbit surfactant pool – PHL and 0.5 mg protein/ml. The pool was isolated as described recently [3].

Lung Mechanics

At the end of the experiment the animals were anesthetized, tracheotomized and intubated. After killing the animals with an i.v. overdose of Thiopental (Sanabo, Vienna, Austria) the chest was opened by midline incision and the lungs were inflated once with approximately twice the tidal volume. The tracheal tube was then connected to a gas-tight syringe (Hamilton, Bonaduz, Switzerland), driven by a pump (Harvard Apparatus, 907A, Millis, Mass.). The pressure transducer (Gould Electronics, TH1017, Bilthoven, Netherlands) was adjusted to zero before closing the system. The lungs were inflated in 15-ml increments, allowing 1-min equilibrium pauses until a stable plateau pressure of 20 mm Hg (= 27.2 cm H_2O) was reached. This volume was defined as vital capacity (VC). Deflation was performed in the same rhythm. Static compliance was calculated as the steepest slope from at least 3 consecutive data points of the deflation limb. All data were taken from the first curve.

Statistic Analysis

ANOVA and group comparisons were performed by linear contrast according to Linder [4].

Differences were accepted to be significant at $p < 0.05$. Results are expressed as means ± SD of the mean.

Results

Figure 1 (compliance) and figure 2 (vital capacity) clearly show the beneficial effects of NS and SAS, while AS turned out to be ineffective. Actually, 4 of the 8 animals treated with AS died within the observation period.

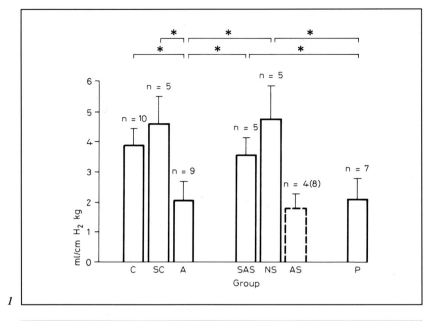

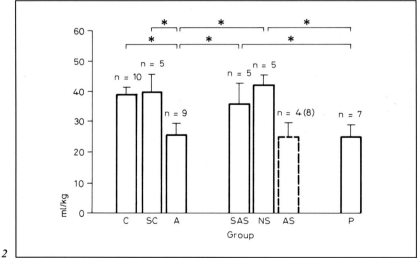

Fig. 1. Static compliance. Note: the dotted bar (group AS) represents the 4 surviving animals. * p < 0.05.

Fig. 2. Vital capacity. Note: the dotted bar (group AS) represents the 4 surviving animals. * p < 0.05.

Discussion

Both types of exogenous surfactant, which are able to compensate for the effects of acid/pepsin instillation by re-establishing normal lung mechanic properties contain the PHL-associated protein fraction (NS, SAS). The therapeutic inefficacy of pure PHL (AS) may be – at least in part – due to the high viscosity of the suspension, which disables the rabbits to 'inhale' the material.

The beneficial effect of SAS is lower than that of NS, although the PHL dosage per kg BW was fourfold. From these data it cannot be elucidated clearly, which factors are responsible for this discrepancy in vivo, because the in vitro properties of SAS are superior to those of NS with regard to their readsorption behavior after extensive cycling on a Wilhelmy balance (Biegler, Vienna, Austria).

References

1   Lachmann, B.; Danzmann, E.: In Robertson, VanGolde, Batenburg, Adult respiratory distress syndrome, pp. 505–548 (Elsevier, Amsterdam 1984).
2   Strohmaier, W.; Redl, H.; Schlag, G.: Exogenous surfactant in experimental aspiration trauma; in Schlag, Redl, Progress in clinical and biological research. Subseries: Vienna Shock Forum (Liss, New York 1989).
3   Redl, H.; Strohmaier, W.; Stachelberger, H.; Schlag, G.: Morphological appearance (REM & TEM) of surfactant derived from lavage, from lamellar body fraction and from lung tissue in dogs. Prog. resp. Res. *18:* 64–67 (1984).
4   Linder, A.: Trennverfahren bei qualitativen Merkmalen. Metrika *6:* 76–83 (1963).

W. Strohmaier, MD, Ludwig Boltzmann Institute for Experimental and Clinical Traumatology, A–1200 Vienna (Austria)

Wichert P von, Müller B (eds): Basic Research on Lung Surfactant.
Prog Respir Res. Basel, Karger, 1990, vol 25, pp 204–208

The Infected Preterm Rabbit Lung

A Model to Test the Effect of Surfactant Replacement on Lung Host Defenses[1]

Michael P. Sherman[a], *Laurie A. Campbell*[a], *T. Allen Merritt*[b],
Donald L. Shapiro[c], *Walker A. Long*[d], *J. Harry Gunkel*[e]

[a] Department of Pediatrics, University of California, Los Angeles, Calif., USA;
[b] Department of Pediatrics, University of California, San Diego, Calif., USA;
[c] Department of Pediatrics, University of Rochester, Rochester, N.Y., USA;
[d] Burroughs Wellcome Co., Research Triangle Park, N.C., USA;
[e] Ross Laboratories, Columbus, Ohio, USA

Surfactant replacement has been advocated immediately after delivery of very preterm infants [1]. Among those infants exhibiting symptoms of respiratory distress at birth, approximately 5–10% will have congenital bacterial pneumonia [2]. We have developed a preterm rabbit model of bacterial pneumonia that recreates the pulmonary environment seen in human newborns with intrauterine lung infection. Clinically available surfactant preparations were instilled into the lungs of these infected animals to determine whether surfactant prophylaxis might have an adverse effect on pulmonary host defenses against bacteria.

Methods

Experimental Model
Pregnant rabbits were delivered by hysterotomy at 28 ± 0.5 days of gestation. These pups were placed in warm (37 °C), humidified (R.H. = 75%) incubators containing 60% inspired oxygen. At 1 h of age, preterm rabbit litters were infected with concentrated aerosols of *Streptococcus agalactiae* type Ia (GBS) [3]. The infected pups were returned to the incubators for 4 h, an interval that allowed an influx of polymorphonuclear leukocytes

[1] This work was supported by grant HL 40675 from the US Public Health Service.

(PMNL) and mononuclear leukocytes (MNL) into the alveoli. In separate studies, lung lavage effluents were analyzed to quantitate the influx of PMN and MNL into the air spaces of preterm rabbits after GBS aerosol infection.

Surfactant Therapy and Intrapulmonary Bacterial Inactivation

Four hours after aerosol infection, litters were randomly divided into a 0-hour group and a surfactant therapy group. The 0-hour animals were immediately sacrificed, the left lung aseptically removed, homogenized and the numbers of viable GBS determined by standard pour plate techniques. Simultaneous with the 0-hour time point, the remaining littermates had their tracheas isolated surgically and 60 mg/kg of surfactant phospholipid were instilled by direct needle puncture [3]. The following surfactant preparations were used in these studies: (1) human amniotic fluid derived surfactant (prepared by T.A. Merritt); (2) calf lung surfactant extract (supplied by D.L. Shapiro); (3) Exosurf (supplied by W.A. Long), and (4) Surfactant TA (supplied by J.H. Gunkel). Infected control animals for these studies included: (1) preterm rabbits given intratracheal rabbit surfactant (60 mg/kg) prepared from adult rabbit lung lavage [3]; (2) preterm rabbits treated with intratracheal sterile saline at a volume equal to the largest amount of instilled surfactant, and (3) untreated preterm rabbits. Preterm animals were also compared to untreated term (< 1-day-old) rabbits. Four hours later, the left lungs of the treated and control animals were removed and cultured as outlined above. The inactivation of GBS in the lung was determined by subtracting the $\log_{10}$ bacterial colony-forming units (cfu)/lung of an individual rabbit pup studied at 4 h from the group mean $\log_{10}$ cfu of the lungs of littermates tested at 0 h. Statistical comparisons were made using methods described in our previous studies [3].

In vitro Growth of Group B Streptococci in Surfactant

Individual surfactant preparations, at a dose of 15 mg/ml of surfactant phospholipid, were incubated with an intrapulmonary colonizing dose of GBS (ca. 10^4 cfu). Sodium, potassium, chloride, bicarbonate, dextrose (final concentration = 4 mM) and protein (final concentration = 0.03% Bactopeptone, Difco, Detroit, Mich.) were added to this bacterial and surfactant mixture so that electrolyte, sugar and protein concentrations of alveolar lining fluid were reproduced [3]. The mixture was shaken at 37 °C in 7.3% CO_2. The bicarbonate/CO_2 system maintained the pH of the suspension at 6.9–7.0 throughout the assay, thereby reproducing the $[H^+]$ concentration of alveolar lining fluid [4]. After 1 and 4 h of incubation, samples were removed and cultured on Todd-Hewitt agar.

Results

Streptococci proliferated in the lungs of preterm rabbits, while term rabbit lungs inactivated GBS (fig. 1). Surfactant therapy did not affect the streptococcal growth seen in the preterm lung as treated pups had numbers of intrapulmonary GBS that were equal to that noted in control animals. One surfactant, Exosurf, actually restricted the intrapulmonary multiplication of GBS when contrasted to the other preterm groups.

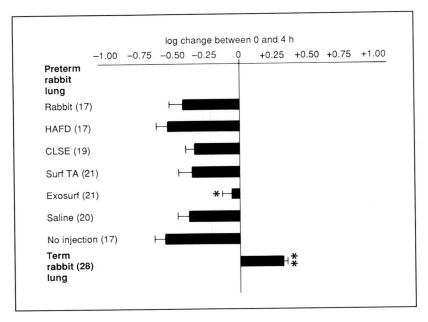

Fig. 1. Effect of surfactant therapy on the intrapulmonary inactivation of group B streptococci by the preterm rabbit lung. The histograms represent mean ± SEM; the numbers of animals studied at 0 and 4 h are given within the parentheses to the right of a designated group. A negative log change to the left of zero means that bacterial proliferation took place in the lung and a positive log change to the right of zero indicates that bacterial killing occurred in the lung. * $p < 0.01$ for Exosurf versus the other preterm groups; ** $p < 0.001$ for term rabbits versus the preterm groups. Rabbit = rabbit surfactant; HAFD = human amniotic fluid derived surfactant; CLSE = calf lung surfactant extract; Surf TA = Surfactant TA; saline = intratracheal saline.

Streptococcal proliferation was likely to occur in the preterm rabbit lung because it had fewer MNL at the time of surfactant therapy (7,492 ± 1,362/g BW) when compared to term rabbits (22,062 ± 2,025, mean ± SEM, n ≥ 12 animals, $p < 0.005$). Lung lavage analyses revealed, however, that similar numbers of PMNL were present in preterm and term lungs at the time of therapy (5,400 ± 987 vs. 4,365 ± 1,138/g BW, respectively). The fact that Exosurf restricted the growth of GBS in the preterm lung was unexpected. However, this finding was consistent with our in vitro studies of GBS growth in surfactant. Those studies showed that Exosurf caused death of GBS relative to the control conditions, while rabbit and human

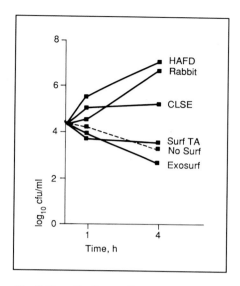

Fig. 2. Growth of group B streptococci in clinically relevant surfactant preparations. Each data point symbolizes the mean of two or more experiments. Abbreviations as in figure 1. No Surf = A reaction mixture without surfactant (control).

amniotic fluid derived surfactants promoted GBS proliferation, and calf lung surfactant extract and Surfactant TA supported minimal streptococcal growth (fig. 2).

Discussion

We developed an in vivo model to evaluate whether surfactant replacement influences lung host defenses. By infecting the preterm rabbit lung, we created a fibrinous alveolar exudate that included MNL, PMNL, and streptococci. This histologic picture is similar to that seen in low birth weight humans who have GBS pneumonia [5]. Over a 4-hour period of observation, clinically relevant surfactants did not affect streptococcal growth beyond that noted in control animals. Although our findings may be reassuring to physicians administering surfactant in the delivery room before a definitive diagnosis is available, clinical studies are still needed to determine the outcome of such infected, preterm human infants who receive replacement therapy at birth. Our most interesting observation was that Exosurf killed GBS during in vitro studies and also restricted GBS growth in the preterm lung. Ordinarily, bacteria are removed from the lungs via phagocytes, extraphagocytic antibacterial effectors, and the mucociliary escalator. The relative importance of phagocytic mechanisms is illustrated by GBS proliferation in the lungs of our preterm rabbits where alveolar macrophages were found in limited numbers. In a lung with a relative deficiency of phagocytes, Exosurf may have its adverse effects on GBS multiplication via an extraphagocytic bactericidal mechanism. However, we cannot exclude that sublethal damage was not induced in the GBS by Exosurf, which in turn augmented phagocytic killing. With the caveat that a surfactant preparation be nontoxic to bronchoalveolar cells, Exosurf serves as an example of a surfactant that has antibacterial effects while retaining surface tension lowering characteristics. Such a concept may be desirable in designing future preparations for surfactant replacement.

References

1   Lawson, E.E.: Exogenous surfactant therapy to prevent respiratory distress syndrome. J. Pediat. *110:* 492–493 (1987).
2   Sherman, M.P.; Goetzman, B.P.; Ahlfors, C.E.; Wennberg, R.P.: Tracheal aspiration and its clinical correlates in the diagnosis of congenital pneumonia. Pediatrics, Springfield *65:* 258–263 (1980).
3   Sherman, M.P.; D'Ambola, J.B.; Aeberhard, E.E.; Barrett, C.T.: Surfactant therapy of newborn rabbits impairs lung macrophage bactericidal activity. J. appl. Physiol. *65:* 137–145 (1988).
4   Neilson, D.W.; Goerke, J.; Clements, J.A.: Alveolar subphase pH in the lungs of anesthetized rabbits. Proc. natn. Acad. Sci. USA *78:* 7119–7123 (1981).
5   Katzenstein, A.-L.; Davis, C.; Braude, A.: Pulmonary changes in neonatal sepsis due to group B β-hemolytic *Streptococcus.* Relation to hyaline membrane formation. J. infect. Dis. *133:* 430–435 (1976).

Michael P. Sherman, MD, UCLA Medical Center, Room B2-325, MDCC,
Los Angeles, CA 90024-1752 (USA)

Wichert P von, Müller B (eds): Basic Research on Lung Surfactant.
Prog Respir Res. Basel, Karger, 1990, vol 25, pp 209–214

Alterations of Surfactant Homeostasis in Diseased Lungs[1]

*Bernd Müller, Heike Hasche, Gudrun Hohorst, Christiane Skurk,
Annette Püchner, Wolfgang Bernhard, Peter von Wichert*

Department of Internal Medicine, Medical Policlinic, Pulmonary Research Unit,
Philipps University, Marburg, FRG

In order to get more insight into the systems that are involved in the pathogenesis of respiration disorders several approaches have been used. Among them analysis of the bronchoalveolar lavage material occupies a prominent position as it yields particularly valuable information. Many components of the bronchoalveolar lavage such as cells, phospholipids, fatty acids, and proteins have been studied often for diagnostic reasons. Taking into account that the lavage material from patients is influenced by a number of factors such as limited standardization of the washing procedure, age of patient, variability and degree of lung injury, and medication at the time of lung wash, studies concerning the basic events of respiration pathology should not be limited to the analyses of bronchoalveolar lavage material. There is also a need for studies on tissue and homogenates from diseased animal and human lungs. In addition, studies are required concerning type II pneumocytes and their role in the homeostasis of lung surfactant in the diseased lung.

The purpose of the present study was to analyze the bronchoalveolar lavage parameters and the metabolism of type II pneumocytes that were isolated from diseased rat lungs. As surfactant homeostasis, we investigated the composition of the bronchoalveolar lavage material, synthesis of phospholipids, and secretion.

[1] This work was supported in part by the Deutsche Forschungsgemeinschaft Wi 359/8 and by PEF 85/007/1C.

Materials and Methods

Animals: For all experiments male Wistar rats (160–180 g) were used.

Experimental groups: Development of acute respiratory distress syndrome (ARDS) was induced by a single administration of lipopolysaccharide from *Salmonella abortus equi* (4 mg/kg). The animals were sacrificed after 4 h of shock lung development, the lungs were excised, perfused, lavaged extracorporally and the type II cells isolated as described below. Exposure to NO_2 was performed in a tight exposure chamber with a gas concentration of 10 ppm over a period of 72 h; lung lavage and cell isolation was the same as for the ARDS group. As control groups animals received either saline solution only or were exposed to an atmosphere of normal air.

Lung lavage was assayed for total protein and phospholipid as well as for phospholipid subclass composition and fatty acids of the phosphatidylcholine species. The methods used were identical to those cited elsewhere [7].

Type II cell isolation and secretion experiments were done according to Dobbs et al. [1].

Cytosol and microsome preparation and Enzyme assays were performed as reported by Haagsman et al. [3].

Results and Discussion

As all the animals used in the experiments had about the same body weight and the recovery of the bronchoalveolar lavage fluid was in the same range (about 95%), we calculated the total washed out protein and phospholipid material and compared it to the experimental groups. We found an increase of total protein and total phospholipid both in the endotoxin-treated and in the NO_2-exposed rats, although the effect was most striking in the NO_2-treated animals (table 1). Analyses of the individual phospholipid classes only showed a decrease for the phosphatidylcholine species (77.2 ± 4.8% of total phospholipids in controls; 65.1 ± 6.5% for the endotoxin group and 61.2 ± 4.7% for the NO_2-exposed rats). Further analyses of the phosphatidylcholines concerning their fatty acid composition only showed slight differences: whereas the proportion of palmitic acid was unchanged in the experimental groups, the contents of arachidonic acid increased from 0.9 ± 0.7% in the controls to 2.3 ± 2.2% in the endotoxin and to 2.5 ± 1.0% in the NO_2 group. Additional physiological measurements, such as determination of the surface activity in the bronchoalveolar lavage should be done to see if the change in phospholipid composition is accompanied by alterations in physiological activity.

After analyzing the bronchoalveolar lavage parameters, cellular surfactant homeostasis was studied. In view of the fact that in the lung other

Table 1. Protein and phospholipid contents in bronchoalveolar lavage of endotoxin-treated and NO_2-exposed rats

	mg/lavage	
	protein	phospholipid
Control	3.45 ± 1.40	0.85 ± 0.50
Endotoxin (20 mg/kg, 4 h)	5.77 ± 2.22^a	1.61 ± 0.51^a
NO_2 (10 ppm, 72 h)	32.45 ± 19.23^b	3.11 ± 1.41^a

Significantly different from control: [a] $p < 0.01$; [b] $p < 0.001$.

cell types may influence the type II pneumocytes, whole animals rather than isolated type II cells were exposed to the treatments with NO_2 lipopolysaccharide. This approach was also preferred to studies with type II cell co-cultures as hereby the influence of only one additional cell type on type II pneumocytes could be examined. It is, however, possible that injured type II cells may recover during the isolation and incubation period.

Since it was observed that there was an increase in total protein and phospholipid in the bronchoalveolar lavage material the question arose as to whether this was due to an increased secretory activity or just to the material that passively transudated into the alveolar lumen. The basic secretion of isolated type II pneumocytes ranged from 2.6 to 3.1%, and no differences were observed between controls and the experimental groups (table 2). However, tetra-phorbol-acetate (TPA)-stimulated secretion showed a decreased rate for the endotoxin and the NO_2 groups when compared to the control. Inhibition of TPA-stimulated secretion with isolated surfactant apoprotein-A (SP-A) [2, 4] was the same for all groups. The reason for not seeing any differences in the basic secretion rate among the three groups could be due to the small amounts that were secreted without stimulation by TPA.

In the secretion experiments it was observed that during the synthesis phase there was a difference in the ^{14}C-choline label per 500,000 type II pneumocytes. When the ^{14}C-choline content was set at 100% for the control group, the endotoxin group contained $131 \pm 37\%$ and the NO_2 group $217 \pm 35\%$. Whether this observation was due to an experimental cell damage in that way that a passive accumulation of the choline took place in the cells, or to an active increase in phospholipid synthesis was tested in synthesis experiments. For this purpose the CDP-choline pathway was chosen. We analyzed the enzymes choline kinase and CTP:choline-

Table 2. Secretion rate of phosphatidylcholine in cultures of type II pneumocytes from endotoxin-treated and NO_2-exposed rats

	^{14}C-labeled phosphatidylcholine, % secreted in 3 h[1]		
	control	endotoxin (20 mg/kg, 4 h)	NO_2 (10 ppm, 72 h)
Saline	2.9 ± 0.3	3.1 ± 0.6	2.6 ± 0.6
DMSO	2.4 ± 0.6	2.6 ± 0.3	2.0 ± 0.2
SP-A (1 µg/ml)	1.8 ± 0.3	1.3 ± 0.3	1.6 ± 0.5
TPA (10^{-8} M)	9.6 ± 1.5	5.2 ± 0.9[a]	5.8 ± 1.3[a]
TPA + SP-A	1.9 ± 0.5	1.8 ± 0.2	2.0 ± 0.5

[a] Significantly different from control $p < 0.01$.
[1] After treatment with endotoxin or NO_2 the cells were incubated for 21 h with [^{14}C]choline. After washing secretion was stimulated by addition of various substances for 3 h after which the [^{14}C]phosphatidylcholine was measured in the cells and the medium.

Table 3. Enzyme activities in type II pneumocytes of endotoxin-treated and NO_2-exposed rats

	Choline kinase activity in cytosol nmol/mg protein min	CTP: choline-phosphate cytidylyltransferase activity, nmol/mg protein min		
		cytosol – PG	cytosol + PG	microsomal
Control	$4.1 (\pm 0.6)$	$0.7 (\pm 0.4)$	$2.1 (\pm 1.2)$	$2.3 (\pm 0.9)$
Endotoxin (20 mg/kg, 4 h)	$5.2 (\pm 0.1)$[a]	$0.4 (\pm 0.2)$	$3.6 (\pm 1.4)$	$2.1 (\pm 0.4)$
NO_2 (10 ppm, 72 h)	$7.0 (\pm 0.6)$[b]	$1.6 (\pm 0.3)$[a]	$4.9 (\pm 1.2)$[a]	$2.2 (\pm 0.6)$

After treatment with endotoxin or NO_2 type II cells were isolated and the enzyme measured in subcellular fractions.
PG = Phosphatidylglycerol.
Significantly different from control: [a] $p < 0.01$; [b] $p < 0.001$.

phosphate cytidylyltransferase and the synthesized product phosphatidylcholine. Regarding the specific activity of the choline kinase we observed an increase for the endotoxin-treated and the NO_2-exposed rats (table 3). Regarding the CTP:choline-phosphate cytidylyltransferase that is claimed to be the rate-limiting enzyme in the CDP-choline pathway [5], the observations were focused on both the microsomal and the cytosolic part of this enzyme. Whereas the microsomal part, that is thought to be the active form of the enzyme [6], was not changed in the experimental groups, the cytosolic form showed an increase in the endotoxin and the NO_2 experimental groups. Using phosphatidylglycerol as activator the same effects were observed (table 3). We have not yet determined whether these changes in enzyme activities are also reflected in alterations of the pool sizes of the intermediates of the CDP-choline pathway.

The additional determination of the newly synthesized phosphatidylcholines, however, encouraged us to hypothesize that an increased specific activity of both enzymes tested also led to an increased synthesis of the final metabolic product after NO_2 exposure. Surprisingly, in these experiments the rate of phosphatidylcholine synthesis from labelled choline seemed lower in the endotoxin-treated animals than the controls, despite the higher specific activity of choline kinase and CTP: choline-phosphate cytidylyltransferase in the endotoxin-treated group (table 4).

Table 4. [^{14}C]choline incorporation in type II pneumocytes in the synthesis experiments of endotoxin-treated and NO_2-exposed rats

Hours of synthesis	Contents, dpm/500,000 cells		
	control	endotoxin (20 mg/kg, 4 h)	NO_2 (10 ppm, 72 h)
1.5	41,892 ($\pm$ 1,463)	39,195[a] ($\pm$ 1,433)	62,068[b] ($\pm$ 680)
3.0	75,081 ($\pm$ 6,408)	61,781[c] ($\pm$ 4,376)	92,633[a] ($\pm$ 9,600)

After treatment with endotoxin or NO_2 the cells were incubated for 21 h with [^{14}C]choline. Following a washing procedure the [^{14}C]choline incorporation was measured within the cells.
Significantly different from control: [a] $p < 0.1$; [b] $p < 0.01$; [c] $p < 0.001$.

In conclusion, our study shows that endotoxin treatment and, particularly, NO_2 exposure lead to enhanced protein and phospholipid contents in lavage. The rate of synthesis of PC from choline in type II cells appears to be increased after NO_2 exposure but not by endotoxin treatment. The unstimulated secretion is not different between the control and the experimental groups, but the stimulated secretion is only half of that observed for the control.

References

1   Dobbs, L.G.; Gonzalez, R.; Williams, M.C.: An improved method for isolating type II cells in high yield and purity. Am. Rev. resp. Dis. *134:* 141–145 (1986).

2   Dobbs, L.G.; Wright, J.R.; Hawgood, S.; Gonzalez, R.; Venstrom, K.; Nellenbogen, J.: Pulmonary surfactant and its components inhibit secretion of phosphatidylcholine from cultured rat alveolar type II cells. Proc. natn. Acad. Sci. USA *84:* 1010–1014 (1987).

3   Haagsman, H.P.; Schuurmans, E.A.J.M.; Batenburg, J.J.; van Golde, L.M.G.: Synthesis of phosphatidylcholines in ozone-exposed alveolar type II cells isolated from adult rat lung: is glycerolphosphate acyltransferase a rate-limiting enzyme? Exp. Lung Res. *14:* 1–17 (1988).

4   Rice, W.R.; Ross, G.F.; Singleton, F.M.; Dingle, S.; Whittsett, J.A.: Surfactant-associated protein inhibits phospholipid secretion from type II cells. J. appl. Physiol. *63:* 692–298 (1979).

5   Vance, D.E.; Choy, P.C.: How is the phosphatidylcholine biosynthesis regulated? Trends Biochem. Sci. *4:* 145–148 (1979).

6   Vance, D.E.; Pelech, S.L.: Enzyme translocation in the regulation of phosphatidylcholine biosynthesis. Trends Biochem. Sci. *4:* 17–20 (1984).

7   Wichert, P. von; Temmesfeld, M.; Meyer, W.: Influence of septic shock upon phosphatidylcholine remodeling mechanism in rat lung. Biochim. biophys. Acta *664:* 487–497 (1981).

Bernd Müller, PhD, Department of Internal Medicine, Medical Policlinic,
Pulmonary Research Unit, Philipps University, Baldingerstrasse,
D–3550 Marburg (FRG)

Wichert P von, Müller B (eds): Basic Research on Lung Surfactant.
Prog Respir Res. Basel, Karger, 1990, vol 25, pp 215–223

Alterations in Alveolar Surfactant following Severe Multiple Trauma

*W. Seeger, U. Pison, R. Buchhorn, T. Joka, M. Brand, U. Obertacke,
H. Neuhof, K.-P. Schmit-Neuerburg*

Division of Clinical Pathophysiology, Department of Internal Medicine,
Justus-Liebig University, Giessen, FRG; Department of Trauma Surgery,
University of Essen, FRG

Lack of surfactant material is the underlying abnormality of the respiratory distress syndrome in preterm infants [1–3]. Surfactant abnormalities have also been implicated in the development of acute respiratory distress in adults (ARDS) [4]. This view is supported by similarities in pulmonary failure between patients with ARDS and preterm babies with respiratory distress syndrome, as well as diffuse alveolar collapse suggesting altered surfactant properties in ARDS. Mechanisms of surfactant alteration in the course of ARDS could not only include absolute lack of surface-active compounds, but also alterations in its phospholipid or protein composition [5, 6]. There are several lines of evidence that increased lung endothelial and epithelial permeability is the decisive step in the development of ARDS or different etiology [7–11]. Subsequent leakage of protein-rich edema fluid into the interstitial and the alveolar space has been implicated in alterations of surfactant properties due to protein surfactant interaction [12–16].

There are, however, only very few data on surfactant properties of patients suffering from ARDS and being artificially ventilated. Von Wichert and Kohl [17] described a decreased dipalmitoyl-phosphatidylcholine content in lung specimens from patients dying of ARDS. Petty et al. [18] observed abnormalities in lung elastic properties and in surface tension behavior of alveolar surfactant, obtained by postmortem lung washing in patients who succumbed because of this syndrome [18]. Hallman et al. [19], who studied the lavage phospholipid profile in patients with respiratory failure of different etiology, noted a marked decrease in phosphatidylcholine (PC) as well as phosphatidylglycerol (PG), accompanied by a relative increase in the other phospholipid species in the lavage material. Moreover, they found some decrease in the minimal surface tension of surfactant material isolated from lavage fluid of 3 of these patients.

We performed a prospective study, designed to evaluate surfactant abnormalities in severely injured patients during the course of posttraumatic pulmonary dysfunction. Trauma patients were selected because of the homogeneity of the primary event (underlying disease of ARDS) as well as the consistency in time relationship between the insult and the development of respiratory failure in several of these patients. It was the aim of the study to describe the time course of surfactant abnormalities after onset of the initial event, to find correlations between biophysical (surface tension behavior) and biochemical (phospholipids, protein extravasation) alterations of the surfactant material and to correlate changes in surfactant properties with the severity of respiratory failure developing in these patients.

Methods

The alveolar phospholipid composition, surface tension characteristics, as well as protein and neutrophil content were determined in serially obtained bronchoalveolar lavage fluids. For this reason fiberoptic bronchoscopy was carried out every second day, beginning within the first 24 h after trauma and ending on the 14th day after the insult (lavage volume 100 ml 0.15 M sodium chloride with an overall recovery of 60–70% by volume). After spinning, the supernatant was analyzed for protein content (staining with Coomassie brilliant blue) [21], organic phospholipids after extraction by the Folch procedure [22, 23], and hemoglobin by standard techniques. High-performance liquid chromatography was used to separate different surfactant phospholipids including PC, PG, phosphatidylethanolamine (PE), phosphatidylinositol (PI), phosphatidic acid (PA), sphingomyelin (SPH), phosphatidylserine (PS) and lysophosphatidylcholine (LPC) [24]. Surface tension was measured in a modified Langmuir-Wilhelmy surface balance [25, 26], using techniques previously described in detail [16]. As the lavage material obtained from these severely injured patients is highly diluted, it was approximately fivefold concentrated by positive pressure ultrafiltration [27, 28]. 50 µg surfactant phospholipids in a maximum volume of 500 µl were then applied to the Wilhelmy balance by using Trurnit's method for surface spreading [29, 30]. The reproducibility of this technique was found to be very high for all surface tension parameters (standard deviation less then 5%). The coordinates of the sixth surface tension area diagram were evaluated for hysteresis area, minimal surface tension and stability index (example given in fig. 1).

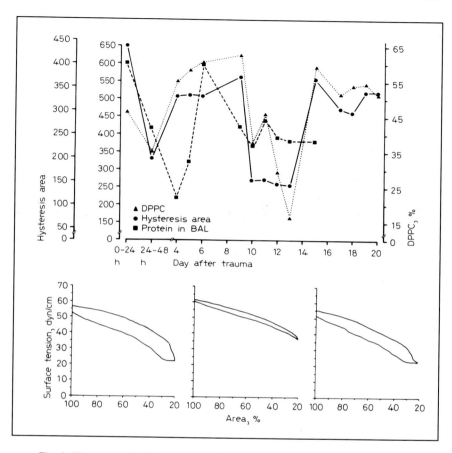

Fig. 1. Upper panel: Time course of hysteresis area, lavage protein content and percentage of dipalmitoyl-phosphatidylcholine in lavage phospholipids (% DPPC) of a patient with high overall ARDS score. Lower panel: Examples of surface tension isotherms (6th cycle) obtained from lavage samples of this patient taken at days 1 (left), 11 (middle) and 16 (right) after trauma.

The study included multiply traumatized patients surpassing 40 points of a standardized injury severity score [31], the time between the trauma and arrival at the hospital being less than 2 h. All patients required controlled mechanical ventilation (with Peep application depending on lung function) and were treated by a standardized regimen. The assessment of respiratory failure was based on a composite scoring system that combined separate evaluation of chest radiographs (range 0–4), oxygenation index ($p(Aa)O_2/pAO_2$), mean pulmonary arterial pressure, and respiratory system compliance [32]. The variables

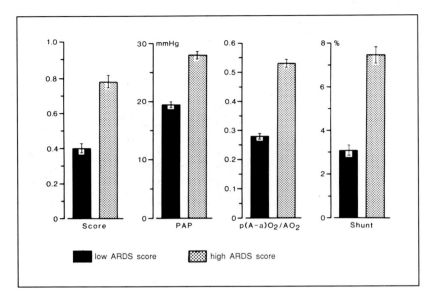

Fig. 2. ARDS score, mean pulmonary arterial pressure (PAP), oxygenation index (p(A-a)O_2/pAO_2) and shunt flow of the severely traumatized patients, divided in a group with overall high ARDS score and a group with overall low ARDS score. Mean values ± SE calculated for the whole observation periods of all patients are given (17 traumatized patients, 10 with low ARDS score and 7 with high ARDS score).

of this composite respiratory failure score are weighted by using different multiplicators. Patients were scored for respiratory failure every 6 h within the first 48 h after trauma and subsequently every 24 h. Based on these sequential respiratory failure scores, a mean score was calculated for the whole observation period. Using a computerized regression model the whole group of patients was thus divided in two subgroups, one with a high overall ARDS score, and one with low overall ARDS score (mild pulmonary dysfunction) (see fig. 2, which also includes basic clinical data of the two groups of patients). The groups did not differ significantly in age or sex, and the overall survival rate was 40% in the patients with low ARDS score and 28% in the patients with high ARDS score (in total 17 traumatized patients).

As a control group, 10 healthy volunteers (all engaged in the study) underwent three bronchoalveolar lavages each by the corresponding procedures.

The lavage protein content was markedly increased in the patients with a high overall ARDS score, compared to traumatized patients with low ARDS score and compared to healthy volunteers (fig. 3). Analysis of

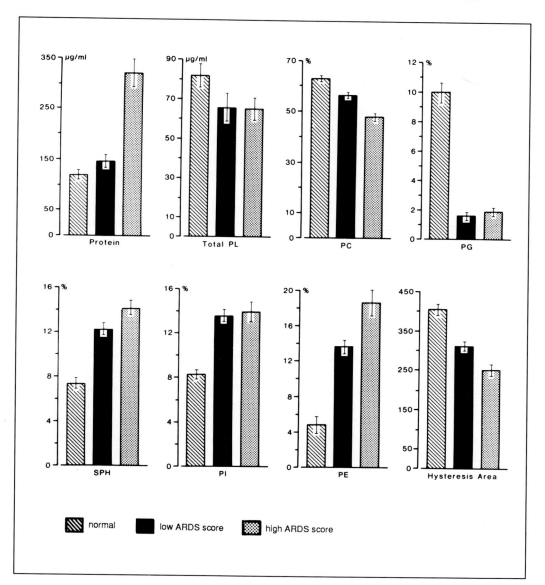

Fig. 3. Lavage variables of severely traumatized patients (divided in those with low and high ARDS score) compared to healthy volunteers. Mean values calculated over the whole observation period of the traumatized patients/volunteers are indicated (mean ± SE). Details are given in the text. All phospholipid (PL) species are given in % of the total lavage phospholipid content; hysteresis area is given in arbitrary units.

time course showed that the significantly higher alveolar protein load could also be demonstrated within the first 24 h after trauma, often already in the first lavage taken within 6 h (see example in fig. 1). In many patients, this protein leakage anteceded alterations in surfactant biophysical properties and phospholipid composition (see below). Correlations between these parameters (for instance hysteresis area and % PC) and the alveolar protein content, calculated for the mean values of each patient, ranged between r = 0.4 and r = 0.6. Hemoglobin was very low in all lavage samples (lavages were performed in nontraumatized areas) and did not differ significantly between the two ARDS severity groups or the healthy controls.

There was a marked influx of granulocytes into the alveolar space in all traumatized patients ($198 \pm 58 \times 10^3$ PMN/ml in all injured patients versus $13.4 \pm 3.4 \times 10^3$ PMN/ml in the healthy volunteers). However, there was no significant difference in the absolute or relative alveolar PMN content between the patients with high and low ARDS, and there was no significant correlation between PMN efflux and alterations in surfactant properties.

The total phospholipid content in the lavage samples displayed quite large variability, but no significant difference between the two ARDS severity groups or the whole group of multiply injured patients and the control subjects was noted. The phospholipid profile in the lavage fluid was, however, markedly altered (fig. 3). There was a marked decrease in % PC in all injured patients, significantly more pronounced in patients with high overall ARDS score than those with mild pulmonary dysfunction. Studies of time course revealed that this decrease occurred progressively, with a minimum after 10–14 days in the patients who developed severe ARDS. PG was very low in all traumatized subjects (with similar progressive decrease), but no significant difference between the two ARDS severity groups was observed. In contrast, % SPH, % PI, % PE and % PA were increased in all multiply injured patients. When calculated for all patients, there was a significant correlation between these alterations in phospholipid profile and ARDS severity score (r = 0.77, calculated for the decreasing PC). The changes in the alveolar phospholipid compositions might be caused by contamination with plasma phospholipids as well as phospholipids from migrated inflammatory cells or injured lung cells. However, the progressive time course more likely suggests that these alterations reflect alveolar type II cell injury in the course of posttraumatic pulmonary insufficiency.

Among the parameters of surfactant function in vitro, hysteresis area was markedly altered in all traumatized patients compared to healthy volunteers, and it efficiently discriminated between the patients with low and high ARDS score. In most patients with severe respiratory failure, there was an initial drop in hysteresis area, followed by a more progressive decline with a minimum after 10–14 days. There was a significant correlation between the decrease in hysteresis area and the lavage phospholipid profile ($r = 0.62$ for % PC). Stability index and minimal surface tension revealed marked alterations in magnitude within single patients; however, the mean values showed no significant difference between the multiply injured patients and healthy controls.

Conclusion

Surfactant abnormalities can be demonstrated in severely traumatized patients and they are significantly correlated with the development of respiratory failure. The main features are a rapidly occurring alveolar protein load (demonstrated within the first 48 h) and a progressive alteration in the alveolar phospholipid composition. The latter include a decrease in PC and PG, accompanied by a relative increase in the other phospholipid species. Alterations in the surfactant biophysical activity are noted, which are correlated with these two events (protein leakage and progressive PC decline) and with the development of respiratory failure.

References

1   Farrell PM, Avery ME: Hyaline membrane disease. Am Rev Respir Dis 1975;111: 657–688.
2   Avery ME, Mead J: Surface properties in relation to atelectasis and hyaline membrane disease. Am J Dis Child 1959;97:517–523.
3   Notter RH, Shapiro DL: Lung surfactant in an era of replacement therapy. Pediatrics 1981;68:781–789.
4   Ashbaugh DG, Bigelow DB, Petty TL, et al: Acute respiratory distress in adults. Lancet 1967;ii:319–323.
5   Kulovich MV, Gluck L: The lung profile. II. Complicated pregnancy. Am J Obstet Gynec 1979;135:64–70.
6   Jackson JC, Palmer S, Truog WE, et al: Surfactant quantity and composition during recovery from hyaline membrane disease. Pediat Res 1986;20:1243–1247.
7   Brigham KL: Metabolites of arachidonic acid in experimental lung vascular injury. Fed Proc 1985;44:43–45.

8   Traber DL: Pulmonary microvascular dysfunction during shock; in Janssen HF, Barnes GD (eds): Circulatory Shock: Basic and Clinical Implications. New York, Academic Press, 1985, pp 23–46.

9   Brigham KL, Meyrick B: Interactions of granulocytes with the lungs. Circ Res 1984; 54:623–634.

10 Repine JE, Tate RM: Oxygen radicals and lung edema. Physiologist 1983;26:171–181.

11 Seeger W, Suttorp N: Role of membrane lipids in the pulmonary vascular abnormalities caused by bacterial toxins. Am Rev Respir Dis 1987;136:462–466.

12 Taylor FB Jr, Abrams ME: Effect of surface active lipoprotein on clotting and fibrinolysis, and of fibrinogen on surface tension of surface active lipoprotein. Am J Med 1966;40:346–350.

13 Tierney DF, Johnson RP: Altered surface tension of lung extracts and lung mechanics. J Appl Physiol 1965;20:1253–1260.

14 Jobe A, Ikegami M, Jacobs H, et al: Permeability of premature lamb lungs to protein and the effect of surfactant on that permeability. J Appl Physiol 1983;55:169–176.

15 Ikegami M, Jobe A, Jacobs H, et al: A protein from airways of premature lambs that inhibits surfactant function. J Appl Physiol 1984;57:1134–1142.

16 Seeger W, Stöhr G, Wolf HRD, et al: Alteration of surfactant function due to protein leakage: special interaction with fibrin monomer. J Appl Physiol 1985;58:326–338.

17 Von Wichert P, Kohl FV: Decreased dipalmitoyl lecithin content found in lung specimens from patients with so-called shock-lung. Int Care Med 1977;3:27–30.

18 Petty TL, Silvers GW, Paul GW, et al: Abnormalities in lung elastic properties and surfactant function in adult respiratory distress syndrome. Chest 1979;75:571–574.

19 Hallman M, Spragg R, Harrell JH, et al: Evidence of lung surfactant abnormality in respiratory failure. Study of bronchoalveolar lavage phospholipids, surface activity, phospholipase activity, and plasma myoinositol. J Clin Invest 1982;70: 673–683.

20 Pison U, Seeger W, Buchhorn R, et al: Surfactant abnormalities in patients with respiratory failure following multiple trauma. Am Rev Respir Dis (in press, 1989).

21 Bradford MM: A rapid and sensitive method for the quantitation of microgram quantities of protein utilizing the principle of protein-dye binding. Anal Biochem 1976;72:248–254.

22 Folch J, Lees M, Stanley GHS: A simple method for the isolation and purification of total lipids from animal tissue. J Biol Chem 1957;226:497–509.

23 Henry RJ: Journal of Clinical Chemistry. New York, Harper & Row, 1964.

24 Pison U, Gono E, Joka T, et al: High performance liquid chromatography of adult human bronchoalveolar lavage. Assay for phospholipid lung profile. J Chromatogr 1986;377:79–89.

25 Clements JA, Hustead RF, Johnson RP, et al: Pulmonary surface tension and alveolar stability. J Appl Physiol 1961;16:444–450.

26 Hildebran JN, Goerke J, Clements JA: Pulmonary surface film stability and composition. J Appl Physiol 1979;47:604–611.

27   Parsons PE, Fowler AA, Hyers TM, et al: Chemotactic activity in bronchoalveolar lavage fluid from patients with adult respiratory distress syndrome. Am Rev Respir Dis 1985;132:490–493.

28   Afford SC, Stockley RA, Kramps JA, et al: Concentration of bronchoalveolar lavage fluid by ultrafiltration. Evidence of differential protein loss and functional inactivation of proteinase inhibitors. Anal Biochem 1985;151:125–130.

29   Trurnit HJ: A theory and method for the spreading of protein monolayers. J Coll Sci 1966;15:1–13.

30   Abrams ME: Isolation and quantitative estimation of pulmonary surface-active lipoprotein. J Appl Physiol 1966;21:718–720.

31   Baker SP, O'Neill B, Haddon W, et al: The injury severity score. A method for describing patients with multiple injuries and evaluating emergency care. J Trauma 1984;14:187–196.

32   Obertacke U, Kalotai J, Coenen T, et al: Ein linearer ARDS-Schweregradscore (0–1). Submitted.

W. Seeger, MD, Division of Clinical Pathophysiology, Department of Internal Medicine, Justus-Liebig-University, D–6300 Giessen (FRG)

Wichert P von, Müller B (eds): Basic Research on Lung Surfactant.
Prog Respir Res. Basel, Karger, 1990, vol 25, pp 224–228

Effects of in vivo Dexamethasone Administration on Surfactant Protein Expression in the Rat[1]

David S. Phelps, Joanna Floros[2]

Department of Pediatrics, Harvard Medical School, Boston, Mass., USA

Twenty years ago Liggins [1] published his findings that lung maturation could be accelerated by glucocorticoid treatment. This led to many studies on the effects of glucocorticoids on the pulmonary surfactant system. These studies generally showed that glucocorticoids have a stimulatory effect on the synthesis of surfactant lipids. With the surge in recent years of research on the surfactant-associated proteins, studies have begun to assess the effects of glucocorticoids on these proteins [2–5]. Most of the regulatory work with surfactant proteins so far has involved SP-A, the 35,000-dalton surfactant glycoprotein. Almost all of these studies have been performed in various culture systems. These studies have shown that glucocorticoids have stimulatory effects, inhibitory effects or both stimulatory and inhibitory effects on SP-A expression depending on variables such as hormone dosage, factors present in the medium, incubation of tissue or cells in serum-containing or serum-free medium [3–5].

We have focused our own investigations on the effect of dexamethasone on SP-A and its mRNA in the intact animal. In our earlier studies we examined fetal rat lung following dexamethasone treatment of the mother [2]. We found a consistent stimulation of SP-A by glucocorticoid treat-

[1] This work was supported by grants HL 38288, HL 34788 and HL 34616 from the National Heart, Lung and Blood Institute.
[2] We would like to thank Heather Harding and Rodney Lequillo for excellent technical assistance.

ment. Since lung development continues postnatally in the rat, we were interested in studying whether glucocorticoids continue to enhance SP-A synthesis postnatally. Towards this goal, we treated animals with glucocorticoids at various ages up to adulthood. The results from this study are discussed here.

Materials and Methods

Rats were injected intraperitoneally with various doses of dexamethasone or with vehicle alone and sacrificed by decapitation 24 h later. The lungs were dissected out and either frozen immediately at −70 °C for ELISA or for RNA preparation or a portion of them was minced and incubated in culture medium containing [^{35}S]methionine for metabolic labeling. To determine the content of SP-A in the tissue by ELISA, the lung tissue was sonicated, extracted with detergent, centrifuged to remove insoluble material and an aliquot was analyzed by ELISA using a competitive ELISA protocol. Metabolic labeling, immunoprecipitation and two-dimensional gel electrophoresis was performed as described previously [2]. RNA was prepared as described elsewhere [6] and size-fractionated on formaldehyde-agarose gels, transferred to nitrocellulose, hybridized with radiolabeled cDNA probes, subjected to autoradiography and then to laser densitometry.

Results and Discussion

The results of a dose-response study using ELISA are shown in table 1. A stimulatory effect due to dexamethasone treatment is clearly seen at most doses. This analysis examines the tissue content of SP-A, but cannot

Table 1. Dexamethasone dose-response curve: SP-A levels have been determined by ELISA and expressed as a ratio using the protein content of the homogenate as a denominator (6 rats were analyzed at each point)

Dexamethasone dose	SP-A/protein
0	6.94 ± 1.96
2 µg/kg	6.49 ± 1.11
20 µg/kg	9.81 ± 1.37
200 µg/kg	10.58 ± 1.69
2 mg/kg	13.23 ± 2.25
20 mg/kg	12.25 ± 2.80

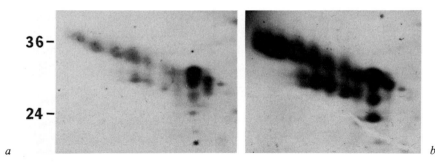

36 –

24 –

a   *b*

Fig. 1. Two-dimensional SDS-gel electrophoresis of [³⁵S]methionine-labeled SP-A immunoprecipitated from lung tissue of 46-day-old rats. *a* Control tissue. *b* Dexamethasone-treated tissue. Molecular mass in kilodaltons is indicated.

differentiate newly synthesized protein from old, or stored material from secreted.

To examine the glucocorticoid effect on newly synthesized SP-A we injected rats with 200 µg/kg body weight as we did in our earlier studies [2]. The tissue was incubated for 6 h in [³⁵S]methionine, disrupted by sonication and the SP-A isolated by immunoprecipitation with a specific antiserum. The labeled protein was then subjected to two-dimensional gel electrophoresis, autoradiography and quantitation by laser densitometry. Labeled SP-A in the culture medium was isolated and analyzed similarly. At every age tested, glucocorticoid treatment resulted in a marked increase in newly synthesized SP-A from both tissues and medium. Figure 1 shows representative autoradiographs from rat lung tissue at day 46. There is about a 2.5-fold difference between dexamethasone-treated animals (b) and controls (a) in these samples. Similar increases are seen throughout development and in the adult. These are summarized in table 2. Control levels of intracellular SP-A and secreted SP-A have been set equal to 100% and the percent change of SP-A with glucocorticoid treatment is shown.

In order to gain further insight into the mechanism by which the glucocorticoids exert their effect, levels of SP-A mRNA were measured as well. To analyze SP-A mRNA, total mRNA was isolated, subjected to agarose gel electrophoresis, blotting and hybridization with a [³²P]-labeled rat SP-A cDNA probe, followed by autoradiography and densitometry. These

Table 2. Summary of dexamethasone effects on SP-A metabolic labeling and SP-A mRNA: the nature of each type of sample, the number of pairs of samples analyzed, percent increase after dexamethasone treatment and significance values are indicated

Sample	n	% increase	Significance
Intracellular SP-A	21	211 ± 11	$p < 0.001$
Secreted SP-A	16	233 ± 16	$p < 0.003$
SP-A mRNA	33	144 ± 5	$p < 0.001$

data are also summarized in table 2. To examine changes in SP-A mRNA levels as a function of dexamethasone concentrations, we performed a dose-response experiment using dexamethasone doses as high as 20 mg/kg body weight. The SP-A mRNA levels were increased at all doses used (data not shown).

The results of this study suggest that glucocorticoid treatment not only accelerates lung maturation but also that this treatment continues to stimulate SP-A production throughout life. Glucocorticoid-mediated inhibition of SP-A secretion, SP-A synthesis, or SP-A mRNA is not detected in these in vivo studies, despite the use of doses of hormones as high as 20 mg/kg body weight. The studies presented here demonstrate that SP-A synthesis continues to be enhanced by glucocorticoid treatment throughout the postnatal life in the rat. Since animals at these ages presumably have sufficient amounts of surfactant to maintain normal respiratory function one might speculate that SP-A and its stimulation by glucocorticoids has some other purpose. The response of other surfactant proteins such as SP-B and SP-C remain to be determined in this system.

References

1   Liggins, G.C.: Premature delivery of foetal lambs infused with glucocorticoids. J. Endocr. *45:* 515–523 (1969).
2   Phelps, D.S.; Church, S.; Kourembanas, S.; Taeusch, H.W.; Floros, J.: Increases in the 35 kDa surfactant-associated protein and its mRNA following in vivo dexamethasone treatment of fetal and neonatal rats. Electrophoresis *8:* 235–238 (1987).
3   Mendelson, C.R.; Chen, C.; Boggaram, V.; Zacharias, C.; Snyder, J.M.: Regulation of the synthesis of the major surfactant apoprotein in fetal rabbit lung tissue. J. biol. Chem. *261:* 9938–9943 (1986).

4   Ballard, P.L.; Hawgood, S.; Liley, H.; Wellenstein, G.; Gonzales, L.W.; Benson, B.;
     Cordell, B.; White, R.T.: Regulation of pulmonary surfactant apoprotein SP 28-36
     gene in fetal human lung. Proc. natn. Acad. Sci. USA *83:* 9527–9531 (1986).
5   Whitsett, J.A.; Pilot, T.; Clark, J.C.; Weaver, T.E.: Induction of surfactant protein in
     fetal lung. J. biol. Chem. *262:* 5256–5261 (1987).
6   Floros, J.; Phelps, D.S.; Kourembanas, S.; Taeusch, H.W.: Primary translation prod-
     ucts and tissue specificity of the major surfactant protein in rat. J. biol. Chem. *261:*
     828–831 (1986).

David S. Phelps, PhD, Harvard Medical School, Richardson Fuller Bldg., A5,
221 Longwood Avenue, Boston, MA 02115 (USA)

Wichert P von, Müller B (eds): Basic Research on Lung Surfactant.
Prog Respir Res. Basel, Karger, 1990, vol 25, pp 229–230

Effect of Calcium, and Lanthanum and Other Calcium Antagonists on the Lung Phospholipid Secretion Produced by Lung Distension in Newborn Rabbits

Anthony Corbet, Robert Voelker, Frederick Murphy

Baylor College of Medicine, Houston, Tex., USA

Abstract. We examined the hypothesis that surfactant secretion is controlled at the apical membrane of granular pneumocytes by an influx of calcium from the alveolar lining liquid. Lung distension and stimulation of secretion was produced by saline lavage in freshly killed newborn rabbit pups of 29.5 days' gestation. Each wash was two-thirds total lung capacity.

When calcium was added to saline lavage fluid the fractional stimulation of phospholipid secretion was 1.11 ± 0.16 for 0.0 mM calcium (NS), 1.94 ± 0.28 for 1.0 mM calcium (p < 0.001) and 1.69 ± 0.34 for 3.0 mM calcium (p < 0.05).

When the calcium transport inhibitor lanthanum was added to lavage fluid and the lungs lavaged 10 times, there was a dose-related decrease of phospholipid secretion, the fractional inhibition increasing from 1.01 ± 0.23 at $10^{-5}\,M$ (NS), to 0.43 ± 0.19 at $0.5 \times 10^{-4}\,M$, 0.22 ± 0.03 at $10^{-4}\,M$ and 0.19 ± 0.05 at $10^{-3}\,M$ (p < 0.001). This inhibition by lanthanum was significantly reversed by the addition of 10 mM calcium, the pooled fractional inhibition decreasing from 0.25 ± 0.04 to 0.50 ± 0.08 (p < 0.02).

In this rabbit pup model, by thin-layer chromatography, lavage phospholipid was $97 \pm 1\%$ phosphatidylcholine and $3 \pm 1\%$ phosphatidylinositol, while approximately 50% of phosphatidylcholine was disatured. Monitoring for LDH and DNA showed no significant changes with calcium or lanthanum. Measurements of ATP showed that oxidative metabolism was well maintained.

Examination of thin lung sections by X-ray elemental analysis suggested that lanthanum, easily demonstrated outside the plasma membrane, did not penetrate inside the alveolar epithelial cells.

To examine whether lanthanum caused phospholipid to be bound at the cell surface, tritiated surfactant phospholipid was prepared in 3-day-old rabbits, and known amounts of radioactivity were lavaged into the lungs of premature pups. Then recovery was examined using 9 further lavages with either saline or lanthanum. Recovery was 0.90 ± 0.04 with saline and 0.74 ± 0.02 with $10^{-3}\,M$ lanthanum. If the data for phospholipid secretion with 10^{-3} M lanthanum are corrected for this recovery, there remained a large suppression of phospholipid secretion, the corrected fractional inhibition being 0.23.

Blockers of L-type calcium channels such as nifedipine and verapamil did not inhibit phospholipid secretion, nor did putative intracellular calcium blockers such as trimethoxybenzoate or tetracaine. On the other hand, phospholipid secretion was significantly blocked by 10^{-2} M nickel (p < 0.001) and partially by 10^{-2} M cadmium.

The data suggest that a pool of calcium on the outside of the plasma membrane is partly responsible for secretion, and that calcium enters through a calcium channel sensitive to inorganic, but not organic blockers. This pool is very difficult to chelate. When lungs were lavaged with 15 mM EGTA and 5 mM BAPTA in HEPES buffer at pH 7.4, the fractional inhibition was 0.81 $\pm$ 0.08 (p < 0.01). This result with calcium chelation confirms that extracellular calcium modulates surfactant secretion.

Anthony Corbet, MD, Baylor College of Medicine, Houston, TX 77030 (USA)

Wichert P von, Müller B (eds): Basic Research on Lung Surfactant.
Prog Respir Res. Basel, Karger, 1990, vol 25, pp 231–236

Surfactant Proteins and Respiratory Distress Syndrome

Joanna Floros[a], *David S. Phelps*[b], *Daphne E. deMello*[c],
Jeff Longmate[a], *Heather Harding*[b]

[a] Department of Pediatrics, Harvard Medical School, Boston, Mass.; and
[b] Department of Biostatistics, Harvard School of Public Health;
[c] Department of Pathology and Laboratory Medicine,
Cardinal Glennon Children's Hospital, St. Louis, Mo., USA

Pulmonary surfactant is present in at least three distinct morphological forms. These are lamellar bodies (LB) found inside the epithelial type II cells and within the alveolar space, an ordered lattice-like structure called tubular myelin (TM), which is derived from lamellar bodies extracellularly and which gives rise to the third form, a surface-active monolayer [1].

To date there are two groups of nonserum surfactant-associated proteins that have been studied extensively. One of these is a 30- to 40-kdalton sialoglycoprotein (SP-A) [2–5]. This protein has been localized within type II cells and within TM complexes [6, 7]. The other group of surfactant proteins consists of low molecular weight hydrophobic proteins (SP-B and SP-C) [8]. Although the precise role of each of these surfactant proteins in the metabolism and function of pulmonary surfactant is not clearly understood, numerous studies have suggested that the surfactant proteins may play a key role in both surfactant function and metabolism. A deficiency of pulmonary surfactant in prematurely born infants can result in respiratory distress syndrome (RDS), a condition that leads to death in about 15% of the cases and morbidity in many others.

Recently, an electron-microscopic study of lungs of infants that died from RDS and infants that died from other causes showed an absence of TM only in RDS samples even though both groups contained structurally similar LB intracellularly and in the alveolar spaces [9]. We were able to obtain some of the samples from that study and re-examine them with an antiserum to SP-A and an antiserum recognizing SP-B and SP-C. The data from this retrospective study are presented and discussed in this report. Furthermore, to better understand the potential role of the surfactant proteins in RDS we are in the process of expanding the observations from the retrospective study by examining prospectively the expression of the surfactant proteins at the protein and mRNA levels in lung tissue of infants dying from RDS or from other nonrespiratory causes. The lung tissue we will use in these studies is obtained at autopsy and in this report we present preliminary data supporting in part the utility of this postmortem material for RNA studies.

Methods and Materials

Immunostaining of lung tissue sections was performed using standard histological techniques. RNA preparation and analysis was done by employing routine techniques. The cDNA probes (SP-A, SP-B, SP-C, actin) were [^{32}P]-labeled. The autoradiograms were quantitated using a soft laser densitometer. Statistical analysis was performed as follows: for the slot blots, each sample was analyzed by using four dilutions each with concentrations of 0.25, 0.5, 1.0 and 2.0 μg of total RNA. The optical density from each dilution was adjusted by dividing by its relative concentration to obtain comparable measurements. Logarithms were taken to stabilize variance, and the average of the replicate logs of adjusted optical densities was used as the single measure for each sample. Regional comparisons are based on t statistics computed on the log scale and back-transformed. Half-lives are based on a linear regression of mean logs on time.

Results and Discussion

Expression of the Surfactant Proteins in Lung Tissues Obtained from Patients Dying from RDS and Patients Dying from Other Causes

Tissues from 10 control patients showed abundant staining within the alveolar epithelial cells and macrophages with both SP-A and SP-BC antisera. Samples of lung tissue from RDS patients showed absent to minimal staining with antisera to surfactant proteins. In all RDS cases examined, the immunostaining for SP-A was consistently less than that for SP-BC and it never reached the levels seen in control cases. An example of the immunohistochemistry from this study is shown in figure 1. Since these RDS tissues were shown to lack TM [9], we suggest that decreased levels of surfactant proteins in RDS tissues correlates with the absence of TM.

To better understand the role of the surfactant proteins in the pathogenesis of RDS we are expanding this study and continuing to address

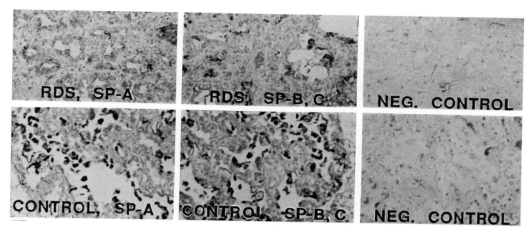

Fig. 1. Sections from formalin-fixed and paraffin-embedded tissue were immuno-stained with antisera to SP-A and SP-BC. The gestational ages for the RDS and control tissues are 30 and 35 weeks, respectively. The postnatal ages are 4.5 h (RDS) and 5 weeks (control). Nonimmune rabbit serum was used instead of the primary antiserum for the negative controls.

questions relating to the expression of these proteins in control and RDS cases by examining their expression at both protein and RNA levels. To carry out these studies we will employ immunohistochemistry (IMH), in situ hybridization (ISH) and RNA blotting using lung tissue obtained post-mortem from control and RDS cases. To first evaluate the utility of post-mortem lung tissue for RNA studies we are carrying out preliminary stud-ies utilizing IMH, ISH, and RNA blotting. In the following section of this report we discuss some of this background work using RNA blotting.

Characterization of Postmortem Lung Tissue for Its Utility in RNA Studies Using RNA Blotting: Regional Differences and Half-Lives

Using RNA blotting we have examined the levels of surfactant protein mRNAs in the upper and lower left lobes from the same individual to determine whether regional differences in the expression of surfactant pro-teins exist and therefore whether it is important to maintain consistency in our sampling procedure. The probes used in this study were [^{32}P]-labeled cDNAs for each of the surfactant proteins. The autoradiograms from the RNA blots were quantitated using a laser densitometer. The data suggested that there may be slight differences between upper and lower lobes, particularly for SP-C. Therefore, sampling will remain consistent.

One of the techniques, in situ hybridization, that we will use in our studies requires an initial fixation step prior to freezing the tissue for storage and subsequent analysis. To determine the stability of the surfactant protein mRNAs at 4 °C in the event that initial fixation could not be performed immediately after the autopsy, lung tissue was removed from 4 patients with postmortem intervals of 4, 5, 9 and 10 h. A portion of each sample was processed immediately to determine initial surfactant protein mRNA content in each individual. This initial value was of importance since the question of interest was to determine the decline in surfactant protein mRNA during tissue refrigeration as compared to the initial surfactant protein value. The remaining tissue was stored at 4 °C and at varying intervals (up to 34 h after acquisition) a portion of each tissue was removed and analyzed to determine surfactant protein mRNA content using slot blots. The surfactant protein mRNA levels were assessed as described under 'Methods and Materials'. The estimated half-lives for SP-A, SP-B, and SP-C were 14, 22 and 21 h, respectively, suggesting that tissue refrigeration for a few hours does not result in significant losses of the surfactant protein mRNA.

Expression of Surfactant Proteins in Normal Adult Lung Tissue:
Variability and Correlation of Expression

To facilitate comparison between experiments, adult lung RNA is being used as a standard so that data can be expressed as a percent of adult RNA values. For the final data interpretation it was important, however, to conduct studies in adult lung tissue obtained from surgical biopsies (rather than postmortem) to determine the variation in the levels of expression of the surfactant proteins in the adult population. In addition, we compared the relative expression of the different surfactant proteins in each of these individuals.

Lung biopsies from 11 patients were analyzed for their surfactant protein mRNA content via blotting, autoradiography and densitometry. Because densitometric values are a relative measure, the coefficient of variation (CV = standard deviation/mean) was used to assess variability. SP-A had a coefficient of variation of 0.53 and SP-B of 0.44. These values were more variable than either SP-C (CV = 0.27) or actin (CV = 0.21), a nonsurfactant protein included for comparison purposes. The tendency of the proteins to vary in concert among individuals was also examined by computing correlation coefficients (CC). The expression of SP-A mRNA was highly correlated with that of SP-B (CC = 0.90; p = 0.002). The expression of SP-C, on the other hand, did not correlate either with the expression of SP-A (CC = 0.02; p = 0.96) or with the expression of SP-B (CC = 0.3; p = 0.39).

Summary

The retrospective study performed here showed that absence or decreased levels of surfactant proteins in lung tissues from prematurely born infants correlates with the incidence of RDS and the absence of tubular myelin. It appears that the SP-A level is much lower (or undetectable) than the levels of SP-BC in these individuals. Preliminary data from studies to establish the utility of postmortem tissue for RNA studies have suggested that: (i) there might be small regional differences in the expression of surfactant proteins in postmortem tissue as determined by RNA blotting; (ii) postmortem lung tissue can be kept at 4 °C for a few hours without major losses in surfactant protein mRNAs; (iii) expression of surfactant proteins in normal lung tissue is more variable for SP-A and SP-B and less so for SP-C among different individuals and (iv) the expression of SP-A correlates with that of SP-B, but the expression of SP-C does not correlate with either the expression of SP-A or SP-B. The results discussed here have provided us with some information about the feasibility of using postmortem material for RNA experiments. Studies are currently underway using this combination of molecular biological and histological techniques to more completely define the levels of the surfactant proteins and their mRNAs in RDS and normal development.

Acknowledgements

The authors would like to thank Drs. B. Benson and R.T. White from California Biotechnology, Inc. for making available the SP-C cDNA probe and Rodney Lequillo for excellent technical assistance. Supported by NIH HL34788, HL34616, Genentech/ALA Career Investigator Award to J.F., and the Fleur-de-Lis Fund of Cardinal Glennon Children's Hospital.

References

1   Wright, J.R.; Clements, J.A.: Metabolism and turnover of lung surfactant. Am. Rev. resp. Dis. 135: 426–444 (1987).

2   Phelps, D.S.; Floros, J.; Taeusch, H.W.: Post-translational modification of the major human surfactant-associated proteins. Biochem. J. 237: 373–377 (1986).

3   Floros, H.; Phelps, D.S.; Taeusch, H.W.: Biosynthesis and in vitro translation of the major surfactant-associated protein from human lung. J. biol. Chem. 260: 495–500 (1985).

4   Phelps, D.S.; Floros, J.: Localization of surfactant protein synthesis in human lung by in situ hybridization. Am. Rev. resp. Dis. *137:* 939–942 (1988).

5   Floros, J.; Steinbrink, R.; Jacobs, K.; Phelps, D.S.; Kriz, R.; Recny, M.; Sultzman, L.; Jones, S.; Taeusch, H.W.; Frank, H.A.; Fritsch, E.F.: Isolation and characterization of cDNA clones for the 35kDa pulmonary surfactant associated protein (PSP-A). J. biol. Chem. *261:* 9029–9033 (1986).

6   Walker, S.R.; Williams, M.C.; Benson, B.: Immunocytochemical localization of the major surfactant apoproteins in type II cells, Clara cells, and alveolar macrophages of rat lung. J. Histochem. Cytochem. *34:* 1137–1148 (1986).

7   Coalson, J.J.; Winter, V.T.; Martin, H.M.; King, R.J.: Colloidal gold immunoultrastructural localization of rat surfactant. Am. Rev. resp. Dis. *133:* 230–237 (1986).

8   Phelps, D.S.; Smith, L.M.; Taeusch, H.W.: Characterization and partial amino acid sequence of a low molecular weight surfactant protein. Am. Rev. resp. Dis. *135:* 1112–1117 (1987).

9   deMello, D.E.; Chi, E.Y.; Doo, E.; Lagunoff, D.: Absence of tubular myelin in lungs of infants dying with hyaline membrane disease. Am. J. Pathol. *127:* 131–139 (1987).

Joanna Floros, PhD, Harvard Medical School, Richardson-Fuller Building A-5, 221 Longwood Avenue, Boston, MA 02115 (USA)

Wichert P von, Müller B (eds): Basic Research on Lung Surfactant.
Prog Respir Res. Basel, Karger, 1990, vol 25, pp 237–246

Structural and Functional Characterization of Porcine Surfactant Isolated by Liquid-Gel Chromatography[1]

Bengt Robertson[a], *Tore Curstedt*[b], *Jan Johansson*[c], *Hans Jörnvall*[c], *Tsutomu Kobayashi*[d]

[a] Department of Pediatrics, St. Göran's Children's Hospital;
[b] Department of Clinical Chemistry, Karolinska Hospital, and
[c] Department of Physiological Chemistry I, Karolinska Institutet, Stockholm, Sweden;
[d] Department of Anesthesiology, Kanazawa University, Kanazawa, Japan

Surfactant substitutes should be carefully characterized by a number of biochemical, biophysical, physiological, bacteriological, and immunological methods before being used in clinical practice. This is essential for our understanding of the mechanisms by which this new therapy prevents or mitigates respiratory distress syndrome (RDS) and, hence, for the safety of the treated patients. Since the properties of the surfactant material may vary with the methods used for purification [1], separate standards have to be set for different forms of exogenous surfactant. In this paper, we characterize a new type of porcine surfactant ('Curosurf'), successfully used in a recent large international clinical trial [2].

Composition of Curosurf

This surfactant is isolated from minced lungs by a combination of washing, centrifugation, chloroform:methanol extraction, and liquid-gel chromatography, as previously described [3]. Curosurf contains approximately 99% polar lipids, mainly phospholipids, and 1% hydrophobic proteins (SP-B and SP-C, in approximate molar proportions 1:3), essential for

[1] This work was supported by The Swedish Medical Research Council (Projects 3351 and 3532), The Swedish National Association against Heart and Chest Diseases, and Axel Tielmans Minnesfond.

the physiological activity [4–7]. Phosphatidylcholine is the predominant phospholipid (about 75 mol%) and dipalmitoylphosphatidylcholine constitutes about 46% of this fraction. The level of phosphatidylglycerol is approximately 4 mol%. Neutral lipids (including triglycerides, cholesterol, and cholesteryl esters) and the 28- to 36-kdalton hydrophilic apoprotein (SP-A) are absent. We believe there is a certain tolerance as to both the lipid and protein concentration of an effective surfactant preparation, because a film or surfactant lipids tends to refine spontaneously during cyclic compression [8].

We have determined the complete amino acid sequence of porcine SP-B (8.7 kdaltons) by analysis at the protein level [9]. The primary structure consists of 79 residues with 8 half-cystine residues and a total of 39% branched-chain hydrophobic residues, mainly leucine and valine. The N-terminal part of porcine SP-B shows some heterogeneity, with Phe-Pro-Ile-Pro- as the predominant structure (a truncated form, two residues shorter, also exists). The structure of the C-terminus was identified as -Arg-Cys-Ser-Ser. The hydrophobic nature of the protein complicated purification, proteolytic cleavages, and sequence analysis. Complete release of branched-chain residues required acid hydrolysis at 110 °C for no less than 7 days [9].

Direct protein analysis of porcine SP-C (3.7 kdaltons) revealed a major sequence of 35 amino acid residues, with a total of 13 Val and 8 Leu and a middle region consisting entirely of branched-chain hydrophobic residues [10]. The N-terminal structure is heterogeneous, with Leu-Arg-Ile-Pro- representing the longest form of the molecule. The C-terminal region contains the residues -Leu-Met-Gly-Leu. Because of its extreme hydrophobicity, SP-C is resistant to cleavage with proteolytic enzymes. Exceptional conditions (150 °C, 72 h) are also required for release of all residues by acid hydrolysis [10].

Surface Properties of Curosurf in Relation to Hypophase Concentration and Temperature

We have investigated the surface properties of Curosurf with pulsating bubble [11]. Maximum and minimum surface tension were recorded at a pulsation rate of 38/min (50% surface compression). Recordings were made at various concentrations and temperature, in order to define the conditions required for optimal physiological activity (maximum surface

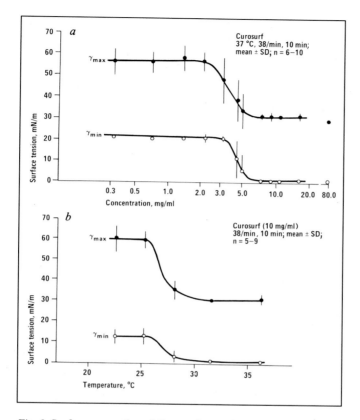

Fig. 1. Surface properties of Curosurf at various hypophase concentrations (*a*) and temperatures (*b*), recorded with pulsating bubble. Values for maximum and minimum surface tension (γ_{max}, γ_{min}) were obtained after pulsation for 10 min. Other experimental conditions are specified in the diagrams.

tension close to 30 mN/m, minimum surface tension close to 0 mN/m). At 37 °C, there seems to be a critical concentration range (3–5 mg/ml) above which optimal activity is consistently recorded, and below which surfactant activity is unsatisfactory (maximum surface tension > 50 mN/m, minimum surface tension about 20 mN/m) (fig. 1a). Also temperature is critical. At a concentration of 10 mg/ml, optimal surface properties are lost at temperatures below 30 °C (fig. 1b).

The concept of a critical concentration of surfactant, essential for adequate stabilization of the airspaces at birth, may provide a guideline to the dosage of surfactant administered for prevention of RDS. In babies receiving prophylactic treatment in the delivery room, the exogenous material becomes mixed with the fetal lung liquid before (or soon after) the onset of ventilation. It seems likely that the concentration of surfactant in this liquid, which at birth has a volume of about 30 ml/kg [12] must exceed the critical level referred to above, to ensure proper lung function.

Physiological Properties of Curosurf Evaluated in Newborn Rabbits

Instillation of Curosurf into the airways of immature newborn rabbits leads to a dramatic improvement of lung-thorax compliance [3]. The dose required for a satisfactory response in surfactant-deficient rabbit neonates, delivered on day 27 of gestation, amounts to about 160 mg/kg. Assuming complete mixing, this would correspond to a concentration of exogenous surfactant in fetal lung liquid of about 6 mg/ml, a figure close to the critical concentration level, above which optimal surface properties are recorded with pulsating bubble (see above).

In earlier studies, we have evaluated the effect of surfactant replacement therapy in immature newborn rabbits by recording tidal volumes during artificial ventilation with a sequence of standardized insufflation pressures [3]. This method is adequate for screening purposes; however, it tends to overestimate the difference in lung-thorax compliance between surfactant-treated and control animals, since the lungs of the latter may remain essentially nonexpanded during the course of the experiment.

In more recent experiments we have ventilated surfactant-treated newborn rabbits and control animals with individualized insufflation pressure, as suggested by Ikegami et al. [13]. Using a standardized tidal volume of 10–12 ml/kg, we found that at least some immature newborn rabbits with a gestational age of 26.5 days can be ventilated for 60 min with adequate PCO_2. Average values for lung-thorax compliance in Curosurf-treated animals and controls were 0.55 and 0.35 ml/cm $H_2O \cdot kg$, respectively. In the same series of experiments, we could document that treatment of mature newborn animals with Curosurf did not cause any deterioration of compliance [Sun et al., unpubl. observations].

All batches of Curosurf to be used in clinical trials are routinely tested on ventilated immature newborn rabbits. The evaluation may also include quantification of alveolar expansion in histological sections with manual morphometry or image analysis [14]. As expected from the in vitro data discussed above, treatment with an adequate dose of effective surfactant promotes uniform alveolar air expansion in the immature neonatal lung [14].

Inactivation of Curosurf by Serum Proteins

Lung surfactant can become inactivated by fibrin monomer [15], by the 110-kdalton serum protein isolated and characterized by Ikegami et al. [16, 17] and by other serum proteins [18]. Protein leakage into the airspaces is probably an important factor in the pathogenesis of both neonatal and adult RDS, and may interfere with the therapeutic effect of surfactant replacement therapy. Since surfactant preparations vary in their resistance to inhibitory proteins [1], this particular aspect deserves special attention. Serum proteins deteriorate the in vitro activity of Curosurf (10 mg/ml) at protein concentrations > 0.3 mg/ml, as indicated by increasing adsorption time (fig. 2a). At protein concentrations > 1 mg/ml, minimum surface tension remains > 20 mN/m, implying loss of film stability during compression (fig. 2b). Curosurf is also inhibited by lung edema fluid from animals subjected to prolonged hyperoxia [19].

In general, Surfactant-TA (a mixture of bovine surfactant and synthetic lipids) is more resistant to inhibitory proteins than are other surfactants, at least under in vitro conditions [1, 18]. The physiological significance of this intriguing observation remains unclear, as other natural surfactant preparations with lower resistance to inhibitory proteins were superior to Surfactant-TA in their capacity to improve gas exchange and lung compliance in prematurely delivered lambs [1]. Also, the inhibition is probably a concentration-dependent phenomenon which can be overcome by increasing the pool size of surfactant in the airspaces. This would be in keeping with the clinical observation that large doses of surfactant are required to obtain a therapeutic effect in patients with adult RDS (ARDS), a disease characterized by flooding of the airspaces with proteinaceous edema [20].

Immunogenicity of Surfactant-Associated Proteins

Although the surfactant-associated proteins are immunogenic, serum antibodies have not been detected in surviving babies treated with Curosurf in the neonatal period [3]. Circulating surfactant-antisurfactant im-

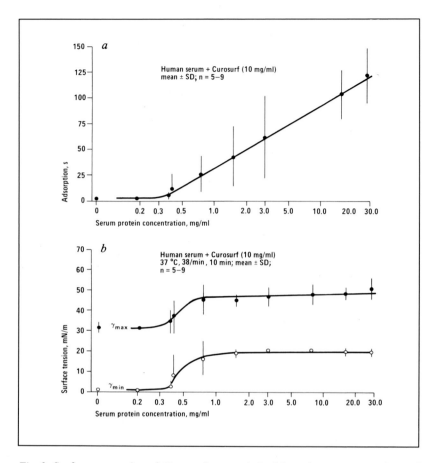

Fig. 2. Surface properties of Curosurf suspended with various concentrations of human serum proteins. Adsorption (*a*) was measured after aspiration of the surface, and maximum and minimum surface tension (γ_{max}, γ_{min}) (*b*) were assessed with pulsating bubble after 10 min of area oscillation. Other experimental conditions are specified in the diagrams.

mune complexes appear during the course of neonatal RDS, irrespective of whether or not the patient has been treated with exogenous surfactant. This is probably secondary to a disturbance of alveolar permeability allowing leakage of surfactant-specific proteins from the airspaces into the interstitial compartment. The immune complexes do not seem to have any harmful effects [21].

Monoclonal antibodies to SP-B inhibit the in vitro activity of Curosurf [Suzuki et al., unpubl. observations], and intraperitoneal inoculation with a hybridoma making antibodies to the same protein results in respiratory failure probably triggered by damage to the alveolar epithelium and flooding of the airspaces with proteinaceous edema including large amounts of the specific antibody [22]. These findings illustrate indirectly the functional significance of SP-B in the surfactant system, without discouraging the use of heterologous surfactant for treatment of neonatal and adult RDS.

Intradermal injection of Curosurf in normal adults causes only a non-specific skin reaction. Curosurf does not elicit an in vitro lymphocyte proliferation response or release of interleukin-2, nor does it inhibit the positive control proliferation responses to the mitogens concanavallin A and phytohemagglutinin. The latter toxicological studies were made prior to the first clinical trial of Curosurf for treatment of ARDS [23].

Curosurf and the Lung Defence System

Some babies treated with surfactant for neonatal RDS may suffer from a combination of surfactant deficiency and infection. It is therefore important that the exogenous surfactant does not promote bacterial growth or interfere with cellular mechanisms in the pulmonary defence system. From recent in vitro studies it seems that Curosurf (15 mg/ml) may have a bacteriostatic effect, at least against *Escherichia coli* [24]. We have shown that short-term (30 min) incubation with low concentrations of Curosurf stimulates phagocytosis and enhances the metabolic burst of blood monocytes (recruitment of these cells is part of a nonspecific inflammatory response), whereas long-term (24 h) incubation seems to decrease the functional activity of the phagocytic cells [25]. Data from experiments on full-term newborn rabbits furthermore indicate that tracheal instillation of small amounts of natural surfactant (5 mg/kg) reduces the capacity of alveolar macrophages to kill group B streptococci [26].

Clearly, the interaction of exogenous surfactant and phagocytic cells is an important target for future research. Analysis of bacteriocidal capacity and possible stimulation of interference with macrophage function should be part of the experimental evaluation of any surfactant preparation designed for clinical use.

Concluding Remarks

Several clinical trials have now documented an impressive effect of surfactant replacement therapy in babies with RDS [2, 27, 28], and it seems likely that this new therapeutic approach will become part of routine treatment, at least in severely ill patients. Very recently, significant progress has also been made in research concerning the structure and functional significance of the surfactant-associated proteins, especially the small hydrophobic ones. Surfactant preparations with nearly optimal physical and physiological properties can be reconstituted from isolated hydrophobic proteins and synthetic phospholipids [7, 29–31] and will probably soon be available for clinical trials. It is important that these new substitutes are subjected to rigorous experimental evaluation along the lines suggested above, before being instilled into the airways of babies or other patients with respiratory failure due to surfactant dysfunction.

Summary

Surfactant isolated from minced porcine lungs by a combination of washing, centrifugation, chloroform-methanol extraction, and liquid-gel chromatography contains 99% polar lipids and 1% hydrophobic proteins. The latter have been identified as SP-B (8.7 kdaltons) and SP-C (3.7 kdaltons), present in approximate molar proportions 1:3. When examined with pulsating bubble at 37 °C, this surfactant reduces minimum surface tension to near zero at hypophase concentrations > 5 mg/ml. When instilled into the airways of immature newborn rabbits ventilated with a tidal volume of 10–12 ml/kg, it improves lung-thorax compliance by $> 50\%$. The same surfactant is inactivated by serum proteins at a protein to phospholipid weight ratio of $> 10\%$. Although the surfactant-associated hydrophobic proteins are potentially immunogenic, no circulating antibodies have been detected in surviving infants treated with porcine surfactant in the neonatal period. Short-term incubation with porcine surfactant initially stimulates the phagocytic activity of human monocytes, but prolonged incubation suppresses macrophage function. These various aspects should be taken into account in experimental evaluation of surfactant preparations for clinical use.

References

1   Ikegami, M.; Agata, Y.; Elkady, T.; Hallman, M.; Berry, D.; Jobe, A.: Comparison of four surfactants. In vitro surface properties and responses of preterm lambs to treatment at birth. Pediatrics, Springfield 79: 38–46 (1987).
2   Collaborative European Multicenter Study Group: Surfactant replacement therapy for severe neonatal respiratory distress syndrome. An international randomized clinical trial. Pediatrics, Springfield 82: 683–691 (1988).

3 Noack, G.; Berggren, P.; Curstedt, T.; Grossmann, G.; Herin, P.; Mortensson, W.; Nilsson, R.; Robertson, B.: Severe neonatal respiratory distress syndrome treated with the isolated phospholipid fraction of natural surfactant. Acta paediat. scand. *76:* 697–705 (1987).

4 Takahashi, A.; Fujiwara, T.: Proteolipid in bovine lung surfactant. Its role in surfactant function. Biochem. biophys. Res. Commun. *135:* 527–532 (1986).

5 Whitsett, J.A.; Ohning, B.L.; Ross, G.; Meuth, J.; Weaver, T.; Holm, B.A.; Shapiro, D.L.; Notter, R.H.: Hydrophobic surfactant-associated protein in whole lung surfactant and its importance for biophysical activity in lung surfactant extracts used for replacement therapy. Pediat. Res. *20:* 460–467 (1986).

6 Yu, S.-H.; Possmayer, F.: Reconstitution of surfactant activity by using the 6 kDa apoprotein associated with pulmonary surfactant. Biochem. J. *236:* 85–89 (1986).

7 Curstedt, T.; Jörnvall, H.; Robertson, B.; Bergman, T.; Berggren, P.: Two hydrophobic low-molecular-mass protein fractions of pulmonary surfactant. Characterization and biophysical activity. Eur. J. Biochem. *168:* 255–262 (1987).

8 Morley, C.; Bangham, A.: Physical properties of surfactant under compression; in von Wichert, Clinical importance of surfactant defects. Prog. respir. Res., vol. 15, pp. 188–193 (Karger, Basel 1981).

9 Curstedt, T.; Johansson, J.; Barros-Söderling, J.; Robertson, B.; Nilsson, G.; Westberg, M.; Jörnvall, H.: Low-molecular-mass surfactant protein type 1. The primary structure of a hydrophobic 8-kDa polypeptide with eight half-cystine residues. Eur. J. Biochem. *172:* 521–525 (1988).

10 Johansson, J.; Curstedt, T.; Robertson, B.; Jörnvall, H.: Size and structure of the hydrophobic low molecular weight surfactant-associated polypeptide. Biochemistry *27:* 3544–3547 (1988).

11 Enhörning, G.: Pulsating bubble technique for evaluating pulmonary surfactant. J. appl. Physiol. *43:* 198–203 (1977).

12 Scarpelli, E.M.: Fetal pulmonary fluid; in Scarpelli, Cosmi, Rev. perinat. Med., vol. 1, pp. 49–106 (University Park Press, Baltimore 1976).

13 Ikegami, M.; Berry, D.; ElKady, T.; Pettenazzo, A.; Seidner, S.; Jobe, A.: Corticosteroids and surfactant change lung function and protein leaks in the lungs of ventilated premature rabbits. Clin. Invest. *79:* 1371–1378 (1987).

14 Robertson, B.; Lachmann, B.: Experimental evaluation of surfactants for replacement therapy. Exp. Lung Res. *14:* 279–310 (1988).

15 Seeger, W.; Stöhr, G.; Wolf, H.R.D.; Neuhof, H.: Alteration of surfactant function due to protein leakage. Special interaction with fibrin monomer. J. appl. Physiol. *58:* 326–338 (1985).

16 Ikegami, M.; Jobe, A.; Jacobs, H.; Lam, R.: A protein from airways of premature lambs that inhibits surfactant function. J. appl. Physiol. *57:* 1134–1142 (1984).

17 Ikegami, M.; Jobe, A.; Berry, D.: A protein that inhibits surfactant in respiratory distress syndrome. Biol. Neonate *50:* 121–129 (1986).

18 Fuchimukai, T.; Fujiwara, T.; Takahashi, A.; Enhorning, G.: Artificial pulmonary surfactant inhibited by proteins. J. appl. Physiol. *62:* 429–437 (1987).

19 Ennema, J.J.; Kobayashi, T.; Robertson, B.; Curstedt, T.: Inactivation of exogenous surfactant in experimental respiratory failure induced by hyperoxia. Acta anaesth. scand. *32:* 665–671 (1988).

20 Lachmann, B.: Animal models and clinical pilot studies of surfactant replacement in adult respiratory distress syndrome. Eur. resp. J. *2* (suppl. 3): 98s–103s (1989).

21 Strayer, D.S.; Merritt, T.A.; Lwebuga-Musaka, J.; Hallman, M.: Surfactant-anti-surfactant immune complexes in infants with respiratory distress syndrome. Am. J. Path. *122:* 353–362 (1986).

22 Suzuki, Y.; Robertson, B.; Fujita, Y.; Grossmann, G.: Respiratory failure in mice caused by a hybridoma making antibodies to the 15 kDa surfactant apoprotein. Acta anaesth. scand. *32:* 283–289 (1988).

23 Richman, P.S.; Spragg, R.G.; Robertson, B.; Merritt, T.A.; Curstedt, T.: The adult respiratory distress syndrome. First trials with surfactant replacement. Eur. resp. J. *2* (suppl. 3): 109s–111s (1989).

24 Sherman, M.P.; Campbell, L.A.; Merritt, T.A.; Shapiro, D.L.; Long, W.A.; Gunkel, J.H.; Robertson, B.A.: In vitro growth of bacteria in clinically relevant surfactant preparations. 3rd Int. Symp., Basic Research on Lung Surfactant, Marburg 1988, abstr. book, p. 71.

25 Wiernik, A.; Curstedt, T.; Johansson, A.; Jarstrand, C.; Robertson, B.: Morphology and function of blood monocytes after incubation with lung surfactant. Eur. J. resp. Dis. *71:* 410–418 (1987).

26 Sherman, M.P.; D'Ambola, J.B.; Aeberhard, E.E.; Barrett, C.T.: Surfactant therapy of newborn rabbits impairs lung macrophage bactericidal activity. J. appl. Physiol. *65:* 137–145 (1988).

27 Hallman, M.; Merritt, T.A.; Jarvenpaa, A.-L.; Boynton, B.; Mannino, F.; Gluck, L.; Moore, T.; Edwards, D.: Exogenous human surfactant for treatment of severe respiratory distress syndrome. A randomized prospective clinical trial. J. Pediat. *106:* 963–969 (1985).

28 Fujiwara, T.; Konishi, M.; Nanbu, H.; Ogawa, Y.; Niitsu, N.; Naito, T.; Akamatsu, H.; Tada, H.; Okuyama, K.; Nishida, H.; Imura, S.; Takeuchi, Y.; Goto, A.; Shimura, K.; Kito, H.; Kuroyanagi, M.; Ogino, T.; Fujimura, M.; Nakamura, H.; Takemine, H.; Nakata, E.; Hashimoto, T.: Surfactant replacement for respiratory distress syndrome (RDS). A multicenter clinical trial (in Japanese). Jap. J. Pediat. *40:* 549–568 (1987).

29 Tanaka, Y.; Takei, T.; Aiba, T.; Masuda, K.; Kiuchi, A.; Fujiwara, T.: Development of synthetic lung surfactants. J. Lipid Res. *27:* 475–485 (1986).

30 Yu, S.-H.; Wallace, D.; Bhavnani, B.; Enhorning, G.; Harding, P.G.R.; Possmayer, F.: Effect of reconstituted pulmonary surfactant containing the 6000-dalton hydrophobic protein on lung compliance of prematurely delivered rabbit fetuses. Pediat. Res. *23:* 23–30 (1988).

31 Revak, S.D.; Merritt, T.A.; Degryse, E.; Stefani, L.; Courtney, M.; Hallman, M.; Cochrane, G.C.: Use of human surfactant low molecular weight apoproteins in the reconstitution of surfactant biologic activity. J. clin. Invest. *81:* 826–833 (1988).

Bengt Robertson, MD, Research Unit for Experimental Perinatal Pathology, St. Göran's Hospital, S–112 81 Stockholm (Sweden)

Wichert P von, Müller B (eds): Basic Research on Lung Surfactant.
Prog Respir Res. Basel, Karger, 1990, vol 25, pp 247–252

Results with Artificial Surfactant: Aspects of Morphological Appearance, Surface Activity, and in vivo Activity of Artificial Surfactant[1]

M. Obladen, H. Schwarz, E. Kattner, P. Stevens

Department of Neonatology, University Children's Hospital, Berlin;
Max Planck Institute for Biology, Tübingen, FRG

A major goal in preparing artificial surfactant is to achieve rapid adsorption and spreading, at the air-water interface, of properties inherent to natural surfactant. In vitro behavior of exogenous phospholipids is dependent on the preparation technique and temperature. Large unilamellar vesicles can adsorb more rapidly than small ones, presumably because they can donate material to the surface film with minimal energy requirement [5, 9]. In this study we examined the relationship between the morphology and in vitro and in vivo activity of 15 surfactant specimens used in a clinical trial.

Patients and Methods

Entry Criteria. Infants of 26–29 weeks of gestation, birth weight 810–1,260 g, requiring artificial ventilation for respiratory distress syndrome grade 2 or more.

Compliance. Dynamic compliance of the respiratory system (C_{rs}) was measured using a 00 pneumotachograph inserted into the airway and connected to a model 6 Buxco lung function analyzer as previously described [6].

Artificial Surfactant. Large unilamellar vesicles containing 80% dipalmitoylphosphatidylcholine, 20% egg phosphatidylglycerol and no protein were studied. The material was obtained in a freeze-dried form from Nattermann, Köln, and was dissolved in phos-

[1] Supported by Deutsche Forschungsgemeinschaft, grant No. Ob 43/5-1.

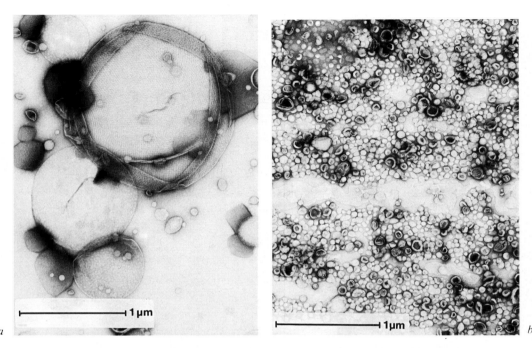

Fig. 1. a Large unilamellar vesicles containing 80% dipalmitoylphosphatidylcholine and 20% phosphatidylglycerol. Negative staining shows lattice-like structures at the vesicle surface. *b* After sonication for 1 min, the same material is transformed into small unilamellar vesicles which have completely lost surface activity.

phate buffer by vortexing. No ultrasound was applied as this transforms the material into small vesicles without surface activity, as shown in figure 1b. A single dose of 50 mg/kg was instilled into the endotracheal tube. The clinical results obtained with this preparation have been published [10].

Surface Tension Studies. In a modified Wilhelmy balance (film surface 392 cm²), all samples adsorbed in less than a second from the 0.15 M NaCl subphase at 37 °C. They formed a stable film, which reduced the surface tension to less than 1 mN/m at compression to 20% and which displayed fast respreading after overcompression. In all samples the stability index was 1.98.

Electron Microscopy. For negative staining, vesicles were adsorbed to carbon-coated grids and stained with 1% uranyl acetate. For thin sectioning, samples were fixed with 1.25% glutaraldehyde in 0.05 M cacodylate buffer, and plastic embedded in Epon. Sections were stained with lead citrate. Specimens were observed at 60 kV with a Philips EM 201 electron miscroscope.

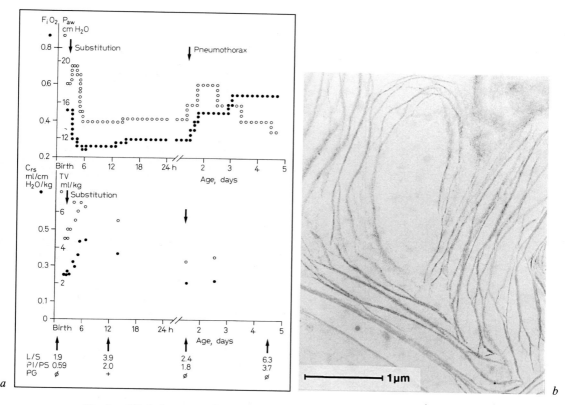

Fig. 2. a Clinical course of a 1,010-gram girl of 28 weeks' gestation with grade 2 RDS, who showed rapid improvement of oxygenation and marked increase of compliance after substitution with artificial surfactant. *b* Electron-micrograph of the sample used in this infant. P_{aw} = Peak inspiratory pressure; C_{rs} = dynamic compliance of the respiratory system; TV = tidal volume; L/S = lecithin/sphingomyelin ratio; PI/PS = phosphatidylinositol/phosphatidylserine ratio; PG = phosphatidylglycerol in tracheal aspirate.

Results

After substitution, most infants had a dissociated response: oxygenation improved immediately, F_iO_2 could be lowered by 19% within 90 min. Improvement in dynamic compliance, however, was delayed. From an initial value of 0.25 ml/cm H_2O/kg, it only started to increase 4 h after substitution to a mean of 0.34 ml/cm H_2O/kg at 12 h. The infants' response

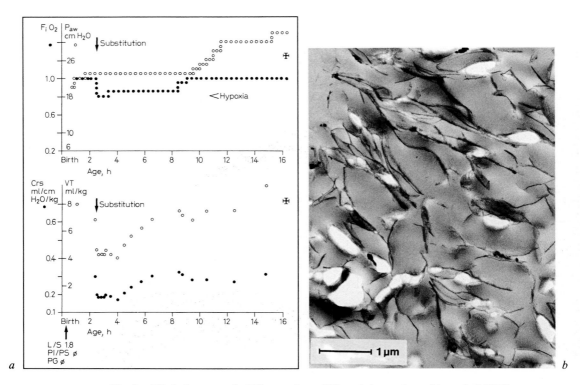

Fig. 3. a Clinical course of a 810-gram boy of 27 weeks' gestation with grade 3 RDS, who showed no response of oxygenation after substitution and died from hypoxemia despite a slight increase in compliance from age 4–6 h. Postmortem examination showed extensive hyaline membranes and intraalveolar hemorrhage. *b* Electron micrograph of the sample used in this infant. For abbreviations, see figure 2.

was temporary, in most of them relapse occurred 10–24 h after substitution. Identical phospholipids, buffer solution and preparation technique were used in all cases. Electron-microscopic analysis of the samples, however, showed considerable variation in vesicle size and type of aggregation. Fourteen of 15 samples studied consisted exclusively of large unilamellar vesicles with a mean diameter of 1.2 μm as shown in figures 1a and 2b. In 1 infant whose oxygenation did not respond to substitution, the vesicles looked quite different, as shown in figure 3b: they were smaller and more densely aggregated. Some parts of the material had an amorphous aspect.

Discussion

Nonresponders have been described in several trials, even with highly effective exogenous surfactants [2, 3]. They have been explained by inadequate dosage [7] and by preparation differences [4]. For surfactant TA, morphologic heterogeneity of vesicles, lamellar sheets, and amorphous material or individual samples were observed and have been related to the presence of calcium and other ions in the solvent [8, 11]. Our data suggest an imperfect sample preparation as another cause of therapeutic failure. We would therefore like to encourage researchers testing exogenous surfactants to study the morphology of their substitutes.

In our study, oxygenation improved well before compliance. The reasons for this are unclear. A sudden increase in lung volume upon substitution, as suggested by Davis et al. [1], could have gone undetected in our set-up. The reduction in right to left shunt resulting from the increase in residual capacity and capillary filling could at least partially explain the improved oxygenation.

References

1   Davis, J.M.; Veness-Neehan, K.; Notter, R.H.; Bhutani, V.K.; Kendig, J.W.; Shapiro, D.L.: Changes in pulmonary mechanics after the administration of surfactant to infants with respiratory distress syndrome. New Engl. J. Med. *319:* 476–479 (1988).

2   Fujiwara, T.: Surfactant replacement in neonatal RDS; in Robertson, Van Golde, Batenburg, Pulmonary surfactant, pp. 479–503 (Elsevier, Amsterdam 1984).

3   Hallman, M.; Merritt, T.A.; Schneider, H.; Epstein, B.L.; Mannino, F.; Edwards, D.K.; Gluck, L.: Isolation of human surfactant from amniotic fluid and a pilot study of its efficiency in respiratory distress syndrome. Pediatrics, Springfield *71:* 473–482 (1983).

4   Ikegami, M.; Agata, Y.; Elkady, T.; Hallman, M.; Berry, B.; Jobe, A.: Comparison of four surfactants. In vitro surface properties and responses of preterm lambs to treatment at birth. Pediatrics, Springfield *79:* 38–46 (1987).

5   Jähnig, F.; Obladen, M.: The physical principles for developing a synthetic lung surfactant. Nuovo cimento *3:* 211–218 (1984).

6   Kattner, E.; Kosack, K.; Obladen, M.: Pneumotachographic measurements of lung function in preterm infants with RDS; in Lachmann, Surfactant replacement therapy, pp. 77–87 (Springer, Heidelberg 1989).

7   Konishi, M.; Fujiwara, T.; Naito, T.; Takeuchi, Y.; Ogawa, Y.; Inukai, K.; Fujimura, M.; Nakamura, H.; Hashimoto, T.: Surfactant replacement therapy in neonatal respiratory distress syndrome. A multi-centre randomized clinical trial. Comparison of high versus low dose of surfactant-TA. Eur. J. Pediat. *147:* 20–25 (1988).

8   Notter, R.H.; Penney, D.P.; Finckelstein, J.N.; Shapiro, D.L.: Adsorption of natural lung surfactant and phospholipid extracts related to tubular myelin formation. Pediat. Res. *20:* 97–101 (1986).

9   Obladen, M.; Popp, D.; Schöll, C.; Schwarz, H.; Jähnig, F.: Studies on lung surfactant replacement in respiratory distress syndrome. Rapid film formation from binary mixed liposomes. Biochim. biophys. Acta *735:* 215–224 (1983).

10   Obladen, M.; Stevens, P.; Kattner, E.: Rapid response of oxygenation, slow response of compliance after liposomal phospholipid substitution in RDS; in Lachmann, Surfactant replacement therapy, pp. 168–180 (Springer, Heidelberg 1989).

11   Taeusch, H.W.; Keough, K.M.W.; Williams, M.; Slavin, R.; Steele, E.; Lee, A.S.; Phelps, D.; Kariel, N.; Floros, J.; Avery, M.E.: Characterization of bovine surfactant for infants with respiratory distress syndrome. Pediatrics, Springfield *77:* 572–581 (1986).

M. Obladen, MD, Department of Neonatology, University Children's Hospital, D–1000 Berlin 19

Wichert P von, Müller B (eds): Basic Research on Lung Surfactant.
Prog Respir Res. Basel, Karger, 1990, vol 25, pp 253–255

Surfactant and Lung Hypoplasia

C.M. Hill[a], *B.D. Brown*[a], *C.J. Morley*[a], *J.A. Davis*[a], *A.J. Barson*[b, 1]

[a] Department of Paediatrics, University of Cambridge, UK;
[b] Department of Pathology, St Mary's Hospital, Manchester, UK

Lung hypoplasia is a condition arising from a failure to develop the appropriate number of generations of bronchioles. Surfactant in lung tissue samples from babies dying with lung hypoplasia resembled that of babies with hyaline membrane disease [1]. As part of a study of lung surfactant in premature and mature babies, lung wash specimens were collected from babies with various types of hypoplasia.

Materials and Methods

Postmortem lavage samples were obtained by washing the lungs with saline (10 ml/kg body weight). These were centrifuged at 300 g and the supernatant stored at $-20\,^{\circ}C$. Lavage samples from the babies with unilateral hypoplasia were from the hypoplastic lung only.

Extraction of samples with solvents, determination of total phospholipid-phosphorus, percentage phospholipid composition and percentage phosphatidylcholine (PC) fraction fatty acid composition (for dipalmitoyl PC; DSPC) were as reported [2]. Statistical comparisons were by Student's t test.

Results and Discussion

The clinical background of the groups examined is shown in table 1. Surfactant of babies with bilateral lung hypoplasia was similar in composition to that of babies with HMD, except that it contained a significantly

[1] C.M.H. would like to acknowledge the assistance of Action Research and currently that of the Children's Research Fund. The authors would like to thank M. Negus for many of the analyses carried out.

higher percentage of DSPC ($p < 0.02$) (table 2). It was different to that of the mature babies in that it contained significantly lower percentages of PC ($p < 0.001$) and DSPC ($p < 0.004$) and a significantly higher percentage of sphingomyelin ($p < 0.001$). Surfactants of babies and infants with bilateral lung hypoplasia were similar. The surfactant composition of babies with

Table 1. Clinical data (mean values)

	HMD	BLH1	BLH2	ULH	MAT
Number	26	19	13	10	15
Gestation, weeks	28.2	32.5	38.0[1]	37.3	38.1
Survival, days	1.0	0.7	140.0	1.7	4.0
LW/BW ratio[2]	2.9	1.2	1.4	1.1	2.3

HMD = Premature babies with hyaline membrane disease, dying within 2 days of birth; BLH1 = babies with bilateral hypoplasia dying within 2 days of birth; BLH2 = infants with bilateral hypoplasia dying between 2 days and 1 year of age; ULH = babies with congenital diaphragmatic hernias and unilateral hypoplasia; MAT = mature babies dying within 14 days of birth, 7 from asphyxia and 8 from congenital heart disease.

[1] Four values only.

[2] Percentage combined lung weight/body weight ratio.

Table 2. Phospholipid composition (mean ± SEM)

	HMD	BLH1	BLH2	ULH	MAT
	n = 26	n = 19	n = 13	n = 10	n = 15
%PC	58.2 ± 2.2	63.2 ± 2.8	63.4 ± 3.2	74.0 ± 1.6	76.1 ± 1.5
%SM	12.4 ± 1.2	8.9 ± 1.2	8.5 ± 1.6	4.9 ± 0.8	3.1 ± 0.5
%LPC	2.0 ± 0.4	3.8 ± 0.9	4.7 ± 1.8	1.2 ± 0.3	2.4 ± 0.4
%PG	1.8 ± 0.4	2.6 ± 0.7	3.5 ± 1.0	1.3 ± 0.4	3.6 ± 0.6
%PI + PS	11.4 ± 1.4	10.5 ± 1.0	8.4 ± 0.9	9.5 ± 0.5	7.1 ± 0.8
%PE	13.2 ± 0.7	9.1 ± 1.5	10.8 ± 1.4	7.6 ± 0.6	6.2 ± 1.1
	n = 8	n = 8	n = 7	n = 6	n = 10
%DSPC	15.4 ± 2.5	34.3 ± 5.8	37.0 ± 5.0	30.5 ± 5.4	51.3 ± 2.7

SM = Sphingomyelin; LPC = lysophosphatidylcholine; PI = phosphatidylinositol; PS = phosphatidylserine; PE = phosphatidylethanolamine.

unilateral hypoplasia was similar to that of the mature babies, except that it contained significantly lower percentages of phosphatidylglycerol (PG; $p < 0.007$) and DSPC ($p < 0.009$).

Infants with bilateral lung hypoplasia probably survived longer than babies with bilateral hypoplasia because their lungs were larger rather than owing to differences in gestational age. The low survival of the babies with unilateral lung hypoplasia is probably because both lungs are relatively small. Lungs with hypoplasia are still capable of undergoing biochemical maturation, i.e. producing surfactant with higher proportions of DSPC and PG.

References

1   Wigglesworth, J.S.; Desai, R.; Guerrini, P.: Fetal lung hypoplasia: biochemical and structural variations and their possible significance. Archs Dis. Childh. *56:* 606–615 (1981).
2   Hill, C.M.; Brown, B.D.; Morley, C.J.; Davis, J.A.; Barson, A.J.: Pulmonary surfactant. I. In immature and mature babies. Early hum. Dev. *16:* 143–151 (1988).

C.M. Hill, PhD, Department of Paediatrics, Cambridge University Clinical School, Addenbrooke's Hospital, Hills Road, Cambridge CB2 2QQ (UK)

Wichert P von, Müller B (eds): Basic Research on Lung Surfactant.
Prog Respir Res. Basel, Karger, 1990, vol 25, pp 256–264

Is It Possible to Enhance the Therapeutic Effect of Exogenous Natural Surfactant

Mikko Hallman[a], *T. Allen Merritt*[b], *Toyoaki Akino*[c], *David Strayer*[d], *Pirkko Arjomaa*[a]

[a] Departments of Pediatrics, University of Helsinki, Finland;
[b] University of California, San Diego, Calif., USA;
[c] Department of Medical Chemistry, Sapporo Medical College, Sapporo, Japan;
[d] Department of Pathology, University of Texas, Houston, Tex., USA

Several questions regarding which surfactants should be used and how they should be used to treat respiratory distress syndrome (RDS) still require comprehensive studies and demonstration of both short- and long-term efficacy and safety for reversing respiratory failure. These open issues include: (1) What is the appropriate patient selection? (2) What is the proper dose and time of administration? (3) What are the pharmacodynamics of exogenous and endogenous surfactant after treatment? (4) What is the best measure of efficacy and safety of exogenous surfactant? (5) What is the proper management of patients undergoing surfactant substitution? In this brief review we discuss these questions in the light of our most recent findings.

Patients and Methods

Amniotic fluid from 85 fetuses of less than 32 weeks of gestation was analyzed within 48 h from delivery. Altogether 255 airway specimens were recovered from 48 intubated small preterm infants (< 30 gestational weeks at birth) who were treated using mechanical ventilation. All patients were cared for in the University Central Hospital in Helsinki.

Human surfactant was recovered from amniotic fluid during elective cesarean sections, isolated and tested for surface activity and lack of microbial contamination, as described previously [1; Hallman and Merritt, unpubl.]. Phosphatidylcholine (PC) was analyzed using an enzymatic method [2], and SP A quantified by ELISA technique using two monoclonal antibodies [3].

The concentrations of PC in epithelial lining fluid (PC$_{alv}$) were calculated as follows:

$$PC_{alv} = PC_{airway\ specimen} \times urea_{serum}/urea_{airway}.$$

The concentrations of the protein in epithelial lining fluid were calculated analogously, using the supernatant of the airway specimen, centrifuged at 15,000 g for 2 h. The present method was verified in rabbits. Briefly, varying amounts of urea in saline was given in order to maintain serum urea at an elevated but constant level for 2 h. Thereafter, the animals were tracheostomized, and a small quantity of normal saline (1–3 ml/kg BW) was introduced into the airways. Within < 20 s, any liquid suctioned from the airways was recovered. Fetal animals were paralyzed by intraperitoneal Pavulon. While the umbilical circulation was maintained, the trachea was cannulated. Any fluid that was recovered from the airways was collected. Thereafter, saline (0.2 ml) was introduced, followed by suctioning of the airways.

The lipid extract of natural surfactant, isolated by lung lavage, was added to Hanks' solution at a final concentration of 1.5 mg/ml. After the addition of SP A, the mixture was incubated at 37 °C for 15 min and the surface activity was measured using a pulsating bubble surfactometer, essentially as described by Enhörning [4a, b].

Results and Discussion

Appropriate Patient Selection

Numerous surveys have confirmed that only 30–70 % of infants of less than 32 weeks of gestation actually develop RDS. While the presently available tests for fetal lung maturity predict mature lungs with close to 100% accuracy, the figures for prediction of lung immaturity are lower. As shown in table 1, SP A accurately predicted lung immaturity among the small preterm infants, and altogether 63 % of infants with RDS had SP A of lower than 0.6 µg/ml [5]. We therefore believe that measurement of SP A in amniotic fluid helps to define the risks of RDS in the fetus and that an immature SP A is useful in defining the recipients of prophylactic surfactant who most likely would have developed RDS. These results demonstrate that a greater specificity can be achieved in prediction of fetal lung immaturity and offer a more rational approach for treatment at birth using 'prophylactic' surfactant administration.

Proper Dose and Time of Administration

The doses of surfactant substitution in RDS have ranged from 40 to 200 mg/kg. It has become clear that the maintenance of surfactant function until endogenous secretion and synthesis is sufficient requires a higher dose than that giving a maximal acute effect in immature animals [6].

Table 1. Value of surfactant indices of predicting RDS in small preterm fetuses (< 32 weeks)

	Predictive value of 'immature' result		Predictive value of 'mature' result		Overall accuracy %
	n	%	n	%	
PG (< 1 % of phospholipids)	67	58	18	89	65
L/S ratio (< 1)	19	100	67	67	75
L/S ratio (< 2)	59	61	26	81	67
SP A (≤ 0.6 µg/ml)	26	100	59	75	82
SP A (≤ 3 µg/ml)	74	55	11	100	61

However, there are no controlled trials to demonstrate whether it is beneficial to give a single large dose [7] or one to several smaller doses, whenever the respiratory failure is severe enough to warrant retreatment [8]. Since RDS is a spectrum of disease with varying degrees of surfactant deficiency it is prudent to individualize the surfactant doses, particularly since an excess of surfactant might be harmful. On the other hand, frequently the requirement for exogenous surfactant retreatment depends upon the management of RDS.

Controlled randomized studies are underway addressing the question whether there are differences between the efficacy of prophylactic over rescue administration of surfactant. This question may seem to be trivial since according to some experimental studies prophylactic surfactant is more efficacious in alleviating respiratory failure. If indeed the same conclusion is not borne out by the ongoing trials (as it may well be), this would indicate the complexity of the clinical treatment and clinical picture of RDS in small preterm infants, rather than the untenability of the Enhörning's concept [9].

Pharmacodynamics of Surfactant in Respiratory Distress Syndrome

The pharmacodynamics of surfactant have been clarified in experimental animals during acute experiments [10]. The majority of exogenous surfactant is rapidly cleared from the airways, but the lavageable surfactant remains larger than in untreated controls. There has been no evidence that feedback inhibition of endogenous synthesis occurs with exogenous treatment; instead, a significant fraction of surfactant is reutilized. These con-

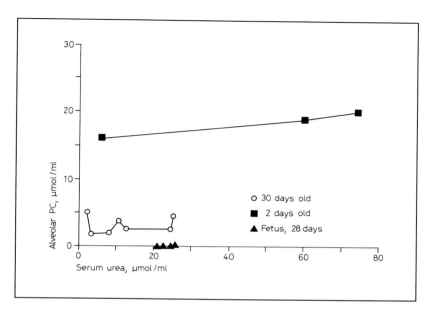

Fig. 1. Effect of serum urea on concentrations of PC in epithelial lining fluid from developing rabbits. For experimental details, see 'Patients and Methods'.

cepts of surfactant clearance and reutilization are dependent on the degree of maturity, and, insofar as the ontogeny of human surfactant possess a unique timing and variability, it is important to understand the surfactant kinetics in human RDS, especially the long-term consequences of exogenous surfactant.

Surfactant pools have been estimated on the basis of the Fick principle by means of a marker in exogenous surfactant. According to these data exogenous surfactant acutely increased the apparent pool size by 7- to 20-fold. The fractional turnover rate of exogenous phosphatidylglycerol (PG) was $2.3 \pm 0.2\%$/h, and PG was replaced by phosphatidylinositol that increased concomitant with the decrease in PG [11].

In order to better understand surfactant function and the role of surfactant inhibitors, the concentrations in epithelial lining fluid were calculated using urea as a marker of the extracellular space. As shown in figure 1, the 'alveolar' concentration of PC was independent of the urea concentration. Actually, there has been an excellent agreement of the direct measurement of PC concentration and the estimate of the PC concentration in fetal lung fluid from preterm rabbits. According to preliminary evidence the concentration of PC in epithelial lining fluid following exogenous surfactant is significantly higher than in placebo (air)-treated infants. However, beyond 7 days of age there were no more detectable differences in the concentrations of PC as compared between the various treatment groups (prophylactic vs. rescue vs. placebo). There were no consistent differences in the alveolar concentrations of nonsedimentable protein as compared between the surfactant- and placebo-treated infants [12].

The Quality of Exogenous Surfactant

A crucial consideration for the requirement of exogenous surfactant includes not only its efficacy but also the safety of these agents. According to present evidence all three surfactant proteins are highly immunogenic. We have recently found that rat antibodies against human, bovine, and porcine surfactant strikingly inhibit surfactant function and that there is cross-reactivity among all three species. Thus far, there are no reports demonstrating antibodies against exogenous surfactant in humans, although a transient increase in human surfactant-specific immunocomplexes were noted among small preterm infants with RDS. This phenomenon was neither associated with complement activation nor symptoms indicative of immunocomplex-mediated disease, and there was no detectable prediction of the immunocomplexes amongst the recipients of exogenous surfactant. However, investigators using any exogenous surfactant, especially multiple doses of animal preparations, need to further address the question of potential immunogenic sensitization and its consequences [13]. A further question of importance is whether the preparations possess any microbiological or pharmacological toxicity.

A critical determinant of any surfactant preparation is the resistance to inhibitors of surfactant function. According to the current concept, the proteinaceous inhibitor(s), derived from the blood as a result of the high permeability lung edema, are important in determining the surfactant function and the alveolar stability in disease [10]. In addition, there are lipids and membranous components that are inhibitory [14]. Proteolytic cleavage of surfactant components as a result of either plasma or inflammatory cell-derived enzyme activity in the alveolar space [15, 16] has been demonstrated and has emerged in pathophysiologic investigations.

We have recently found evidence of surfactant inhibitors that interfere with the function of surfactant proteolipid-lipid complexes [17]. However,

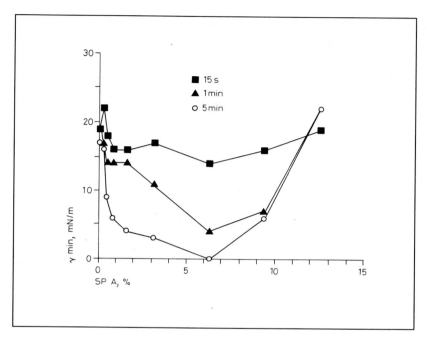

Fig. 2. Effect of SP A on the minimum surface tension of the surfactant lipid extract in Hanks' solution. The results are the means of 3 experiments. For experimental details see 'Patients and Methods'.

when a defined concentration of SP A is added to the system the inhibition disappears (fig. 2). The biochemical characterization of these inhibitors is far from complete. Some amino acids (arginine, lysine), plasmin (splits arginine residues from proteins), Ca^{++}, and some monoclonal antibodies against surfactant proteolipid inhibit surfactant function, whereas addition of SP A restores the surfactant function in vitro, suggesting that SP A 'protects' or shields surfactant against these inhibitors.

Using multiple regression analysis we found a significant negative correlation between the degree of respiratory failure and the SP A/PC ratio in RDS, which implies that a function of SP A is in the shielding against some surfactant inhibitors. Furthermore, these findings indicate that SP A cannot be deleted from the list of components that improve surface properties, and, hence, may be a critical component of exogenous surfactant in lung diseases demonstrating lack of endogenous SP A.

Management of Infants Undergoing Surfactant Substitution

The management of both fetuses and infants may profoundly influence the therapeutic efficacy, since surfactant does not 'cure' the immaturity that is associated with a myriad of progressive, life-threatening problems. An improvement in the surfactant function induces rapid circulatory changes – especially the increase in the left-to-right shunt through the patent ductus arteriosus – or necessitates changes in ventilatory settings to avoid overdistension of alveoli. Failure to appreciate these interacting factors can have grave consequences, and does not reduce the requirement for intensive care. We have closely observed about 250 infants participating in trials of surfactant substitution. This has allowed us to appreciate several problems in management.

According to our early experience, in very small preterm infants (< 26 weeks of gestation), a striking surfactant-induced improvement resulting in a decrease in O_2 requirements and ventilator settings was followed by a shock-like state at the age of 8–24 h (n = 8). This condition rapidly led to demise in most cases (75%). The modification of treatment has been associated with considerably lower mortality (20%); we believe that early red blood cell infusions to correct the iatrogenic loss of the blood volume and redistribution of the circulation, early treatment of the patent ductus arteriosus by slow intravenous infusions of indomethacin, and the 'no-touch' instructions have been instrumental in explaining the decrease of the mortality among these extremely preterm, surfactant-treated infants.

The 'nonresponders' constitute a group of infants who, according to our criteria, demonstrate no improvement that would last longer than 1 h. In the Helsinki trial, such infants include 7% (n = 7) of all infants treated with surfactant. This heterogenous group of babies had a more severe respiratory failure and a higher postnatal age than the responders. The fact that 3 of these infants were offspring of the mothers with severe early-onset preeclampsia (23% of the surfactant-treated offspring of preeclamptic mothers), led us to investigate this group of small preterm infants. There was a significant positive correlation between the fluid intake during the first 24–48 h and the severity of RDS at the age of 2 days. This result was not explained by severity of RDS at birth or by the length of the treatment of maternal preeclampsia. We feel that it is mainly the interstitial lung edema caused by the fluid overload of these cardiovascularly compromized infants that hampers the surfactant effect. This supposition is strengthened by our recent observation that a severe fluid restriction and ductal closure can restore the responsiveness to exogenous surfactant.

It is our belief, borne out by recent experience, that surfactant substitution will virtually eliminate the need for supplemental oxygen and high airway pressures during the early neonatal period. Whether this, together with the elimination of lung ruptures, actually decreases the chronic morbidity that affects the growth and neurodevelopment, remains a challenge for future research.

Acknowledgements

This investigation was supported by grants from the Sigrid Juselius Foundation, the Finnish Academy, the Finnish Foundation for Pediatric Research (M.H.), and from the NIH (HL-35036, to T.A.M., M.H.). The authors are grateful to E. Riihelä, H. Ahola, and S. Tsupari for their skillful technical assistance, and to the nurses of the Newborn Intensive Care Unit, Childrens' Hospital, University of Helsinki, for collection of the airway specimens. We thank Byk Gulden Pharmazeutica for a generous gift of recombinantly produced SP-A.

References

1   Hallman, M.; Merritt, T.A.; Schneider, H.; Epstein, B.L.; Mannino, F.; Edwards, D.K.; Gluck, L.: Isolation of human surfactant from amniotic fluid and pilot study of its efficacy in respiratory distress syndrome. Pediatrics, Springfield *71:* 473–482 (1983).
2   Muneshige, A.; Okazaki, T.; Quirk, J.G.; MacDonald, P.C.; Nozaki, M.; Johnston, J.M.: A rapid and specific enzymatic method for the quantification of phosphatidylcholine, disatured phosphatidylcholine, and phosphatidylglycerol in amniotic fluid. Am. J. Obstet. Gynec. *145:* 474–480 (1983).
3   Kuroki, Y.; Takahashi, H.; Fukada, Y.; Mikawa, M.; Inagawa, A.; Fujimoto, S.; Akino, T.: Two-side 'simultaneous' immunoassay with monoclonal antibodies for the determination of surfactant apoproteins in human amniotic fluid. Pediat. Res. *19:* 1017–1020 (1985).
4a Enhörning, G.: Pulsating bubble technique for evaluating pulmonary surfactant. J. appl. Physiol. *43:* 198–203 (1977).
4b Arjomaa, P.; Hallman, M.: Purification of a hydrophobic surfactant peptide using high-performance liquid chromatography. Analyt. Biochem. *171:* 207–212 (1988).
5   Hallman, M.; Arjomaa, P.; Mizumoto, M.; Akino, T.: Surfactant proteins in diagnosis of fetal lung maturity. I. Predictive accuracy of the 35 kD protein, the L/S ratio, and phosphatidylglycerol. Am. J. Obstet. Gynec. *158:* 531–535 (1988).
6   Ikegami, M.; Adams, F.H.; Towers, B.; et al.: The quantity of natural surfactant necessary to prevent the respiratory distress syndrome in premature lambs. Pediat. Res. *14:* 1082–1085 (1980).
7   Fujiwara, T.; Chida, S.; Watabe, Y.; Maeta, H.; Morita, T.; Abe, T.: Artificial surfactant therapy in hyaline-membrane disease. Lancet *i:* 55–59 (1980).

8   Hallman, M.; Merritt, T.A.; Cochrane, C.G.; Gluck, L.: Human surfactant substitution in severe respiratory distress syndrome. Prog. resp. Res. *18:* 193–203 (1984).

9   Enhörning, G.; Robertson, B.: Lung expansion in the premature rabbit fetus after tracheal deposition of surfactant. Pediatrics, Springfield *50:* 58–66 (1972).

10 Jobe, A.: Metabolism of endogenous surfactant and exogenous surfactants for replacement therapy. Semin. Perinatol. *12:* 231–244 (1988).

11 Hallman, M.; Merritt, T.A.; Pohjavuori, M.; Gluck, L.: Effect of surfactant substitution on lung effluent phospholipids in respiratory distress syndrome. Evaluation of surfactant phospholipid turnover, pool size, and relationship to severity of respiratory failure. Pediat. Res. *20:* 1228–1235 (1986).

12 Hallman, M.; Merritt, T.A.: Estimates of the alveolar concentration of phosphatidylcholine. Clin. Res. *37:* 212A (1989).

13 Merritt, T.A.; Strayer, D.S.; Hallman, M.; Spragg, R.D.; Wozniak, P.: Immunologic consequences of exogenous surfactant administration. Semin. Perinatol. *12:* 221–230 (1988).

14 Rauvala, H.; Hallman, M.: Glycolipid accumulation in bronchoalveolar space in adult respiratory distress syndrome. J. Lipid Res. *25:* 1257–1262 (1984).

15 Hallman, M.; Spragg, R.; Harrell, J.H.; Moser, K.M.; Gluck, L.: Evidence of lung surfactant abnormality in respiratory failure. Study of bronchoalveolar lavage phospholipids, surface activity, phospholipase activity, and plasma myoinositol. J. clin. Invest. *70:* 673–683 (1982).

16 Merritt, T.A.; Revak, S.D.; Hallman, M.; Cochrane, C.G.: Elastolytic degradation of surfactant 35 kD apoprotein. Pediat. Res. *21:* 460A (1987).

17 Hallman, M.; Merritt, T.A.; Akino, T.; Arjomaa, P.: The role of surfactant protein (SP A) in surfactant function. Study of surface activity and airway specimens in RDS. Clin. Res. *37:* 213A (1989).

Mikko Hallman, MD, University of California, Irvine Medical Center,
101 City Dr R 81, Bld 27, Orange, CA 92668 (USA)

Wichert P von, Müller B (eds): Basic Research on Lung Surfactant.
Prog Respir Res. Basel, Karger, 1990, vol 25, pp 265–270

Surfactant Research of Interest to the Clinician

Goran Enhorning

Department of Obstetrics and Gynecology, State University of New York at
Buffalo, N.Y., USA

The etiology of the neonatal respiratory distress syndrome (RDS) is
fairly simple. It is a condition that develops mainly when the baby is born
too early, before surfactant synthesis and release is adequate. For this reason, it seemed logical that RDS should be possible to prevent even when
the delivery is preterm by supplying what is missing: a surfactant preparation with the right properties. Those properties consist of fast adsorption,
low equivalent surface tension, and surface tension reaching zero when
surface area is moderately compressed.

Once the right surfactant suspension was used, it could be shown that
by instilling that preparation into the trachea, or even into the pharynx,
prior to the first breath, the development of RDS could be inhibited [1–3].
If, however, the treatment was delayed or the surfactant was not given in
adequate quantity, its effect was less dramatic and of shorter duration [4,
5]. Investigative work, carried out at the laboratories of Robertson and
Jobe, has made it possible to say today that it is because of damage to the
airway epithelium, that the results of a delayed treatment are inferior [6].
That damage occurs already after a short period of ventilation of a lung
that is surfactant deficient [7]. The damage allows proteins to invade the
airways and a vicious circle has then been started, since the proteins will
inhibit the surfactant even further [8–11]. The inhibition, which seems to
be due to a competition between proteins and surfactant, is an interesting
phenomenon with enormous practical consequences. A simple way of
looking at the problem is the following. Proteins are generally water soluble. Thereby, in the initial adsorption phase, a protein has a better chance
to quickly reach the surface, but with time, surfactant molecules, mainly
consisting of disaturated phosphatidylcholine, will occupy the surface and

squeeze out the protein. Thus, with a surfactant in high concentration and with the required properties, proteins will have a chance to be at the surface for only a very short period of time. They will soon be squeezed out. However, if the proteins are in high concentration, they will, as soon as a new air-liquid interface is produced, cause a delay in the formation of a surfactant film, particularly if the surfactant does not have the right composition.

This problem with proteins inhibiting adsorption of surfactant becomes of importance when treatment of neonatal RDS is delayed. The baby then develops some of the problems of adult RDS (ARDS), but will still benefit from the surfactant treatment. Most likely the typical ARDS would also improve if treated with surfactant supplied in adequate quantity and concentration. Preferably, the treatment should in sequence be given to one section of the lung at a time and not until a lavage has reduced the airway concentration of proteins in that section. That type of therapeutic procedure would yield lavage fluid for cytological, chemical and physical analysis. The surface properties could be studied with the bubble surfactometer but not until the lavage fluid has been concentrated in a standardized fashion. To evaluate the efficacy of the surfactant treatment and to study the surface activity of the lavage fluid will be new areas of research, of considerable interest to the clinician.

As the clinician realizes that prevention and treatment of neonatal RDS is possible with surfactant, he will eagerly ask: Are there other problems which are at least partly due to a surfactant deficiency? The deficiency could result from a poor synthesis and release or an inhibition with proteins or an excessive enzymatic breakdown of the phospholipids. Today, many believe that ARDS is such a condition. But are there others? Perhaps pneumonia? Perhaps asthma? If, indeed, asthma, affecting at least 7 million individuals in the United States [12], is a condition of surfactant deficiency that could be treated by supplying surfactant as an aerosol spray, this would be an area of research of enormous interest which could have far-reaching practical consequences.

Already the fact that some of the most important medication for asthma stimulates synthesis and/or release of surfactant, points to the possibility that asthma is a condition of surfactant deficiency and that the relief the patient encounters when using this medication, might partly be due to this effect on the surfactant system. In the following an attempt will be made to offer reasons why a surfactant deficiency might give symptoms characterizing asthma.

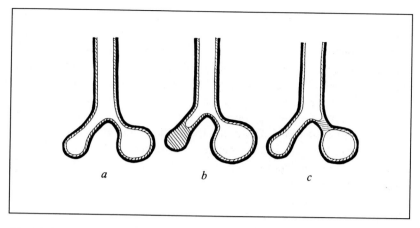

Fig. 1. The surface of the alveolus is conceived as being spherical (*a*) with an outlining layer of liquid (hatched area). A generally accepted concept is that when surface tension is inappropriately high, the small alveolus will collapse and the already large one becomes overexpanded (*b*). There is also a risk that a cylindrical airway leading to an overexpanded alveolus becomes blocked, resulting in trapping of air (*c*). See text for considerations when that might occur.

It is generally accepted that when there is a surfactant deficiency, the lungs are not given normal stability. There is felt to be a risk that during expiration, the small alveolus would collapse and give up its content of air to larger alveoli that would then become overexpanded. However, there might be a greater risk that collapse will occur in the respiratory bronchiole, the cylindrical airway through which air first passes when it is extruded from the alveolus during expiration (fig. 1). For a spherical surface the law of Laplace is $\Delta P = 2\gamma/R$ but for a cylindrical surface it is $\Delta P = \gamma/R$. Consequently, if surface tension, γ has the same value at the two surfaces, the cylinder of the respiratory bronchiole would be the first to collapse if it has a radius, R, less than half that of the spherical alveolus. However, collapse or filling with liquid may occur in the cylinder even though it has a radius which is more than 50% of what it is in the sphere provided its surface tension has a higher value than what is prevailing in the alveolus. Inhibiting factors may be descending from above and since the surfactant originates from the cytoplasm of alveolar cells type II and is delivered to the alveolar space, it is conceivable that if it is produced in subliminal quantities, not enough will be extruded into the respiratory bronchiole. Surface tension is then lowered more in the alveolus than in the adjacent cylindrical airway. It is therefore quite likely that with a surfactant supply which is slightly inadequate, collapse will occur in many respiratory bronchioles, resulting in trapping of air. The closure of many respiratory bronchioles is likely to have a drastic effect on the stability of the bronchioles, the more centripetal airways which are larger, but still lack stabilizing cartilage. A pressure difference, which normally never exists, will have time to build up across the wall of the bronchiolus. This will result in excessive *dilatation* during inspiration and *compression* during expiration. This oscillating expansion and compression may lead to features considered typical for asthma. The excessive dilatation may overdistend the epithelium so that it becomes injured and partly desquamated. This in turn will allow serum proteins and eosinophils to invade the airway. Proteins will have the unfavorable effect of inhibiting surfactant, thereby causing surface tension to increase even further. The proteins will also be part of mucus plugs. The repeated dilatation of the airway may partly be inhibited by its smooth circular muscles which, by being stretched, are perhaps stimulated to go into contraction. When the muscles are repeatedly triggered to contract, they may eventually undergo hypertrophy.

When during expiration, the bronchiolus is *compressed,* its resistance to airflow is raised so that the increased expiratory effort, so troublesome for the patient, may result in wheezing. It is easy to conceive that the increased pressure difference that develops across the airway wall during expiration may result in mucus from glands outside the airway being squeezed through the glands' exit tubes which are traversing the wall (fig. 2). Mucus may thereby be extruded into the airway to form the main part of the mucus plug.

In summary, it is thus possible that relative surfactant deficiency resulting in collapse and closure of many respiratory bronchioles may result in several features characterizing asthma: (a) Air trapping in *overdistended* alveoli. (b) *Hypertrophy* of smooth circular muscles. (c) Formation of *mucus plugs,* which consist of extruded mucus and proteins, and contain desquamated cells. (d) *Increased airflow resistance,* particularly during expiration which causes (e) *wheezing* in airways containing serum and mucus. It is possible that when β-adrenergic agents alleviate the acute symptoms of asthma, they do so not only by relaxing smooth muscles but by releasing surfactant from the cytoplasm of type II cells. From the alveolus the surfactant could be extruded into the respiratory bronchiole and

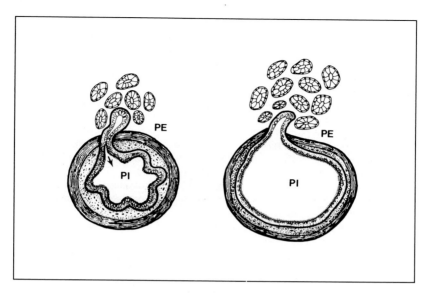

Fig. 2. Schematic drawing showing the conceived effect of a pressure difference developing across the wall of a bronchiolus during an asthma attack. In the expiratory phase (to the left), the exterior pressure, PE, exceeds pressure inside the airway, PI. This would cause an extrusion of mucus from the gland outside the airway. During inspiration, pressure difference is reversed. The higher pressure is now inside the airway, PI. This results in an overstretching of the airway wall including its smooth circular muscles. The muscle being stretched might be stimulated to contract and thereby might eventually be undergoing hypertrophy.

result in an opening of this airway and the acute symptoms of dyspnea may then be alleviated.

Animal experiments are clearly neded to test this hypothesis. If they offer promising results, it is obvious that an area of surfactant research that is of enormous interest to the clinician has opened up.

References

1   Enhorning, G.; Robertson, B.: Lung expansion in the premature rabbit fetus after tracheal deposition of surfactant. Pediatrics, Springfield *50:* 58–66 (1972).
2   Enhorning, G.; Grossmann, G.; Robertson, B.: Pharyngeal deposition of surfactant in the premature rabbit fetus. Biol. Neonate *22:* 126–132 (1973).

3   Enhorning, G.; Shennan, A.; Possmayer, F.; Dunn, M.; Chen, C.P.; Milligan, J.:
     Prevention of neonatal respiratory distress syndrome by tracheal instillation of sur-
     factant. A randomized clinical trial. Pediatrics, Springfield *76:* 145–153 (1985).
4   Jobe, A.; Ikegami, M.; Glatz, T.; Yoshida, Y.; Diakomanolis, E.; Padbury, J.: Dura-
     tion and characteristics of treatment of premature lambs with natural surfactant. J.
     clin. Invest. *67:* 370–375 (1981).
5   Jobe, A.; Ikegami, M.; Jacobs, H.; Jones, S.; Conaway, D.: Permeability of prema-
     ture lamb lungs to protein and the effect of surfactant on that permeability. J. appl.
     Physiol. *55:* 169–176 (1983).
6   Jacobs, H.; Jobe, A.; Ikegami, M.; Glatz, T.; Jones, S.J.; Barajas, L.: Premature
     lambs rescued from respiratory failure with natural surfactant. Clinical and bio-
     physical correlates. Pediat. Res. *16:* 424–429 (1982).
7   Nilsson, R.; Grossmann, G.; Robertson, B.: Lung surfactant and the pathogenesis of
     neonatal bronchiolar lesions induced by artificial ventilation. Pediat. Res. *12:* 249–
     255 (1978).
8   Holm, B.A.; Notter, R.H.; Finkelstein, J.N.: Plasma protein inhibition of lung sur-
     factant activity. Fed. Proc. *44:* 1026 (1985).
9   Holm, B.A.; Notter, R.H.; Finkelstein, J.N.: Surface property changes from interac-
     tions of albumin with natural lung surfactant and extracted lung lipids. Chem. Phys.
     Lipids *38:* 287–298 (1985).
10   Ikegami, M.; Jobe, A.; Jacobs, H.; Lam, R.: A protein from airways of premature
     lambs that inhibits surfactant function. J. appl. Physiol. *57:* 1134–1142 (1984).
11   Fuchimukai, T.; Fujiwara, T.; Takahashi, A.; Enhorning, G.: Artificial pulmonary
     surfactant inhibited by proteins. J. appl. Physiol. *62:* 429–437 (1987).
12   Dodge, R.R.; Burrows, B.: The prevalence and incidence of asthma and asthma-like
     symptoms in a general population sample. Am. Rev. resp. Dis. *122:* 567–575
     (1980).

Goran Enhorning, MD, The Perinatal Center, Children's Hospital of Buffalo,
219 Bryant Street, Buffalo, NY 14222 (USA)

Posters

Wichert P von, Müller B (eds): Basic Research on Lung Surfactant.
Prog Respir Res. Basel, Karger, 1990, vol 25, pp 271–273

Effects of the Surfactant-Associated Proteins, SP-A, SP-B and SP-C, on Phospholipid Surface Film Formation[1]

Ulrich Pison, Kathleen Shiffer, Samuel Hawgood, Jon Goerke

Cardiovascular Research Institute and Departments of Pediatrics and Physiology, University of California at San Francisco, Calif., USA

The alveolar lining layer of mature mammalian lungs contains a surface active material called pulmonary surfactant. This material is a lipoprotein complex composed primarily of phospholipids and three distinct surfactant-associated proteins, SP-A, SP-B, and SP-C [Possmayer, 1988]. Pulmonary surfactant lowers the surface tension at the air-liquid interface of the alveolar space, thereby preventing lung collapse [Clements and Tierney, 1965]. To maintain the interfacial monolayer during breathing, surfactant components have to adsorb rapidly [Goerke and Clements, 1986]. In order to investigate the effects of components of pulmonary surfactant on the rate of surface film formation, multilamellar vesicles (MLV) were prepared from simple phospholipid mixtures and purified surfactant proteins and tested in an assay of surface adsorption in which the change in surface pressure ($\Delta\pi$) is measured with respect to time (Δt).

Material and Methods

Pulmonary surfactant was isolated from lung lavage fluid obtained from adult dogs using differential centrifugation techniques. SP-A and surfactant lipids were prepared from pulmonary surfactant by extraction with butanol. SP-A was further purified by differential solubility in octylglucopyranoside and NaCI. Surfactant lipids contain SP-B and SP-C (1 % by weight). SP-B and SP-C were isolated from surfactant lipids by gel-filtration chromatography using acidified chloroform/methanol as solvent [Hawgood et

[1] Supported in part by the Deutsche Forschungsgemeinschaft grant Pi 165/1-1, and by NIH grants HL-24075 and HL-07275.

al., 1987]. MLV were prepared from synthetic phospholipids and purified SP-B and/or SP-C in various ratios at a temperature above the phase transitions of the synthetic lipid mixtures employed (52 °C) by continuously rotating the aqueous suspension in the presence of a glass bead. The ability of the various MLV preparations to adsorb and spread at an air-liquid interface ($\Delta\pi/\Delta t$) from a stirred subphase was tested at 37 °C. The subphase buffer was 5 mM Tris (pH 6.9), 150 mM NaCl containing 2,5 mM CaCl$_2$ or 1 mM EDTA. Three milliliters of the subphase buffer was added to a circular Teflon chamber 2 cm in diameter, a MLV preparation (100 µl volume; 100 µg total phospholipid) was carefully layered on the bottom of the chamber, stirring was started at 240 rpm, and the surface pressure was measured continuously by a platinum dipping plate suspended from a strain gauge. The signal was amplified, displayed on a chart recorder and expressed as mN/m.

Results

With the method of vesicle preparation used in this study there was no change in the surface pressure with MLV prepared from any of the phospholipids alone in the time frame of these experiments. In most cases the inclusion of SP-B, SP-C, or mixtures of SP-B and SP-C in the MLV caused a biphasic increase in the rate of change in surface pressure: an initial rapid change in $\Delta\pi/\Delta t$ was followed by a much slower phase during which the π very slowly approached the equilibrium π (π_{eq}) expected of a phospholipid film, 45 mN/m. The initial rapid $\Delta\pi/\Delta t$ was promoted by either SP-B or SP-C in a similar dose-dependent fashion with all phospholipids tested (range of phospholipid/protein ratio from 40/1 to 2/1, w/w). The second slower phase was not significantly affected by the protein concentration of SP-B and SP-C. However, by adding SP-A (10 µg) to the chamber during the slower second phase of $\Delta\pi/\Delta t$, 15 min after adsorption had begun, the change in π was accelerated markedly up to the π_{eq} of 45 mN/m for MLV containing both SP-B and SP-C (10 and 5 µg, respectively) or SP-B alone (5–20 µg), but not SP-C alone (5–20 µg). This effect of SP-A on MLV containing SP-B was only observed if negatively charged phospholipids and calcium were also present in the adsorption chamber.

The π at which the change from rapid to slow $\Delta\pi/\Delta t$ occurred was highly reproducible for a given mixture and set of formulation conditions but was found to be dependent on the phospholipid concentration during both sample preparation and adsorption, phospholipid/protein ratio, the phospholipid composition, and the presence of calcium ions. The composition of the actual surface film itself was not assessed in this study, but SP-B or SP-C alone prepared in an identical manner to the samples with lipid caused no change in surface pressure.

Conclusions

The presence of either of the low molecular weight surfactant-associated proteins, SP-B and SP-C, accelerates the formation of a surface film from MLV prepared from different phospholipid mixtures. Not all mixtures, however, adsorb to a π comparable to π_{eq} for a phospholipid film within the time frame of these experiments. The high molecular weight surfactant-associated protein, SP-A, accelerates the formation of a surface film with a π_{eq} of 45 mN/m from MLV provided SP-B and egg phosphatidylglycerol are included in the mixture and calcium ions are present during adsorption. Future experiments will need to assess the properties and composition of the spontaneously adsorbed films in addition to the rate of film formation.

References

Clements, J.A.; Tierney, D.F.: Alveolar instability associated with altered surface tension; in Fenn, Rahn, Handbook of physiology, sect. 3: Respiration, vol. II, chapt. 65, pp. 1565–1583 (American Physiological Society, Washington 1965).

Goerke, J.; Clements, J.A.: Alveolar surface tension and lung surfactant; in Macklem, Mead, Handbook of physiology, sect. 3: The respiratory system, vol. III, chapt. 16, pp. 247–261 (American Physiological Society, Washington 1986).

Hawgood, S.; Benson, B.J.; Schilling, J.; Damm, D.; Clements, J.A.; White, R.T.: Nucleotide and amino acid sequences of pulmonary surfactant protein SP18 and evidence for cooperation between SP 18 and SP 28-36 in surfactant lipid adsorption. Proc. natn. Acad. Sci. USA 84: 66–70 (1987).

Possmayer, F.: Perspective: A proposed nomenclature for pulmonary surfactant proteins. Am. Rev. resp. Dis. (in press, 1988).

Ulrich Pison, MD, Marien-Hospital, Chirurgische Abteilung, Rochusstrasse 2, D–4000 Düsseldorf 30 (FRG)

Wichert P von, Müller B (eds): Basic Research on Lung Surfactant.
Prog Respir Res. Basel, Karger, 1990, vol 25, pp 274–278

Expression of Human Pulmonary Surfactant-Associated Apoprotein B in Chinese Hamster Ovary Cells

Harald R. Eistetter, Tilman Voss, Klaus P. Schäfer, Sabine Löffler, Cirsten Hess, Klaus Melchers

Abteilung für Molekularbiologie, BYK Gulden Pharmazeutika, Konstanz, BRD

Biochemically, pulmonary surfactant is composed primarily of a mixture of lipids and the three surfactant-associated apoproteins SP-A, SP-B, and SP-C, the genes of which have been isolated and characterized recently [1–3]. Structure and surface activity of pulmonary surfactant seem to be crucially influenced by these apoproteins. In particular, the lipophilic low-molecular weight proteins SP-B and SP-C were found to accelerate the spreading kinetics of phospholipids in an air-water interphase [2, 3].

Both proteins are synthesized as water-soluble precursor molecules of 42 kdaltons (SP-B) and 21 kdaltons (SP-C) apparent molecular weights. The mature, highly hydrophobic polypeptides are generated from these precursors by complex proteolytic cleavage mechanisms [2, 3, and own observations]. In addition, both SP-B and SP-C form dimers by disulfide linkages, which can be reduced to monomers of about 8 kdaltons (SP-B) and about 5 kdaltons (SP-C), respectively.

Using SP-B as a working model, we wanted to establish a cell line that was able to express and process the SP-B protein correctly. Therefore, we have transfected the coding sequence for the human pulmonary surfactant-associated apoprotein SP-B into Chinese hamster ovary (CHO) cells. Analyses of genetically stable transgenic cell lines indicate that the recombinant 42-kdalton SP-B precursor molecule is indeed processed into an 8-kdalton polypeptide, which (i) shares the lipophilic properties of the native molecule, and (ii) dimerizes into an 18-kdalton species. These results imply that CHO cells are suitable for both the synthesis of SP-B and the study of its processing.

Materials and Methods

Expression Vector, Cell Culture, Transfection, and Clone Selection

Construction of the mammalian expression plasmid pMT-APO (a kind gift from R.T. White, CalBio Inc., Mountain View, Calif.) carrying the complete functional promoter region of the human metallothionein II (hMTII) gene, the SV 40 enhancer, and the 3′ regulatory sequences of the human apolipoprotein A-I gene will be reported elsewhere. A SP-B cDNA clone [2] (also kindly provided by R.T. White) was subcloned into the vector. Plasmids containing the insert sequences in the correct orientation were selected for transfections using standard procedures [4].

CHO-K1 cells (ATCC) were grown in 50% Dulbecco's modified Eagle's medium/ 50% Ham's F12 medium (Coon's modified) supplemented with 15% fetal calf serum at 37°C in a humidified 10% CO_2 atmosphere. The expression plasmid was co-transfected with pSVtkneo [5] into the cells, as described [6]. Recombinants were selected in medium containing 800 µg/ml neomycin. Resistant colonies were picked and propagated. After growth in medium containing zinc (50 µM), RNA was extracted [7], and analyzed for the presence of SP-B-specific transcripts by Northern blots, as in Eistetter et al. [8]. From SP-B-expressing cell pools, single-cell subclones were derived by limiting dilution in 96-well tissue culture plates.

Partial Purification of SP-B, Gel Electrophoresis and Immunoblotting

Recombinant human SP-B was partially purified from detergent-treated transgenic CHO cells by extraction with $CHCl_3$/methanol (1:1), subsequent chromatography on an LH-60 molecular sieve column [2], and separation by SDS-PAGE [9]. Proteins were electrophoretically transferred onto nitrocellulose. Immunoblots were carried out according to Towbin et al. [10].

Results and Discussion

Expression of SP-B mRNA in transfected CHO:K1 Cells

We have transfected a vector that regulates the expression of the entire precursor molecule of human pulmonary surfactant-associated apoprotein SP-B, under the control of the metal-inducible human methallothionein II gene promoter, into CHO cells. In recombinant single-cell subclones, after zinc induction, polyadenylated SP-B-specific transcripts were found. In figure 1, a Northern blot is shown, using poly(A)$^+$-RNA extracted from cells of a genetically stable SP-B-expressing subclone designated CHO/B1. By hybridization with an SP-B cDNA probe, a transcript of about 1.7 kb in size (as expected from the transcription unit) could be detected at high levels. This implies that the recombinant CHO cells are able to correctly transcribe the introduced gene with high efficiency. Similar findings have

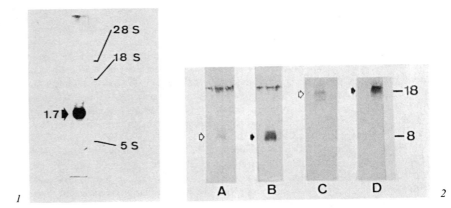

Fig. 1. Hybridization of SP-B cDNA to poly(A)⁺-RNA isolated from recombinant CHO/B1 cells. 5 µg of RNA were loaded. The transcript band is indicated by an arrow. Molecular weights are given in kilobases. Relative mobilities of 28S, 18S, and 5S rRNA are shown.

Fig. 2. Immunoblots of natural SP-B extracted from bovine lung (B, D), and recombinant human SP-B isolated from CHO/B1 cells (A, C). The probes were applied to the gels either in reduced (A, B), or in nonreduced (C, D) conditions. Immunostained bands representing monomeric or dimeric SP-B forms are indicated by black (bovine, natural material) and white (human, recombinant material) arrows. Molecular weights are given in kilodaltons.

been reported earlier, when CHO cells transfected with expression vectors containing the identical promoter were found to produce high levels of human apolipoprotein A-I [12] or human prorenin [13].

Characterization of CHO Cell-Produced Recombinant SP-B

In order to determine whether the CHO/B1 cells were able to synthesize and process the SP-B precursor protein, we have analyzed both serum-free supernatants and cell lysates of zinc-induced cultures. In immunoblots (using a rabbit antiserum directed against its carboxy-terminal 181 amino acids) the 42-kdalton SP-B precursor is found in supernatants as well as in cell lysates (not shown). Moreover, by means of antibodies specific for the mature 8-kdalton SP-B polypeptide [11], proteins were visualized in cell lysates, which have the same apparent molecular weights as natural bovine SP-B, either under reducing or nonreducing conditions (fig. 2). These polypeptides could not be detected in supernatants.

Although the correct amino- and carboxy-terminal amino acid sequences of recombinant SP-B still have to be confirmed, we may conclude that CHO/B1 cells seem to process SP-B into its mature form. In vivo, the (8 kdaltons) mature SP-B molecule is enzymatically cleaved from the precursor by mechanisms which are still not understood in detail. Obviously, the necessary proteolytic enzymes are present in CHO cells, too, and thus are not specific for alveolar type II pneumonocytes, which express SP-B in vivo. This is consistent with the finding that CHO cells convert the recombinant human apolipoprotein A-I precursor molecule into its mature form [12]. In that case, the enzymatic cleavage of the propeptide occurs after a Gln-Gln sequence, which is also found to precede the amino-terminal of mature human SP-B. In addition, several consensus sequences for proline-directed arginine cleavage are found in SP-B. This may imply common proteolytic mechanisms in cells of different germ layer origin.

CHO/B1 cells were shown to synthesize a SP-B dimer. Future experiments have to verify that the recombinant SP-B monomers assemble correctly into this dimeric structure. Since mature SP-B contains 7 cysteine residues, a complex folding pattern into its tertiary structure as well as into the dimer can be anticipated.

The ability to produce considerable amounts of human *pulmonary surfactant*-associated apoprotein SP-B in transgenic mammalian cells permits both an investigation of its physiological role as well as the analysis of its biosynthesis including the study of posttranslational modifications.

References

1   White, R.T.; Damm, D.; Miller, J.; Spratt, K.; Schilling, J.; Hawgood, S.; Benson, B.; Cordell, B.: Isolation and characterization of the human pulmonary surfactant apoprotein gene. Nature, Lond. *317:* 361–363 (1985).
2   Hawgood, S.; Benson, B.J.; Schilling, J.; Damm, D.; Clements, J.A.; White, R.T.: Nucleotide and amino acid sequences of pulmonary surfactant protein SP 18 and evidence for cooperation between SP 18 and SP 28–36 in surfactant lipid adsorption. Proc. natn. Acad. Sci. USA *84:* 66–70 (1987).
3   Warr, R.G.; Hawgood, S.; Buckley, D.I.; Crisp, T.M.; Schilling, J.; Benson, B.J.; Ballard, P.L.; Clements, J.A.; White, R.T.: Low molecular weight human pulmonary surfactant protein (SP 5). Isolation, characterization, and cDNA and amino acid sequences. Proc. natn. Acad. Sci. USA *84:* 7915–7919 (1987).
4   Maniatis, T.; Fritsch, E.F.; Sambrook, K.L.: Molecular cloning. A laboratory manual (Cold Spring Harbor, New York 1982).

5   Nicolas, J.F.; Berg, P.: Regulation of expression of genes transduced into embryonal carcinoma cells; in Cold Spring Harbor Conference on Cell Proliferation, No. 10, pp. 469–485 (Cold Spring Harbor, New York 1983).

6   Wigler, M.; Silverstein, S.; Lee, L.-S.; Pellicer, A.; Cheng, Y.; Axel, R.: Transfer of purified herpes virus thymidine kinase gene to cultured mouse cells. Cell *11:* 223–232 (1977).

7   Edmonds, M.; Vaughn, M.H., Jr.; Nakazato, H.: Polyadrenylic acid sequences in the heterogeneous nuclear RNA and rapidly-labeled polyribosomal RNA of Hela cells. Possible evidence for a precursor relationship. Proc. natn. Acad. Sci. USA *68:* 1336–1339 (1971).

8   Eistetter, H.R.; Melchers, K.; Keller, A.; Schäfer, K.P.; Voss, T.: Expression of human pulmonary surfactant-associated apoproteins in recombinant mammalian cells; in Progress in biotechnology (Elsevier, Cambridge, in press).

9   Schägger, H.; Jagow, G.: Tricine-sodium dodecyl sulfate-polyacrylamide gel electrophoresis for the separation of proteins in the range from 1 to 100 K. Analyt. Biochem. *166:* 368–379 (1987).

10 Towbin, H.; Staehelin, T.; Grodon, J.: Electrophoretic transfer of proteins from acrylamide gels to nitrocellulose sheets: procedure and some applications. Proc. natn. Acad. Sci. USA *76:* 4350–4354 (1979).

11 Eistetter, H.R.; Voss, T.: Characterization of pulmonary surfactant-associated apoproteins by highly specific antibodies (in preparation).

12 Mallory, J.B.; Kushner, P.J.; Protter, A.A.; Cofer, C.L.; Appleby, V.L.; Lau, K.; Schilling, J.W.; Vigne, J.L.: Expression and characterization of human apolipoprotein A-I in Chinese hamster ovary cells. J. biol. Chem. *262:* 4241–4247 (1987).

13 Fritz, L.C.; Arfsten, A.E.; Dzau, V.J.; Atlas, S.A.; Baxter, J.D.; Fiddes, J.C.; Shine, J.; Cofer, C.L.; Kushner, P.; Ponte, P.A.: Characterization of human prorenin expressed in mammalian cells from cloned cDNA. Proc. natn. Acad. Sci. USA *83:* 4114–4118 (1986).

Klaus P. Schäfer, PhD, Abteilung für Molekularbiologie,
BYK Gulden Pharmazeutika, PO Box 6500, D–7750 Konstanz (FRG)

Wichert P von, Müller B (eds): Basic Research on Lung Surfactant.
Prog Respir Res. Basel, Karger, 1990, vol 25, pp 279–284

cDNA and Deduced Amino Sequence of the Rat Hydrophobic Surfactant Associated Protein SP-B

Philip A. Emrie[a–d], *John M. Shannon*[a], *Robert J. Mason*[a], *James H. Fisher*[a, b, d]

[a] Department of Medicine, National Jewish Center for Immunology and Respiratory Medicine; [b] Division of Pulmonary Sciences, Department of Medicine, University of Colorado Health Sciences Center; [c] Eleanor Roosevelt Institute for Cancer Research, and [d] Webb-Waring Institute, Denver, Colo., USA

Pulmonary surfactant prevents collapse of the alveoli of the lung by lowering surface tension at the air-liquid interface. Surfactant is secreted by alveolar type II cells and is a complex mixture of phospholipids, cholesterol and two groups of surfactant-associated proteins.

The most abundant surfactant-associated protein is a glycoprotein of MW 26–38 kdaltons. This protein is now termed SP-A [for a review of the nomenclature and functions of the surfactant-associated proteins, see ref. 1]. SP-A is water soluble, but readily associates with phospholipids. In the presence of calcium, SP-A causes aggregation of phospholipids and increases the rate of adsorption of surfactant lipid to an air-liquid interface. SP-A has also been implicated in the regulation of surfactant secretion and reutilization.

A second group of very hydrophobic surfactant-associated proteins consists of at least two different proteins. One has a nonreduced molecular weight of 18 kdaltons and has been designated SP-B. The second protein has a nonreduced molecular weight of 5–8 kdaltons and has been designated SP-C. These proteins are structurally unrelated to each other and to SP-A. When purified hydrophobic surfactant proteins are mixed with synthetic phospholipids, near-native surfactant properties are restored, including rapid surface adsorption and lowering of surface tension during dynamic compression. Surfactant lipid extracts containing the two hydrophobic proteins as the sole identifiable protein components have been used to treat hyaline membrane disease in premature infants. Hydrophobic surfactant-associated proteins have also been implicated in stimulating the uptake of liposomes by type II cells, inhibiting glycerophospholipid synthesis by type II cells, and along with SP-A may be taken up preferentially in the lungs of intact animals.

cDNA clones for the SP-B protein have been isolated from human [2, 3] and canine [4] sources. In this article we report the isolation and cloning of a full-length cDNA for the rat SP-B protein and the derived amino acid sequence. These data permit comparison of structural similarities between species and may provide insights regarding the function of this protein.

Materials and Methods

A rat lung cDNA library constructed in lambda gt10 was screened on nitrocellulose filters with a ^{32}P-labeled human SP-B cDNA clone as a probe [5]. Positive plaques from the rat lung cDNA library were purified to homogeneity by four successive screenings. Phage DNA was purified on a CsCl gradient [5], digested with EcoR1, and then analyzed by Southern blot hybridization. Purified insert from a positive phage library clone was subcloned into the vector pIBI30 (International Biotechnologies, Inc., New Haven, Conn.). Plasmid containing the rat SP-B cDNA was used to prepare radiolabeled insert for Northern analysis.

Insert from positive cDNA clones was cloned into phage M13 vectors in both orientations. Overlapping ordered deletions were generated using the T4 polymerase method [6]. The nucleotide sequence was established by a modification of the chain-termination method of Sanger [7] using the Sequenase® modified T7 polymerase (US Biochemical Corp., Cleveland, Ohio).

Alveolar type II epithelial cells were purified from 200–250 g Sprague-Dawley rats by the method of Dobbs et al. [8]. Northern analysis was performed after resolution of total cellular RNA on agarose-formaldehyde denaturing gels by standard techniques [5] using ^{32}P-labeled rat or human SP-B insert DNA.

Results and Discussion

When the rat lung cDNA library was screened with the ^{32}P-labeled human SP-B cDNA probe approximately 0.1% of the clones hybridized with the probe. Northern analysis of RNA from rat whole lung and isolated alveolar type II cells with the human cDNA clone for SP-B showed a band of 1.5 kb. A positive clone from the rat lung cDNA library with an insert of 1.5 kb was therefore selected for further characterization. This clone showed specificity for a band of 1.5 kb in Northern analysis of whole lung

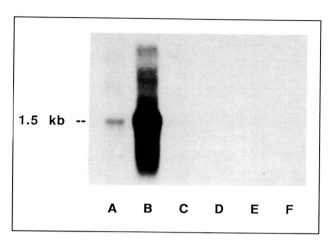

Fig. 1. Northern blot analysis of rat tissue RNA. 5 µg of total cellular RNA from adult whole lung isolated type II cells, liver, spleen, brain and kidney was resolved on a 1% agarose-formaldehyde denaturing gel. RNA was transferred to Nytran® membrane by capillary blotting and hybridized to ^{32}P-labeled rat SP-B cDNA. Hybridization is detected with whole lung RNA (lane A) of 1.5 kilobases and is greatly enriched in isolated type II cells (lane B). No signal was detected for RNA from liver (lane C), kidney (lane D), brain (lane E) or spleen (lane F).

RNA and was greatly enriched in RNA isolated from alveolar type II cells (fig. 1). No hybridization was detected with total cellular RNA from liver, kidney, brain, or spleen. This result implies that rat alveolar type II cells are a source of SP-B, as has been suggested by in situ hybridization [9] and immunohistochemistry [10, 11].

Dideoxynucleotide sequence analysis of the rat SP-B clone was performed. The nucleotide sequence consists of 1,534 bases and encodes an open reading frame of 376 amino acids. The initiating methionine is presented within the context of a Kozak consensus sequence for a translation initiation site. A stretch of more than 40 adenosine residues was found at the 3′ end of the clone. A eukaryotic polyadenylation signal, AATAAA, was present 12 bases upstream from the start of the poly (A) tail. These results imply that this is a full length cDNA clone for rat SP-B. The sequence is shown in figure 2.

The primary translation produce shows extensive homology with the sequences reported for human [2] and canine [4] SP-B (fig. 3). Over the entire protein 69% of the amino acid sequences are conserved between rat

```
        MetAlaLysLeuHisLeuGlnTrpLeuLeuLeuLeuProThrLeuCysSerLeuGlyAlaAlaThr
CCAGCCATGGCCAAGTTACATCTGCAGTGGCTACTGCTCCTTCCTACACTCTGCAGCCTAGGCGCAGCTACA
        12        24        36        48        60        72

GluSerAlaSerSerProAspCysAlaGlnGlyProLysPheTrpCysGlnSerGlnGlnAlaIleGln
GAGTCGGCCTCGTCCCCGGATTGTGCACAAGGCCCTAAATTCTGGTGCCAAAGCCTGGAGCAAGCGATACAG
        84        96        108       120       132       144

CysArgAlaLeuGlyHisCysLeuGlnGluValTrpGlyHisAlaGlyAlaAsnAspLeuCysGlnGluCys
TGCAGAGCCCTGGGGCACTGCCTACAGGAAGTCTGGGGACATGCAGGAGCTAATGACCTGTGCCAAGAGTGT
        156       168       180       192       204       216

GluAspIleValHisLeuThrLysMetThrLysGluAspAlaPheGlnAspThrIleArgLysPheLeu
GAGGATATTGTCCACCTCCTCACAAAGATGACCAAGGAAGACGCTTTCCAGGACACGATCCGGAAGTTCCTG
        228       240       252       264       276       288

GluGlnGluCysAspIleLeuProLeuLysLeuLeuValProArgCysArgGlnValLeuAspValTyrLeu
GAACAAGAATGTGATATCCTACCCTTGAAGCTGCTTGTGCCCCGGTGTCGCCAAGTGCTTGATGTCTACCTG
        300       312       324       336       348       360

ProLeuValIleAspTyrPheGlnHisGlyIleLysProLysAlaIleCysSerHisValGlyLeuCysPro
CCCCTGGTTATCGACTACTTCCAGGGCCAGATTAAACCCAAAGCCATCTGCAGTCATGTGGGCCTGTGCCCA
        372       384       396       408       420       432

LeuGlyGlnThrLysProGluGlnLysProGluMetLeuAspAlaIleProAsnProLeuLeuAsnLysLeu
CTTGGGCAGACTAAGCCAGAGCAGAAGCCCGAGATGCTGGATGCCATTCCAAACCCTCTGCTGAACAAGCTG
        444       456       468       480       492       504

ValLeuProAlaLeuProGlyAlaPheLeuAlaArgProGlyProHisThrGlnProLeuSerGluGlnGln
GTCCTCCCTGCGCTGCCAGGAGCCTTCTTGGCAAGGCCTGGGCCTCACACACAGGACCTCTCTGAGCAACAG
        516       528       540       552       564       576

LeuProIleProLeuProPheCysTrpLeuCysArgThrLeuIleLysArgValGlnAlaValIleProLys
CTCCCCATCCCTCTGCCCTTCTGCTGGCTTTGCAGGACTCTGATCAAGCGGGTCCAAGCTGTGATTCCCAAG
        588       600       612       624       636       648

GlyValLeuAlaValAlaAlaSerGlnValCysHisValValProLeuValValGlyGlyIleCysGlnCys
GGTGTGCTGGCCGTGGCTGTGTCCCAGGTGTGCCACGTGGTACCCCTGGTGGTGGGTGGCATCTGCCAGTGC
        660       672       684       696       708       720

LeuAlaGluArgTyrThrValLeuLeuLeuLeuAspAlaAlaLeuLeuGlyArgValValProGlnLeuValCysGly
CTGGCTGAGCGTTACACAGTACTTCTACTAGATGCACTGCTGGGTCGTGTGGTGCCCCAGCTAGTCTGTGGC
        732       744       756       768       780       792

LeuValLeuArgCysSerThrAlaAspAlaIleGlyProAlaLeuProAlaLeuGluProLeuIleGluLys
CTGGTCCTCCGATGTTCCACTGCAGATGCCATTGGCCCAGCCCTCCCTGCTCTGGAGCCTCTGATTGAAAAA
        804       816       828       840       852       864

TrpProLeuGlnAspThrGluCysHisPheCysLysSerValIleSerAsnGlnAlaTrpAsnThrSerGlu
TGGCCACTGCAAGACACTGAGTGCCATTTCTGCAAGTCTGTGATCAACCAGGCCTGGAACACCAGTGAACAG
        876       888       900       912       924       936

AlaMetProGlnAlaMetHisGlnAlaCysLeuArgPheTrpLeuAspArgGlnLysCysGluGlnProHisVal
GCTATGCCACAGGCAATGCACCAGGCCTGCCTTCGCTTCTGGCTAGACAGGCAAAAGTGTGAGCAGTTTGTG
        948       960       972       984       996       1008

GluGlnHisMetProGlnLeuLeuAlaLeuValProArgSerGlnAsnProAlaHisThrSerCysGlnAlaLeu
GAACAGCACATGCCCCAGCTGCTGGCCCTGGTGCCTAGGAGCCAGGATGCCCACACCTCCTGCCAGGCCCTT
        1020      1032      1044      1056      1068      1080

GlyValCysGluAlaProAlaSerProLeuGlnCysPheGlnThrProHisLeu@@@
GGAGTATGTGAGGCCCCGGCCAGCCCTCTGCAATGCTTCCAAACCCCACACCTCTGAGAACCCAGGCTCCAG
        1092      1104      1116      1128      1140      1152

CAACCTGTACCAAAGGACCACAAGTCTCAGCCTGCACCCCCAGCCCTGTGCCAGCCTGTATGTCCCAGCTCT
        1164      1176      1188      1200      1212      1224

AACCACAGCCCAACAACACCACCAGAGTGGTTTCTGGCTCCTCTGATGATGAGGTGCAGCAGAACCATGTCC
        1236      1248      1260      1272      1284      1296

TCTAGAAGCCTTCAGAGGGGGCTTCAGGCCTTCACCCCCACCAGAGCCAAGCCCAGCTCCAACAGCTCCTAC
        1308      1320      1332      1344      1356      1368

AGCCCCAGGACTGAAAAGGAACTGTCGTGGGTCCACAGAGAGTGGGTAGAAGTGGGGCTAGCCACAGCAACC
        1380      1392      1404      1416      1428      1440

CCTTGCTAACACTTCTGTGAAATTCGAACTTAGATGAATAAACACTGGAAACACAAAAAAAAAAAAAAAAAA
        1452      1464      1476      1488      1500      1512

AAAAAAAAAAAAAAAAAAAAAAAAAA
        1524      1536
```

2

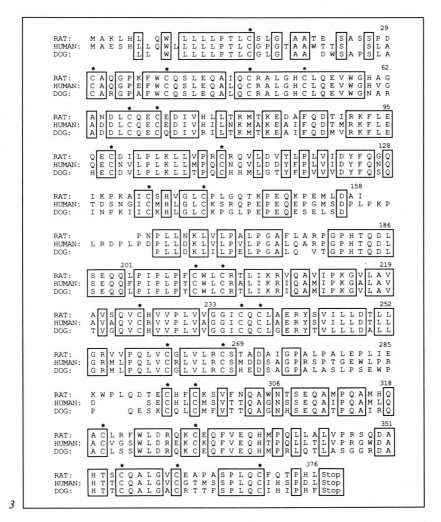

Fig. 2. Sequence of rat SP-B clone 1 and deduced amino acid sequence. The deduced amino acid sequence of SP-B is shown above the cDNA sequence. The proposed amino acid sequence of the mature protein is underlined.

Fig. 3. Comparison of rat, human and canine SP-B sequence. Deduced amino acid sequences of rat, human (31) and canine (33) are shown. Residues which are identical in all three species are boxed to emphasize homology. Gaps have been introduced within the sequences to achieve the greatest homology between all three species. The proposed mature protein is indicated by rat amino acids 201–269. Cysteine residues are marked with an asterisk. An N-linked glycosylation consensus sequence (Asn-X-Ser) is indicated at position 306.

and human SP-B and 67% between rat and dog SP-B. It is known that the primary translation product undergoes NH-2 and COOH terminal proteolysis to yield the native protein found in extracellular fluid. The region corresponding to the mature protein shows that 81% of amino acids are conserved between rat and human SP-B and 89% between rat and dog SP-B. There are 25 cysteine residues present in the proposed primary translation product and their position has been extremely well conserved between all three species. A consensus N-linked glycosylation site in the carboxy region has also been conserved across all three species. The predicted MW of the primary translation product is 42 kdaltons and the predicted MW of the native protein is 8 kdaltons.

References

1   Possmayer, F. Am. Rev. resp. Dis. *138:* 990–998 (1988).
2   Jacobs, K.A.; Phelps, D.S.; Steinbrink, R.; Fisch, J.; Kriz, R.; Mitsock, L.; Dougherty, J.P.; Taeusch, H.W.; Floros, J.: J. biol. Chem. *262:* 9808–9811 (1987).
3   Glasser, S.W.; Kortlagen, T.R.; Weaver, T.; Pilot-Matias, T.; Fox, J.L.; Whitsett, J.A.: Proc. natn. Acad. Sci. USA *84:* 4007–4011 (1987).
4   Hawgood, S.; Benson, B.J.; Schilling, J.; Damm, D.; Clements, J.A.; White, R.T.: Proc. natn. Acad. Sci. USA *84:* 66–70 (1987).
5   Maniatis, T.; Fritsch, E.F.; Sambrook, J.: Molecular cloning: a laboratory manual (Cold Spring Harbor Laboratory, Cold Spring Harbor 1982).
6   Dale, R.M.K.; McClure, B.A.; Houchins, J.P.: Plasmid *13:* 31–40 (1985).
7   Sanger, F.; Nicklen, S.; Coulsen, A.R.: Proc. natn. Acad. Sci. USA *74:* 5463–5469 (1977).
8   Dobbs, L.G.; Geppert, E.F.; Williams, M.C.; Greenleaf, R.D.; Mason, R.J.: Biochim. biophys. Acta *618:* 510–523 (1980).
9   Phelps, D.S.; Floros, J.: Am. Rev. resp. Dis. *137:* 939–942 (1988).
10 Phelps, D.S.; Harding, H.: J. Histochem. Cytochem. *35:* 1339–1342 (1987).
11 Whitsett, J.A.; Weaver, T.E.; Clark, J.C.; Sawtell, N.; Glasser, S.W.; Korfhagen, T.R.; Hull, W.M.: J. biol. Chem. *262:* 15618–15623 (1987).

Philip A. Emrie, MD, Department of Medicine,
National Jewish Center for Immunology and Respiratory Medicine,
1400 Jackson Street, Denver, CO 80206 (USA)

Wichert P von, Müller B (eds): Basic Research on Lung Surfactant.
Prog Respir Res. Basel, Karger, 1990, vol 25, pp 285–287

Stretch-Induced Secretion of Phosphatidylcholine from Rat Alveolar Type II Cells: Inhibition by Surfactant Protein-A[1]

H.R.W. Wirtz, L.G. Dobbs

Cardiovascular Research Institute, San Francisco, Calif., USA

Several stimuli are known to increase the secretion of pulmonary surfactant from alveolar type II cells in primary culture: β-adrenergic agonists, adenosine nucleotides, leukotrienes, calcium ionophores, and phorbol esters. Physiologic correlates for most of the pharmacological agents listed above remain to be elucidated, and the regulation of surfactant secretion in vivo is not well understood.

There is a line of evidence emerging from experiments performed on intact animals as well as excised lungs suggesting that deep inflation of the lung is capable of releasing increased amounts of surfactant phospholipids into the compartment reached by lung lavage [1–4, 6]. It has been hypothesized that mechanical stress acts directly on type II cells. In the lung this type of mechanical stress on type II cells would be created during breathing, with enlarging of the alveolus and thinning of the alveolar walls [5].

Our aim has been to examine whether stretching of alveolar type II cells in vitro results in an increased secretion of surfactant. In one series of experiments, we cultured freshly isolated type II cells from adult Sprague-Dawley rats and plated them on fibronectin-coated elastic silicone membranes in Dulbecco's modified Eagle's medium (H16) supplemented with 10% FCS, gentamicin, penicillin and 1 µCi/ml ^{3}H-choline. After 22 h, cells

[1] Supported in part by NIH grant HL-24075 and DFG grant Wi 824.

were washed, and the membranes were transferred to a stretching device.

In a second series of experiments the type II cells were cultured on membranes in a stretching device; in this case the membranes were not transferred prior to study. In both cases membranes with adherent cells were stretched by applying hydrostatic pressure to the membrane from underneath. A stretch that increased the calculated surface area of the membrane by 21% was performed at the beginning of the experimental period; the pressure was held for 20 s and then released. The cells were allowed to secrete for a subsequent 60 min. Control membranes were treated in the same fashion, but no stretch was performed. ^{3}H-phosphatidylcholine as well as lactic acid dehydrogenase (LDH) was measured in both media and cells. In one series of experiments, we added 1.25 µg/ml of surfactant protein A (SP-A) to the medium shortly before the stretch was exerted.

In the two different series of experiments, we observed similar increases in surfactant secretion. In both series of experiments with a total of 11 different preparations of type II cells, we saw an average increase of 75% over baseline surfactant secretion. When SPA was added shortly before the stretch was performed, the increase in surfactant secretion was abolished and secretion dropped to baseline values. The addition of 12-O-Tetradecanoylphorbol-13-acetate (TPA) in a concentration of $10^{-8}\,M$ to nonstretched membranes resulted in a 100% increase during the 60-min period.

LDH measurements in the medium showed no significant difference between membranes that had been stretched and control membranes and were in general very low (approximately 0.25% of the cell fraction), indicating that stretching did not injure the cells.

Thus, stretching type II cells in vitro stimulates secretion of ^{3}H-phosphatidylcholine, a marker of surfactant. This is not due to injury to the cells. The magnitude of the response to just one stretch is equal to three quarters of the stimulation caused by the continuous presence of $10^{-8}\,M$ TPA. The addition of SPA inhibits the stretch-induced secretion, as it inhibits secretion induced by other factors.

We hypothesize that stretch stimulates surfactant secretion and this provides a means by which production of surfactant could be increased by ventilation.

The mechanism by which a mechanical signal transforms into a secretory response appears to be of great interest in organs that undergo different degrees of distension and deserves further investigation.

References

1   Nicholas, T.E.; Barr, H.A.: Control of release of surfactant phospholipids in the isolated perfused rat lung. J. appl. Physiol. *51:* 90–98 (1981).
2   Nicholas, T.E.; Barr, H.A.: The release of surfactant in rat lung by brief periods of hyperventilation. Resp. Physiol. *52:* 69 (1983).
3   Nicholas, T.E.; Power, H.T.; Barr, H.A.: Surfactant homeostasis in the rat lung during swimming exercise. J appl. Physiol. *53:* 1521–1528 (1982).
4   Oyarzun, M.J.; Clements, J.A.: Control of lung surfactant by ventilation, adrenergic mediators, and prostaglandin in the rabbit. Am. Rev. resp. Dis. *117:* 879–891 (1978).
5   Storey, W.F.; Staub, N.C.: Ventilation of terminal air units. J. appl. Physiol. *17:* 391–397 (1962).
6   Wyszogrodski, I.; Kyei-Aboagye, K.; Taeusch, H.W., Jr.; Avery, M.E.: Surfactant inactivation by hyperventilation. Conservation by end-expiratory pressure. J appl. Physiol. *38:* 461–466 (1975).

Hubert Wirtz, MD, CVRI Box 0130, University of California,
San Francisco, CA 94143 (USA)

Wichert P von, Müller B (eds): Basic Research on Lung Surfactant.
Prog Respir Res. Basel, Karger, 1990, vol 25, pp 288–290

Comparison of Surfactant Biosynthesis in Fetal Rat Alveolar Type II Cells Grown in a Chemically Defined Medium or in the Presence of Serum

Gaëlle Rolland, Jacques Bourbon, Cécile Valenza,
Marie-Elisabeth Dufour, Léa Marin, Michel Rieutort

Centre de Biologie Cellulaire, CNRS, Ivry-sur-Seine, France

Fetal rat lung type II cell cultures represent a convenient model for studying cell differentiation and maturation. The use of serum in culture media, however, raises problems because of its poor chemical definition; the serum also induces a fast cellular growth but dedifferentiation occurs as a counterpart [1, 2].

The aim of our study was to conceive a chemically defined medium, searching for a balance between factors which stimulate cell growth and factors which help maintain a differentiated state.

Material and Methods

Alveolar type II cells were prepared from fetal rat lungs (gestational age 19.5 days) and cultured as described by Bourbon et al. [3] in this issue.

The cells were cultured for 4 days on Gelfoam sponges, then recovered by collagen digestion and plated on multiwell culture dishes (about $3 \cdot 10^5$ cells per well, i.e. 1.9 cm²), and the different media were immediately tested on cell monolayers. Unsupplemented Eagle's MEM was used for control basal conditions. The different supplemented media were MEM containing 2% or 10% whole or charcoal-stripped fetal calf serum (FCS) and the complex defined medium with (DM) or without growth factors. The additives of DM were pyruvate [4, 5], transferrin, putrescine [6], ethanolamine and phosphoethanolamine [7], iron, copper, manganese and zinc, retinoic acid [8], sodium selenite, somatostatin, glycylhistidyl lysine [9], epidermal growth factor (EGF) and insulin-like growth factor 1.

The incorporation af either [^{14}C]-thymidine into cell DNA (0.2 µCi/ml), or of [^{3}H]-choline into surfactant phosphatidylcholine (0.1 µCi/ml) was determined. DNA was extracted and counted according to Rannels et al. [10]. Surfactant fractionation, extraction of lipids and counting were as described by Bourbon et al. [3]; phosphatidylcholine (PC) spots were eluted and counted.

Results and Discussion

The cells survived but did not grow in control MEM; in FCS-supplemented media they grew actively, but flattened and seemed to lose their characteristics. By contrast, cells cultured in DM grew slower but tended to constitute three-dimensional clusters.

The incorporation of [^{14}C]-thymidine into DNA in 48 h was strongly higher in the presence of FCS than in control medium or in DM without growth factors ($4.3 \cdot 10^3$, $0.95 \cdot 10^3$, and $1 \cdot 10^3$dpm/10^5 cells, respectively). Defined medium with growth factors gave intermediary results ($1.8 \cdot 10^3$ dpm/10^5 cells, n = 9).

The incorporation of [^{3}H]-choline into surfactant-fraction PC was the highest in the presence of 2% FCS (99 dpm per 10^5 cells per 72 h), compared to 10% FCS (66 dpm), DM (61 dpm), DM without growth factors (28 dpm) and control MEM (10 dpm, n = 30). It is important to note that increasing FCS from 2 to 10% did not enhance the incorporation of the precursor. The results obtained with 10% FCS are comparable to those obtained with DM. Charcoal stripping greatly reduced the effect of serum (56 dpm per 10^5 cells per 72 h for 2% whole FCS as compared to 9 dpm for 2% stripped FCS, n = 6).

Therefore, the increase in choline incorporation observed with DM was much higher than that induced by charcoal-stripped FCS, which suggests that most of the effects of complete serum was mainly due to hormones likely to be removed by charcoal stripping. Defined medium containing the various additives listed above appears to support growth and maintenance of a differentiated state of fetal alveolar type II cells to a better extent than medium containing charcaol-stripped serum and should be used for culture experiments with type II pneumocytes instead of serum-containing media.

References

1   Cott, G.R.; Walker, S.R.; Mason, R.J.: The effect of substratum and serum on the lipid synthesis and morphology of alveolar type II cells in vitro. Exp. Lung Res. *13:* 427–447 (1987).

2   Scott, J.E.: The role of sera, growth factors, and hormones in the in vitro production of disaturated phospatidylcholine and propagation of undifferentiated type II alveolar cells from fetal rabbit lung. Exp. Lung Res. *12:* 181–194 (1987).

3   Bourbon, J.; Doucet, E.; Rieutort, M.; Marin, L.; Surfactant and non-surfactant phospatidylcholine biosynthesis in cultivated fetal rat alveolar type II cells. Effect of growth factors. Prog. resp. Res., vol. 25, pp. 104–108 (Karger, Basel 1989).

4   Jozwiak, z.; Snyder, C.E.; Murty, V.L.M.; Slomiany, A.; Slomiany, L.B.; Herp, A.: Lipid composition of the secretion from human bronchial explant culture. Biochim. biophys. Acta *802:* 282–286 (1984).

5   Wasilenko, N.J.; Marchok, C.: Pyruvate regulation of growth and differentiation in primary cultures of rat tracheal epithelial cells. Expl Cell Res. *155:* 507–517.

6   Yamane, I.; Kan, M.; Hoshi, H.; Minamoto, Y.: Primary culture of human diploid cells and its long-term transfer in a serum-free medium. Expl Cell Res. *134:* 470–474 (1981).

7   Willey. J.C.; Lechner, J.F.; Harris, C.C.: Bombesin and the C-terminal tetradecapeptide of gastrin-releasing peptide are growth factors normal human bronchial epithelial cells. Expl Cell Res. 153: 245–247 (1984).

8   McDowell, E.M.; Ben, T.; Coleman, B.; Chang, S.; Newkirk, C.; De Luca, L.M.: Effects of retinoic acid on the growth and morphology of hamster tracheal epithelial cells in primary culture. Virchows Arch. B *54:* 38–51

9   Hilfer, S.R.; Schneck, S.L.; Brown, J.W.: The effect of culture conditions on cytodifferentiation of fetal mouse lung respiratory passageways. Exp. Lung Res. *10:* 115–136 (1986).

10 Rannels, S.R.; Yarmell, J.A.; Fisher, C.S.; Fabisiak, J.P.; Rannels, D.E.: Role of laminin in maintenance of type II pneumocyte morphology and function. Am. J. Physiol. *253:* C835–C845 (1987).

G. Rolland, Centre de Biologie Cellulaire, CNRS,
67, rue Maurice-Günsbourg, F–94205 Ivry-sur-Seine (France)

Wichert P von, Müller B (eds): Basic Research on Lung Surfactant.
Prog Respir Res. Basel, Karger, 1990, vol 25, pp 291–293

Effects of Hyperoxia on Type II Cell Surfactant Synthesis

B.A. Holm, S. Matalon

Departments of Obstetrics/Gynecology and Pharmacology,
State University of New York at Buffalo, N.Y.; USA;
Department of Anesthesiology, University of Alabama at Birmingham, Ala., USA

Prolonged exposure to 100% oxygen causes a progressive lung injury which results in pulmonary edema, atelectasis, and death due to respiratory failure [1]. We have previously shown that much of the pathophysiology of hyperoxic lung injury is related to pulmonary surfactant alterations, central to which is a decrease in alveolar phospholipid levels [2–4]. In order to better understand the mechanisms of pulmonary oxygen toxicity, we chose to study type II pneumocyte phospholipid metabolism during in vivo hyperoxic exposure and recovery [5].

Methods

Adult white rabbits were exposed to 100% oxygen and removed to room air for 0, 24, 72 or 200 h. At these time points, the animals were killed and, following lung lavage, type II cells were isolated by the method of Finkelstein and Shapiro [6]. Phospholipid synthesis in type II cell suspensions was assessed by measuring the rate of [Me-^{3}H]choline incorporation into phosphatidylcholine (Pc) as previously described [5, 6]. In addition, the activity of glycerol-3-phosphate acyltransferase was measured in type II cell preparations by the method of Finkelstein and Mavis [7].

Results and Discussion

We have previously shown that rabbits exposed to 100% oxygen for 64 h and removed to room air for periods up to 200 h provides a model for the assessment of hyperoxic lung injury as it progresses through the stages

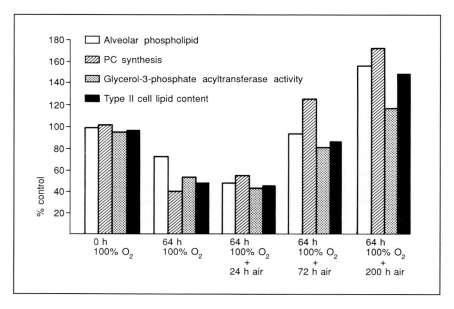

Fig. 1. Comparison of alveolar phospholipid with type II cell lipid content, PC synthesis rate, and glycerol-3-phosphate acyltransferase activity during hyperoxic exposure and recovery. Data are from Holm et al. [5], with permission.

of early development, acute respiratory distress, and eventually recovery [2]. Acute respiratory distress in this model occurs at 24 h postexposure, and recovery takes place after 72 h postexposure. Changes in the pulmonary surfactant system are seen already during their hyperoxic exposure, prior to the development of respiratory distress [2].

Lipid metabolic studies show that the hyperoxic exposure causes a 60% decrease in the type II cell PC synthesis rate which correlates with decreased phospholipid levels both in type II cells and, after a lag period, in the alveolar space. Moreover, the rebound to supranormal alveolar phospholipid levels at 200 h postexposure corresponds to similar increases in type II cell phospholipid content and PC synthesis rates, as shown in figure 1 [5]. The changes in PC synthesis during hyperoxic exposure and recovery also corresponded at all times with similar changes in glycerol-3-phosphate acyltransferase activity. This enzyme is involved in the early stages of phospholipid biosynthesis, and has been previously shown to be sulfhydryl dependent and sensitive to oxidative stress [8]. Although further studies are needed, it seems clear that the detrimental effects of hyperoxia on type II cell metabolism may play a significant role in the development of the pathophysiology of pulmonary oxygen toxicity.

References

1   Clark, J.M.; Lambersten, C.I.: Pulmonary oxygen toxicity. A review. Pharmac. Rev. *23:* 37–133 (1971).
2   Holm, B.A.; Notter, R.H.; Siegle, J.; Matalon, S.: Pulmonary physiological and surfactant changes during injury and recovery from hyperoxia. J. appl. Physiol. *59:* 1402–1409 (1985).
3   Matalon, S.; Holm, B.A.; Notter, R.H.: Mitigation of pulmonary hyperoxic injury by administration of exogenous surfactant. J. appl. Physiol. *62:* 756–761 (1987).
4   Holm, B.A.; Notter, R.H:: Effects of hemoglobin and cell membrane lipids on pulmonary surfactant activity. J. appl. Physiol. *63:* 1434–1442 (1987).
5   Holm, B.A.; Matalon, S.; Finkelstein, J.N.; Notter, R.H.: Type II pneumocyte changes during hyperoxic lung injury and recovery. J. appl. Physiol. *65:* 2672–2678 (1988).
6   Finkelstein, J.N.; Shapiro, D.L.: Isolation of type II alveolar epithelial cells using low protease concentrations. Lung *160:* 85–98 (1982).
7   Finkelstein, J.N.; Mavis, R.D.: Biochemical evidence of internal proteolytic damage during isolation of type II alveolar epithelial cells. Lung *156:* 243–254 (1979).
8   Haagsman, H.P.; Shuurmans, E.A.; Alink, G.M.; Battenburg, J.J.; Golde, L.M.G. van: Effects of ozone on phospholipid synthesis by alveolar type II cells isolated from adult rat lung. Exp. Lung Res. *9:* 67–84 (1985).

Bruce A. Holm, MD, Perinatal Center, Children's Hospital of Buffalo,
Buffalo, NY 14222 (USA)

Wichert P von, Müller B (eds): Basic Research on Lung Surfactant.
Prog Respir Res. Basel, Karger, 1990, vol 25, pp 294–297

Immunohistochemical Verification of Surfactant Replacement in Immature Rabbit Fetuses

K. Albermann[a], U. Bamberger[b], H. Ziegler[a], B. Disse[a], F. Pohlandt[c]

[a]Department of Biological Research and [b]Department of Experimental Pathology,
Dr. K. Thomae GmbH, Biberach, and [c]Department of Neonatology, University
of Ulm, FRG

In the therapy of IRDS by intratracheal instillation of surfactant the distribution of the material and the potential influence of the dose remain an intriguing question. We sought to determine whether exogenous surfactant can be detected and quantified by immunohistochemical methods in sections of rabbit fetus lungs. Additionally, we compared these methods with routine histological phospholipid staining.

Methods

Immature rabbit fetuses of 27 days ± 2 h gestational age were delivered by cesarean section.

They were intubated and treated by intratracheal instillation of 50–180 mg/kg BW SF-RI 1 (Thomae, Biberach, FRG), a purified surfactant extract from bovine lungs containing approximately 1% surfactant-associated proteins. A protein-free phospholipid mixture (PLM) was used for comparison. Untreated littermates served as controls.

The fetuses were ventilated with time-cycled, volume-controlled, pressure-limited ventilators in body plethysmographs. Tidal volume was adjusted at 5–7 ml/kg BW, breath frequency was 60 min^{-1}, I/E ratio was 1:2 and $F_iO_2 = 1.0$. Lung function was tested every 30 min by measuring dynamic lung compliance (C_{dyn}). After 3 h of ventilation the animals were ventilated with nitrogen for 2 min at the same ventilator setting and killed by an overdose of pentobarbital. Only the lungs used for paraffin embedding were inflated to 2.5 kPa and fixed by immersion in 4% phosphate-buffered formaldehyde (pH 7.4) at a deflation pressure of 1.5 kPa for 3 h. For complete fixation, the tracheas were clamped and the lungs were kept in the same fixation medium for at least 72 h. Frozen sections of the lungs were stained with a phosphomolybdic acid reaction [1] or Sudan red 7B [2]. A second set of frozen or formaldehyde-fixed, paraffin-embedded sections were subsequently incubated with a polyclonal rat antibovine surfactant antibody and with a rabbit antirat IgG antibody labelled with alkaline phosphatase [3]. In control sections the surfactant-specific antibody was omitted.

Results and Discussion

In both frozen and paraffin-embedded immunohistochemically stained sections, the exogenous bovine surfactant was visualized. SF-RI 1-treated animals could easily be distinguished from the controls.

The PLM-treated fetuses did not show any positive staining. Conventional phospholipid stains of frozen sections did not reveal distinct staining of PLM or of SF-RI 1.

The best results were achieved with immunohistochemical staining of paraffin-embedded sections: a semiquantitative analysis of the amount of surfactant showed a dose-dependent increase. The upper lung lobes contained less immunoreactive material than the lower lobes. The antigen was not detected in the heart, the blood vessels, the pulmonary interstitium or the esophagus. Sections of mature bovine lung were stained with this method as a positive control: the alveolar type II cells were stained specifically.

Immunohistochemical staining of paraffin sections is appropriate to visualize exogenous surfactant and is a very useful technique to assess the distribution after endotracheal instillation (fig. 1).

It is puzzling that in many lung sections the surfactant can be seen in collapsed but rarely in well-aerated regions. Model calculations indicate

---

(For fig. 1 see page 296)

Fig. 1. Lung sections (paraffin, 5 μm) of premature rabbit fetuses (gestational age: 27 days ± 2 h) after artificial ventilation for 3 h. *a–c* SF-RI 1, 120 mg/kg. *d* Untreated littermate. *a* Fibrinous material (arrows) in the alveoli surrounding the surfactant (*), after AG/AB reaction with alkaline phosphatase. HE × 630, oil immersion. *b* Patchy distribution of surfactant in the lung evaluated by semiquantitative microscopy. Surface-active material (*) surrounded by hyaline plasma exudate. HE × 400. *c* Control section of the same animal as in *(a, b),* where the surfactant-specific antibody was omitted, without any staining or unspecific reaction. HE × 630, oil immersion. *d* Untreated animals even after incubation with surfactant-specific antibody did not show any positive staining either. HE × 630, oil immersion.

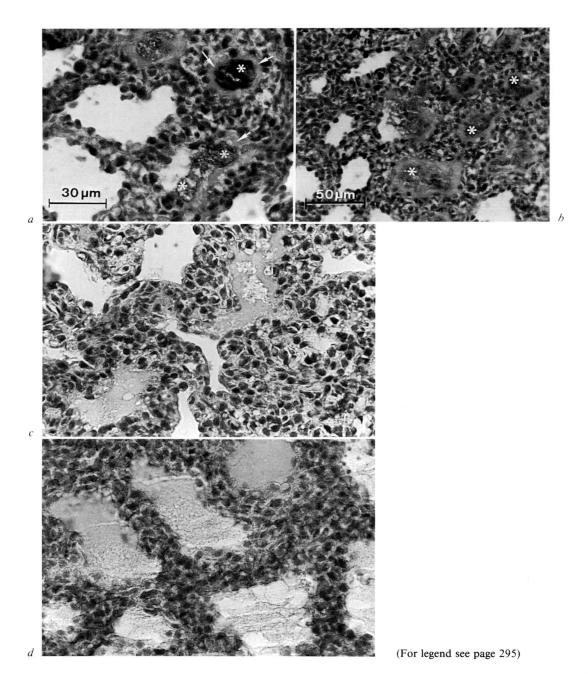

(For legend see page 295)

Table 1. Dynamic lung compliance ($\bar{x}$ $\pm$ SE)of 27-day-old premature rabbit fetuses treated with 180 mg/kg SF-RI 1 and of untreated littermates at V_T = 10% VC (vital capacity) = 0.15 ml

Time on artificial ventilation, min	Surfactant C_{dyn}, ml/kPa/kg	Control C_{dyn}, ml/kPa/kg
5	3.3 ± 0.2* (n = 4)	0.7 ± 0.1 (n = 13)
30	4.6 ± 0.3* (n = 13)	1.7 ± 0.1 (n = 12)
180	7.0 ± 0.9* (n = 10)	1.6 ± 0.2 (n = 5)

Statistically significant difference from control: * $p < 0.05$, Student's t test.

that the higher surfactant content of the lower lung lobes can be explained by their lower grade of aeration. In collapsed alveoli, the instilled surfactant material was concentrated to a relatively small volume. On the contrary, the very thin lining layer (50 Å) in the mostly well-aerated alveoli of the upper lobes cannot be seen by light microscopy. An influence of the dose on the distribution was not observed.

The measurement of dynamic compliance (C_{dyn}) revealed a statistically significant difference ($p < 0.05$) between SF-RI 1-treated animals and untreated littermates (table 1). After 5 min of ventilation C_{dyn} showed a slight dose-dependent increase (data not shown). After 3 h C_{dyn} showed a twofold increase compared to the initial value at 5 min and a fourfold increase compared to control animals, but no longer a difference between the dose groups. However, the amount of stained material in the lung sections was increased in a dose-dependent manner.

References

1   Landing, B.H.; Uzmann, L.L.; Whipple, A.: Phosphomolybdic acid as a staining reagent for lipids. Lab. Invest. *1:* 456–462 (1952).
2   Michaelis, L.: In Romeis, Mikroskopische Technik, p. 256 (Oldenbourg, München 1968).
3   Mason, D.Y.: Immunocytochemical labeling of monoclonal antibodies by the APAAP-immunoalkaline phosphatase technique; in Bullock, Petrusz, Techniques in immunocytochemistry, vol. 3, pp. 25–42 (Academic Press, London 1985).

K. Albermann, MD, Department Biological Research, Dr. K. Thomae GmbH, D–7950 Biberach/Riss (FRG)

Wichert P von, Müller B (eds): Basic Research on Lung Surfactant.
Prog Respir Res. Basel, Karger, 1990, vol 25, pp 298–301

Effect of Maternal Corticosteroid Treatments on the Relationship between Surfactant Pool Size and Compliance in Preterm Ventilated Rabbits

Machiko Ikegami, Alan Jobe

Department of Pediatrics, Harbor-UCLA Medical Center, Torrance, Calif., USA

The alveolar surfactant pool sizes per kilogram body weight (BW) of term newborns of several species exceed that of the healthy adult by about 5- to 10-fold. On the other hand, the surfactant pool sizes per kilogram BW in preterm animals with respiratory failure comparable to that of infants with severe respiratory distress syndrome are not very different from those measured in adult animals [1–3]. We report here the relationship between endogenous surfactant pool sizes, measured as saturated phosphatidylcholine (SPC) in alveolar wash, and lung compliances in premature ventilated rabbits at gestational ages of 27, 28 and 29 days. These gestation ages were chosen to give a range from low to high compliance values. Surfactant treatments of the three different gestational age rabbits improved lung compliances to values equivalent to those for term newborn rabbits [4]. This result indicated that the possible effects of lung maturational factors other than surfactant pool sizes on compliance at 27, 28 and 29 days would be negligible for the purpose of this study. We have described several effects of maternal corticosteroid treatments on lung function of ventilated premature rabbits that are independent of changes in surfactant pool sizes. The treatments decreased protein leaks in ventilated immature lung [5], improved the treatment response of the preterm lung to surfactant [5], increased lung volumes and decreased lung rupture pressures [6] and treatments changed the dose-response relationship of exogenous surfactant in 27-day-old fetal rabbits [7]. Because of the corticosteroid-mediated effects we also asked if maternal corticosteroid (0.1 mg/kg betamethasone) treatments at 48 and 24 h before delivery of the fetuses changed the relationship between endogenous surfactant pool size and lung compliance.

Material and Methods

After an endotracheal tube was placed, the rabbits were ventilated in plethysmographs at tidal volumes of 10–13 ml/kg until sacrifice at 30 min [5]. The surfactant was recovered from the lungs by a thorough lavage procedure, and the amount of SPC was measured after osmium tetroxide treatment of the lipid extracts [5].

Results

The calculated compliance values for the rabbits were graphed versus the SPC recovered by alveolar wash expressed as µmol/kg BW (fig. 1). Surfactant pool sizes below about 0.7 µmol SPC/kg BW did not alter compliance in either the control or the corticosteroid-treated groups. The mean compliance values for the corticosteroid-treated rabbits with pool sizes below 0.7 µmol SPC/kg BW were 41% higher than for control rabbits (p < 0.01). Compliance increased to values comparable to surfactant-treated or term newborn rabbits over a narrow SPC pool size range of 0.7–1.2 µmol SPC/kg for corticosteroid-treated rabbits and 0.9–2.3 µmol SPC/kg for control rabbits (p < 0.01). The maximal compliance achieved in the corticosteroid-treated rabbits exceeded those in the control rabbits by about 50% (p < 0.01). When compared to measurements made with surfactant-treated rabbits [7], these results document that far less endogenous surfactant than exogenous surfactant was needed to alter compliance in the preterm ventilated rabbit.

Summary

This study demonstrated that very little endogenous surfactant was needed to improve lung function in preterm ventilated rabbits and that compliance improved over a narrow range of endogenous surfactant concentrations. Far less endogenous surfactant

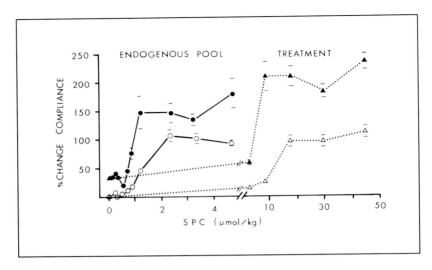

Fig. 1. The data were normalized such that the compliance values at low pool sizes for control rabbits (o) were given a value of 0 and the mean peak compliance value was given as value of 100%. Values for corticosteroid-treated rabbits (●) were expressed relative to the 0% value for the control rabbits. A similar normalization was used for the data from Seidner et al. [7] for control rabbits treated with rabbits surfactant (△) or corticosteroid-treated rabbits treated with surfactant (▲). The X axis gives the amount of SPC recovered by alveolar wash from the rabbits not treated with surfactant or the amount of SPC in the surfactant used to treat the rabbits.

than exogenous surfactant was needed to alter compliance in the preterm ventilated rabbits. The corticosteroid treatments changed the shape of the dose-response curve for compliance versus the surfactant pool size as measured by alveolar wash.

References

1   Prueitt, J.L.; Palmer, S.; Standaert, T.A.; Luchtel, D.L.; Murphy, J.H.; Hodson, W.A.: Lung development in the fetal primate *Macaca nemestrina*. III. HMD. Pediat. Res. *13:* 654–659 (1979).

2   Jackson, J.C.; Palmer, S.; Truog, W.E.; Standaert, T.A.; Murphy, J.H.; Hodson, W.A.: Surfactant quantity and composition during recovery from hyaline membrane disease. Pediat. Res. *20:* 1243–1247 (1986).

3   Jobe, A.H.; Ikegami, M.; Jacobs, H.C.; Jones, S.J.: Surfactant pool sizes and severity of respiratory distress syndrome in prematurely delivered lambs. Am. Rev. resp. Dis. *127:* 751–755 (1983).

4   Ikegami, M.; Jobe, A.; Seidner, S.; Yamada, T.: Effect of corticosteroids and surfactant on lung function and protein leaks in ventilated rabbits at four gestational ages. Am. Rev. resp. Dis. *137:* 18A (1988).

5   Ikegami, M.; Berry, D.; ElKady, T.; Pettenazzo, A.; Seidner, S.; Jobe, A.: Corticosteroids and surfactant change lung function and protein leaks in the lungs of ventilated premature rabbits. J. clin. Invest. *79:* 1371–1378 (1987).

6   ElKady, T.; Jobe, A.: Corticosteroids and surfactant increase lung volumes and decrease rupture pressures of preterm rabbit lungs. J. appl. Physiol. *63:* 1616–1621 (1987).

7   Seidner, S.; Pettenazzo, A.; Ikegami, M.; Jobe, A.: Corticosteroid potentiation of surfactant dose response in preterm rabbits. J. appl. Physiol. *64:* 2366–2371 (1988).

Machiko Ikegami, MD, Department of Pediatrics, Bldg. A-17 Annex,
Harbor-UCLA Medical Center, 1000 W. Carson St., Torrance, CA 90274 (USA)

Wichert P von, Müller B (eds): Basic Research on Lung Surfactant.
Prog Respir Res. Basel, Karger, 1990, vol 25, pp 302–304

Human Surfactant Extracted from Vaginal-Delivered Amniotic Fluid Improves Lung Mechanics as Effectively as Natural Surfactant Derived from Bovine Lungs

D. Gommers[a], *K.L. So*[a], *S. Armbruster*[a], *R. Tenbrinck*[a],
J.L.M. van Remortel[a], *J.E. van Eyk*[b], *B. Lachmann*[a]

Departments of [a]Anesthesiology and [b]Gynaecology, Erasmus University,
Rotterdam, The Netherlands

Previous studies have indicated that human surfactant derived from amniotic fluid (AFS), obtained during cesarean section from full-term pregnancies, is effective in treatment of respiratory distress syndrome [1]. However, the most critical issue concerning this source is the limited number of cesarean sections. This study aimed to investigate the effect of surfactant from vaginally delivered amniotic fluid on lung mechanics, compared to natural surfactant derived from bovine lungs.

Materials and Methods

Amniotic fluid, collected vaginally, was first filtered and then mixed with chloroform (ratio 1:5) for 30 min. In order to remove all chloroform a vacuum rotation evaporator was used. The final extraction was suspended in water; several specimens were taken for phospholipid analyses. Immature rabbit fetuses, with no lung surfactant, were delivered from white New Zealand rabbits by cesarean section on day 27 of pregnancy (term ± 31 days).

Following tracheotomy the animals were transferred to a heated pressure-constant plethysmograph [2], and ventilated with pure O_2 (via a servo ventilator) with a frequency of 30 and peak airway pressure of 25 cm H_2O. After initial recording of thorax-lung compliance (C_{rs}) only animals with a C_{rs} below 0.1 ml/cm $H_2O \cdot kg$ received one of the following administrations (0.15 ml): group A: 7 animals (30 ± 3 g BW) received bovine surfactant (50 mg/ml total phospholipids); group B: 7 animals (28 ± 6 g BW) received

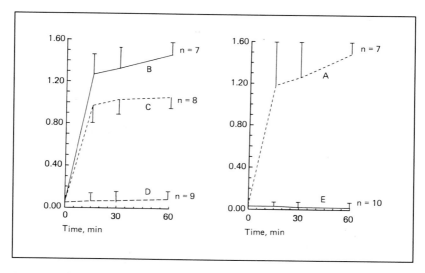

Fig. 1. Thorax-lung compliance in premature rabbit fetuses at various intervals after tracheal instillation of natural lung surfactant (group A), amniotic fluid surfactant (group B), mixture of artificial phospholipids and AFS (group C), pure artificial phospholipids (group D) or no surfactant at all (group E).

pure human surfactant (50 mg/ml total phospholipids); group C: 8 animals (33 ± 4 g BW) received a mixture of artificial phospholipids and AFS in a 1:2 ratio by weight (50 mg/ml total phospholipids); group D: 9 animals (29 ± 5 g BW) received pure artificial phospholipids (DPPC/Ei PG = 7:3) (50 mg/ml total phospholipids); group E: 18 animals (28 ± 4 g BW) with no administration, served as controls. All animals were ventilated for 60 min. Compliance was measured at 15, 30 and 60 min.

Results

Figure 1 shows that surfactant extracted from bovine lungs (group A) dramatically improved thorax-lung compliance within a few minutes. Similar effects were also observed after instillation of pure human surfactant extracted from vaginal-delivered amniotic fluid (group B). The mixture of artificial phospholipids and AFS (group C) also improved C_{rs} highly significantly. The pure artificial phospholipids (group D) led to almost no improvement in lung mechanics.

Discussion

Our results clearly demonstrate that human surfactant extracted from vaginal-delivered amniotic fluid improves lung mechanics as effectively as surfactant extracted from bovine lungs. If one likes to avoid animal surfactant for treatment of neonates with immature lungs, surfactant extracted from amniotic fluid could be a proper and effective alternative. In order to save human surfactant the mixture of amniotic fluid surfactant with artificial phospholipids gives, at least in an immature rabbit fetus model, appropriate results and may therefore also be used for treatment of neonates with respiratory distress syndrome.

References

1   Hallman, M.; Merritt, T.A.; Schneider, H.; Epstein, B.L.; Mannino, F.; Edwards, D.K.; Gluck, L.: Isolation of human surfactant from amniotic fluid and a pilot study of its efficiency in respiratory distress syndrome. Pediatrics, Springfield 71: 473–482 (1983).
2   Lachmann, B.; Grossmann, G.; Freyse, J.; Robertson, B.: Lung-thorax compliance in the artificially ventilated premature rabbit neonate in relation to variations in inspiration/expiration ratio. Pediat. Res. 15: 833–838 (1981).
3   Morley, C.J.; Miller, N.; Bangham, A.D.; et al.: Dry artificial lung surfactant and its effect on very premature babies. Lancet i: 64 (1981).
4   Hallman, M.; Feldman, B.H.; Kirkpatrick, E.; Gluck, L.: Absence of phosphatidylglycerol (PG) in respiratory distress syndrome in the newborn. Study of the minor surfactant phospholipids in newborns. Pediat. Res. 11: 714–720 (1977).

B. Lachmann, MD, PhD, Department of Anesthesiology, Erasmus University, Postbox 1738, NL–3000 DR Rotterdam (The Netherlands)

Wichert P von, Müller B (eds): Basic Research on Lung Surfactant.
Prog Respir Res. Basel, Karger, 1990, vol 25, pp 305–307

Biophysical and Physiological Properties of Synthetic Surfactants

B.A. Holm, R.H. Notter, G. Enhorning

Departments of Obstetrics/Gynecology and Pharmacology, SUNY at Buffalo, N.Y.;
Department of Pediatrics, University of Rochester, Minn., USA

Since the discovery by Avery and Mead [1] that pulmonary surfactant deficiency was involved in neonatal lung disease, investigators have attempted to develop protein-free synthetic surfactants for use in exogenous surfactant replacement therapy [2, 3]. Although some synthetic surfactants have shown promise, their physiological efficacies have been poor. In this study, we attempt to explain apparent discrepancies between the biophysical and physiological efficacies of several simple synthetic surfactant mixtures.

Methods

Calf lung surfactant extract (CLSE) was prepared by the chloroform:methanol extraction of surfactant material obtained from calf lungs by bronchoalveolar lavage, as previously described [4]. Synthetic surfactant mixtures were prepared by combining the components in chloroform, drying the mixture under nitrogen, and resuspending the mixture, by sonication, in 0.15 M NaCl with 5 mM CaCl$_2$. All mixtures were resuspended at 5 mg/ml, unless otherwise stated.

Adsorption and dynamic surface tension measurements were made using an oscillating bubble surfactometer [5]. Physiological efficacy was determined using the surfactant-deficient excised rat lung model of Bermel et al. [6].

Results and Discussion

In these studies, the CLSE mixture used successfully in the treatment of neonatal respiratory distress syndrome (RDS) was used as a positive control, and pure DPPC was used as a negative control. As shown in

Table 1. Biophysical and physiological properties of surfactant mixtures

Surfactant	Equilibrium surface tension, mN/m (adsorption)[1,2]	Dynamic surface tension, mN/m^2	Physiological effectiveness
CLSE	22	0	++
DPPC	68	38	−
DPPC:palmitic acid (9:1)	38	0	−

[1] Adsorption after 10 s.
[2] Measured on an oscillating bubble at 5 mg/ml.

Table 2. Sensitivity to inhibition by albumin

Surfactant	Surfactant concentration mg/ml	Albumin concentration mg/ml	Dynamic surface tension mN/m^2)[1]
CLSE	2	200	0
DPPC:palmitic acid (9:1)	20	2	37

[1] Measured on an oscillating bubble.

table 1, synthetic surfactant mixtures can be prepared which are capable of demonstrating good biophysical properties comparable to those exhibited by CLSE or natural surfactant. However, these mixtures are unimpressive in restoring the lung mechanical properties of surfactant-deficient lungs.

The explanation for the discrepancy between biophysical properties and physiological effectiveness appears to rest in the fact that the biological system drastically alters the synthetic surfactant material. As shown in table 2, the synthetic mixture is at least two orders of magnitude more sensitive to inhibition by plasma proteins than CLSE [4, 7]. Indeed, the lavage material from the surfactant-deficient lungs instilled with synthetic surfactant is biophysically inactive unless the trace levels of protein removed from the lung are extracted, at which point biophysical activity is restored.

These results suggest that inhibition by plasma proteins may explain the ineffectiveness of many synthetic surfactants developed over the past

30 years. Moreover, sensitivity to inhibition should be an important consideration in the development and selection of exogenous surfactants for replacement therapy.

References

1   Avery, M.E.; Mead, J.: Surface properties in relation to atelectasis and hyaline membrane disease. Am. J. Dis. Child. *97:* 517–523 (1959).
2   Notter, R.H.; Finkelstein, J.N.: Pulmonary surfactant. An interdisciplinary approach. J. appl. Physiol. *57:* 1613–1624 (1984).
3   Jobe, A.; Ikegami, M.: State of art. Surfactant for the treatment of respiratory distress syndrome. Am. Rev. resp. Dis. *136:* 1256–1275 (1987).
4   Holm, B.A.; Notter, R.H.; Finkelstein, J.N.: Surface property changes from interactions of albumin with natural lung surfactant and extracted lung lipids. Chem. Phys. Lipids *38:* 287–298 (1985).
5   Enhorning, G.: Pulsating bubble technique for evaluating pulmonary surfactant. J. appl. Physiol. *43:* 198–203 (1977).
6   Bermel, M.S.; McBride, J.T.; Notter, R.H.: Lavaged excised rat lungs as a model of surfactant deficiency. Lung *162:* 99–113 (1984).
7   Holm, B.A.; Notter, R.H.: Effects of hemoglobin and cell membrane lipids on pulmonary surfactant activity. J. appl. Physiol. *63:* 1434–1442 (1987).

Bruce A. Holm, MD, Perinatal Center, Children's Hospital of Buffalo,
Buffalo, NY 14222 (USA)

Wichert P von, Müller B (eds): Basic Research on Lung Surfactant.
Prog Respir Res. Basel, Karger, 1990, vol 25, pp 308–312

Surfactant Kinetics in the Newborn Rabbit[1]

Paul A. Stevens[a], Jo Rae Wright[b], John A. Clements[c]

Departments of [b]Physiology and [c]Pediatrics, [a-c]Cardiovascular Research
Institute, University of California San Francisco, Calif., USA

Surfactant kinetics have been relatively well described in adult animals [2, 3, 5]. In contrast, much less is known about the fate of endogenous and instilled surfactant in the newborn period, mainly because the traditional analyses of the tracer data [1, 6] assumed steady state conditions. The newborn period with its adaptational changes and a rapidly changing alveolar surfactant pool size [4] does not fulfill these requirements.

For a quantitative analysis of surfactant secretion and clearance processes and their phase relationship around birth, transport equations for use with isotope tracers in non-steady state conditions were derived. They were then applied to specific activity values for lamellar body and alveolar phospholipids, obtained from experiments with newborn rabbits.

Methods

Animals

Time-mated pregnant rabbits were injected with 250 μCi (methyl-^{3}H) choline chloride. One group of newborn animals was killed in utero immediately after injection of the dose with the isotope label (designated −8 h). One group was killed in utero 4 h after injection (−4 h). All other groups were delivered by Caesarean section after spinal anaesthesia of the mother. One group of animals was killed in utero by an overdose of pentobarbital for determination of the initial alveolar pool size (birth). The remaining newborns were anaesthetized through the uterine wall with an intraperitoneal injection of

[1] Supported by grants from the American Heart Association, the American Lung Association, and the National Heart, Lung and Blood Institute (HL-24075, HL-30923).

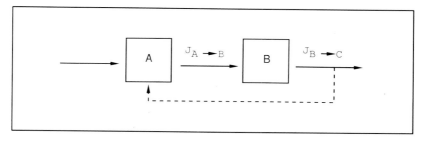

Fig. 1. Model of alveolar surfactant fluxes. A = Component in lamellar bodies of type II cells; B = same component in alveolar surfactant; $J_{A \to B}$ = flux of the component from lamellar bodies to alveolar pool; $J_{B \to C}$ = sum of fluxes of the component out of the alveolar pool. x_B = amount of labeled component in alveolar pool; Q_B = amount of the component in alveolar surfactant, i.e. alveolar pool size; t = time after injection of label; C_A = concentration (i.e. specific activity) of labeled component in pool A; C_B = concentration of label in pool B. Dashed arrow indicates that more or less of the component may recycle to lamellar bodies.

ketamine hydrochloride and acepromazine maleate. Upon delivery, an incision was made through the skin to expose the trachea and 0.1 ml radioactively labeled natural surfactant, harvested by lung lavage from an adult rabbit, was instilled intratracheally.

One group of animals was killed immediately after instillation (5 min), others were permitted to breathe for various lengths of time before sacrifice (0.5, 1, 3, 6, 9, 12 h).

After sacrifice, their lungs were lavaged with ice-cold saline. Lamellar bodies were isolated as described by Baritussio et al. [1]. Phospholipid (PL) and protein content as well as the specific activity of PLs in lavage, lung homogenate and lamellar bodies were determined.

Data Analysis

For the lamellar body fractions, specific activity of light and dense fractions were combined in one lamellar body fraction by calculating a weighted average of the two fractions.

The same number of radioactive counts (7,000 dpm in 100 µg PL) was given to each animal, regardless of weight. To facilitate interpretation, we therefore normalized the data to 7,000 dpm instilled per 50 g of body weight.

Recovery of the instilled label in the whole lung (tissue + lavage) was the same ($\pm 50\%$) at all times. There was a strong linear correlation between specific activities achieved in each pool and the % total recovery. Therefore, the measured specific activities were corrected to values that would be achieved, had the lung received 100% of the instilled label.

The assumptions involved in the model of alveolar surfactant metabolism used (fig. 1) and the derivation of the formulas for the calculation of secretion and clearance rates will be published elsewhere. For calculations we used the following formulas (see explanation of symbols in fig. 1):

Average secretion rate $\bar{J}_{A \to B_{1,2}}$ in theinterval from t_1 to t_2:

$$\bar{J}_{A \to B_{1,2}} = \frac{(Q_{B_1} + Q_{B_2})(C_{B_2} - C_{B_1})}{(C_{A_1} + C_{A_2} - C_{B_1} - C_{B_2})(t_2 - t_1)},$$

where $\bar{J}_{A \to B_{1,2}}$ is average flux for the interval (t_1 to t_2).

Average clearance rate $\bar{J}_{A \to B_{1,2}}$ in the interval from t_1 to t_2:

$$\bar{J}_{B \to C_{1,2}} = \bar{J}_{A \to B_{1,2}} - \frac{Q_{B_2} - Q_{B_1}}{t_2 - t_1}.$$

Results

During the 8-hour period before birth, the PL content of the alveolar lavage of the newborn rabbits did not change significantly. Within 5–10 min after birth, the pool size increased from 10.3 µg PL/g BW before breathing to 14.3 µg after 10 min. Within the first 12 h of life the pool increased almost threefold to 29.2 µg PL/g BW at 12 h (table 1).

Table 1. Alveolar pool size and specific activities of lavage and lamellar body PLs in fetal and newborn rabbits (numbers in parentheses denote SE)

	Time							
	−8 h	−4 h	0 h[1]	0.1 h	3 h	6 h	9 h	12 h
n	6	4	5	8	8	6	6	7
Alveolar PL, µg/g BW	10.3	10.2	10.3	14.3	18.0	23.0	26.1	29.2
	(1.4)	(0.7)	(1.2)	(1.7)	(2.1)	(3.6)	(1.9)	(1.9)
Intratracheal label								
Specific activity lavage	−	−	11.4	6.2	3.2	2.1	2.1	1.5
PL, dpm/µg PL				(0.8)	(0.5)	(0.2)	(0.2)	(0.2)
Specific activity lamellar	−	−	0	1.1	1.6	1.4	1.3	1.2
body PL, dpm/µg PL				(0.3)	(0.2)	(0.2)	(0.2)	(0.1)
Intravenous label								
Specific activity lavage	0	1.1	4.0	4.0	5.2	7.7	8.4	11.0
PL, dpm/µg PL		(0.2)	(0.3)	(0.3)	(0.5)	(0.4)	(0.3)	(1.0)
Specific activity lamellar	0	3.2	5.3	3.2	5.7	8.1	8.0	12.0
body PL, dpm/µg PL		(1.1)	(1.1)	(1.1)	(0.7)	(0.4)	(0.2)	(1.4)

[1] Delivery.

In contrast, the PL content of the whole lung and of the lamellar body fraction did not change significantly with time.

The specific activity of the intratracheally instilled label in the lavage went down from 11.4 dpm/µg PL at instillation to 6.2 dpm/µg PL after 5 min to 1.5 dpm/µg PL at 12 h.

At the same time, the specific activity in lung homogenate rose slightly from 0.1 dpm/µg PL immediately after instillation to 0.2 dpm/µg PL after 30 min, and then stayed elevated for the remaining time.

The specific activity in the lamellar body fraction rose from 1.1 immediately after instillation to a maximum of 1.6 after 3 h. Thereafter, it decreased gradually to 1.2 dpm/µg per µg PL at 12 h of life (table 1). At all times the specific activity in the lamellar bodies was 5–7 times higher than that of the lung homogenate.

The curves of specific activity vs. time for lavage and lamellar body fractions were fit to the sum of two exponentials by curve stripping. The values derived from these curves were used for the calculation of secretion and clearance rates.

For an estimate of surfactant secretion and clearance rates before delivery, the data from the experiments with the intravenous label were used. The curves of specific activity vs. time for lavage and lamellar body fractions were fit empirically to two-term polynomial functions.

Before birth, secretion and clearance rates were equal, and both rose slightly from 1.8 at −6 h to 7.3 µg PL/h/g BW at birth.

Immediately after birth the secretion rate increased rapidly to 37.7 µg PL/h/g BW between 0 and 10 min, and then decreased to 1.8 µg PL/h/g BW between 1.5 and 2 h. It then slowly increased to 6 µg PL/h/g BW between 11 and 12 h.

Clearance rate also increased after birth, but to only 13.6 µg PL/h/g BW between 0 and 10 min, and then decreased further to 24.7 µg PL/h/g BW between 0.5 and 1 h vs. 28.9 mg PL/h/g BW for the secretion rate. It then followed the same pattern as the secretion rate, but its magnitude at all times in the first 12 h after birth was slightly lower than that of the secretion rate.

Discussion

Secretion and clearance rates were substantial even before birth. This indicates that the surfactant system is not dormant and waiting to be activated at birth, but actively and continuously replenishing its components.

The reasons for this continuous renewal before birth are not yet clear. Loss of material up the airways may account for some but not all of the clearance in the fetal animal.

The results of our study raise the question how the alveolar pool is regulated. A simple feedback model with a constant set point is unlikely, since in such a model the increased pool size would have to decrease exponentially with time to the previous value. In the newborn, however, at least in the first 24 h of life, the alveolar pool size continues to move away from its previous set point.

Most likely, a new set point is established because of changing demands on the surfactant system by the onset of breathing, the elimination of substantial clearance up the airways, and removal of alveolar fluid with a resulting concentration of surfactant components in the liquid bathing the type II cells.

References

1   Baritussio, A.G.; Magoon, M.W.; Goerke, J.; Clements, J.A.: Precursor-product relationship between rabbit type II cell lamellar bodies and alveolar surface-active material. Surfactant turnover time. Biochim. biophys. Acta *666:* 382–393 (1981).
2   Hallman, M.; Epstein, B.L.; Gluck, L.: Analysis of labeling and clearance of lung surfactant phospholipid in rabbit. Evidence of bi-directional surfactant flux between lamellar bodies and alveolar lavage. J. clin. Invest. *68:* 742–751 (1981).
3   Jacobs, H.; Jobe, A.; Ikegami, M.; Jones, S.: Surfactant phosphatidylcholine source, fluxes and turnover times in 3-day-old, 10-day-old, and adult rabbits. J. biol. Chem. *257:* 1805–1810 (1982).
4   Stevens, P.A.; Wright, J.R.; Clements, J.A.: Changes in quantity, composition and surface activity of alveolar surfactant at birth. J. appl. Physiol. *63:* 1049–1057 (1987).
5   Wright, J.R.; Wager, R.E.; Hamilton, R.L.; Hwang, M.; Clements, J.A.: Uptake of lung surfactant subfractions into lamellar bodies of adult rabbit lungs. J. appl. Physiol. *60:* 817–825 (1986).
6   Zilversmit, D.B.; Entenman, C.; Fishler, M.C.: On the calculation of 'turnover time' and 'turnover rate' from experiments involving the use of labelling agents. J. gen. Physiol. *26:* 325–331 (1943).

Paul A. Stevens, MD, Freie Universität Berlin, Universitätsklinikum
Rudolf Virchow, Standort Charlottenburg, Heubnerweg 6, D-1000 Berlin 19

Wichert P von, Müller B (eds): Basic Research on Lung Surfactant.
Prog Respir Res. Basel, Karger, 1990, vol 25, pp 313–317

Hormonal Effects on the Synthesis of Phosphatidylcholine Molecular Species by Organ Cultures of Human Fetal Lung[1]

A.D. Postle, P.A. Caesar, I.C.S. Normand

Child Health, Faculty of Medicine, Southampton General Hospital,
Southampton, UK

It is well recognized that immature human fetal lung undergoes a spontaneous apparent morphological maturation when maintained in vitro in an organ culture system [1, 2]. Characteristic changes include the appearance of cytosolic lamellar bodies within presumptive type II alveolar epithelial cells by 4 days in culture, together with the induction of surfactant apoproteins. We have recently demonstrated that morphological criteria of maturation are not necessarily matched in terms of biochemical parameters [3]; the fractional synthesis of dipalmitoylphosphatidylcholine (PC16:0/16:0) declined rather than increased with time in culture. Previous reports have described a stimulation of total culture phosphatidylcholine (PC) synthesis in response to a combination of glucocorticoid and thyroid hormones [4]. This result has been interpreted as evidence for biochemical maturation in culture. Consequently, we have investigated in more detail the effects of dexamethasone and tri-iodothyronine administration to organ cultures of human fetal lung.

Methods

Fetal human lung (15–18 weeks of gestation) from 13 therapeutic terminations was established in organ culture as described previously [3]. Cultures were maintained at 37 °C in a 5% CO_2 atmosphere for 4 days in serum-free RPMI 1640 medium in either the

[1] This work was supported by a project grant from the Children Nationwide Medical Research Fund. The assistance of the Electron Microscopy Unit, Southampton General Hospital with the morphometric analysis is gratefully acknowledged.

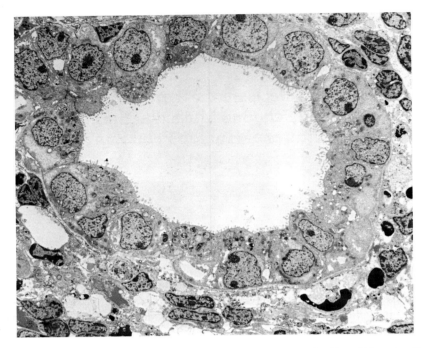

Fig. 1. The appearance of alveolar epithelial cells from human fetal lung (15 weeks of gestation) after serum-free organ culture for 4 days in the absence of hormonal additions. Original magnification $\times$ 870.

presence or absence of $10^{-6}\ M$ dexamethasone and $10^{-7}\ M$ tri-iodothyronine. PC synthesis was monitored by the incorporation of [methyl-^{14}C]choline Cl (400 nCi/ml) over 18 h, followed by chloroform:methanol extraction [5], isolation of a PC fraction on Bondelut NH$_2$ cartridges and HPLC resolution of individual molecular species of PC [3]. Eluted peaks were detected by an on-line dual-channel radioactivity monitor, using [^{3}H]PC16:0/16:0 as an internal standard to assess recovery. DNA was measured by an automated fluorescence-binding method [6]. Surfactant and residual fractions of cultures were prepared for PC analysis by density gradient centrifugation over a layer of 0.75 M sucrose. Sections of cultures were prepared for transmission electron microscopy [3].

Results

The characteristic accumulation of lamellar bodies within epithelial cells of human fetal lung after 4 days in organ culture is shown in figure 1. Exposure to dexamethasone and T$_3$ had no significant effect on this change

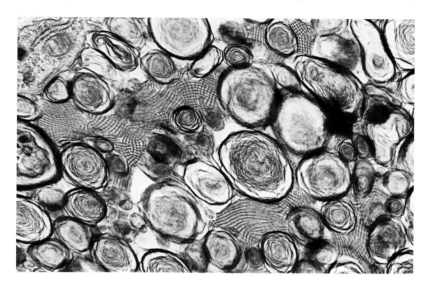

Fig. 2. Secreted lamellar body and tubular myelin material trapped within a lumen of an 'alveolar-like structure' from fetal human lung (17 weeks of gestation) maintained in organ culture for 4 days in the presence of added hormones. Original magnification × 8,700.

in culture. Morphometric analysis showed that there were 21.5 ± 3.0 and 16.4 ± 0.2 lamellar bodies in each type II cell section for control and hormone-treated cultures, respectively (n = 5 cultures). There appeared to be more secreted material within the hormone-treated cultures, but this was not quantified. This secreted material was present both as lamellar bodies and tubular myelin figures (fig. 2).

Hormone exposure for 4 days stimulated total PC synthesis by fetal human lung organ cultures by 346 ± 81% (n = 6) between days 4 and 5 in culture. This hormonal stimulation was not a specific effect on surfactant PC as the increase of [^{14}C]choline incorporation into PC of the surfactant fraction was similar to that of the whole cultures (326 ± 109%).

The distribution of incorporated [^{14}C]choline within individual molecular species of PC is outlined in table 1. There were no significant differences in this distribution between control and hormone-treated cultures. For each case PC16:0/18:1 represented the major synthesized species. The proportion of PC16:0/16:0 represented a considerable decrease from the value in vivo of 32.0 ± 1.2% [3]. Moreover, there was no enrich-

Table 1. Synthesis of molecular species of PC by organ cultures of human fetal lung

Molecular species	Relative retention time	PC synthesis %[^{14}C]choline incorporation (mean $\pm$ SEM)	
		control (n = 6)	hormone-treated (n = 6)
14:0/16:0	0.69	6.4 ± 0.7	9.8 ± 0.8
16:0/16:1	0.75	11.7 ± 0.6	13.6 ± 1.5
16:0/18:2	0.80	16.6 ± 1.4	14.8 ± 1.7
16:0/16:0	1.00	16.6 ± 1.7	18.6 ± 0.7
16:0/18:1	1.07	36.1 ± 3.2	31.3 ± 1.2
18:1/18:1	1.18	12.2 ± 0.4	11.8 ± 1.4

ment of PC16:0/16:0 in the surfactant fraction compared with the whole culture for either control or hormone-treated cultures (results not shown).

Conclusions

Dexamethasone and tri-iodothyronine at the high concentrations used in these experiments produced the expected stimulation over 4 days of total PC synthesis in organ cultures of immature human fetal lung. In contrast to previous reports [4], this hormone treatment caused no apparent enhancement of maturation in vitro in terms of morphological features. The reason for this discrepancy is unclear, but differences in culture medium might be implicated.

The analysis of PC synthesis in terms of individual molecular species shows clearly, however, that this hormonal treatment represented not a specific stimulation of surfactant phospholipid synthesis but a more general increase in total cellular PC. It is important to recognize that, under these experimental conditions in vitro, biochemical criteria of lung maturation do not automatically accompany the expression of morphological markers. This implies that the morphological development and the biochemical maturation of the lung must be controlled by distinct regulatory mechanisms, which are temporally co-ordinated during the phenotypic expression of type II alveolar epithelial cells in vivo.

References

1 Ekelund, L.; Arvidson, G.; Astedt, B.: Cortisol-induced accumulation of phospholipids in organ cultures of human fetal lung. Scand. J. clin. Lab. Invest. *35:* 419–423 (1975).

2 Snyder, J.M.; Johnston, J.M.; Mendelson, C.R.: Differentiation of type II cells of human fetal lung in vitro. Cell Tissue Res. *220:* 17–25 (1981).

3 Caesar, P.A.; Wilson, S.J.; Normand, I.C.S.; Postle, A.D.: A comparison of the specificity of phosphatidylcholine synthesis by human fetal lung maintained in either organ or organotypic culture. Biochem. J. *253:* 451–457 (1988).

4 Gonzales, L.; Ballard, P.L.; Ertsey, R.; Williamson, M.C.: Glucocorticoid and thyroid hormones stimulate biochemical and morphological differentiation of human fetal lung in organ culture. J. clin. endocr. Metab. *62:* 678–691 (1986).

5 Bligh, E.G.; Dyer, W.S.: A rapid method of total lipid extraction and purification. Can. J. Biochem. *37:* 911–923 (1959).

6 Sterzel, W.; Bedford, P.; Eisenbrand, G.: Automated determination of DNA using fluorochrome Hoechst 33258. Analyt. Biochem. *47:* 462–467 (1985).

A.D. Postle, PhD, Child Health, Level G, Centre Block,
Southampton General Hospital, Tremona Road, Southampton SO9 4XY (UK)

Wichert P von, Müller B (eds): Basic Research on Lung Surfactant.
Prog Respir Res. Basel, Karger, 1990, vol 25, pp 318–320

Intratracheal Instillation of Exogenous Surfactant Reduces Pulmonary Hyperoxic Injury[1]

Sadis Matalon[a], *Bruce A. Holm*[b]

[a] Departments of Anesthesiology, Physiology and Biophysics, University of
Alabama at Birmingham, Birmingham, Ala., USA; [b] Departments of Obstetrics
and Gynecology, State University of New York at Buffalo, N.Y., USA

Prolonged exposure to high concentrations of oxygen causes progressive damage to the mammalian blood-gas barrier including the different components of the pulmonary surfactant system. Rabbits returned to air after a sublethal exposure to 100% O_2 developed arterial hypoxemia, abnormal lung mechanics and pulmonary edema [1]. Since these physiological findings were consistent with the existence of a surfactant-deficient state, we investigated whether instillation of lung surfactant may reduce the appearance of lung injury in this model.

Material and Methods

Rabbits were exposed to 100% O_2 for 64 h in environmental chambers as previously described [1]. At this time, they were anesthetized and 125 mg of calf lung surfactant extract (CLSE) dissolved in 8 ml of sterile saline was instilled in their lungs. The rabbits were then returned to room air. Another 125 mg of CLSE were instilled 12 h later. At 24 h postexposure, we measured: (a) arterial blood gases and pH; (b) quasi-static pressure-volume curves; (c) levels of lavagable phospholipids, and (d) minimum surface tension of the bronchoalveolar lavage. All procedures have been previously described in detail [1, 2].

[1] This work was supported by NIH grant HL 31197 and a Career Investigator Award from the American and Alabama Lung Association.

Table 1. Phospholipid and minimum surface tension of bronchoalveoar lavage

	88 h air	64 h in 100% O_2 24 h in air	
	uninstilled	saline	surfactant
PL/BW, µmol/kg	8.7 ± 1.6	$4.3 \pm 0.2^+$	$34 \pm 4^{+*}$
T_{min}, dyn/cm	1 ± 0.2	$26 \pm 2^+$	$1 \pm 0.2^*$

Values are means $\pm$ 1 SE; each group consists of at least 5 animals. PL = Total phospholipid content of the bronchoalveolar lavage; BW = body weight; T_{min} = minimum surface tension of the bronchoalveolar lavage.
Significantly different from 88 h in air (column 1; $^+ p < 0.05$).
Significantly different from the saline value (column 2; $^* p < 0.05$).

Results and Discussion

The distribution of the instilled surfactant was assessed by mixing the instillate with a small amount of Evans blue. In all lobes, the distribution of surfactant was nonuniform with areas closest to the main bronchi receiving the highest concentrations.

Rabbits that breathed room air for 24 h after 64 h in O_2 had significantly lower values of lavagable phospholipids (PL) as compared to air controls (table 1). Furthermore, the minimum surface trension (T_{min}) of cell free-lavage after cycling in an oscillating bubble at 37 °C for 1 h was never lower than 25 dyn/cm. Instillation CLSE increased PL values and returned T_{min} to 1 dyn/cm, which was similar to the values found in air-breathing controls.

Figure 1 shows the deflation limbs of open-chested quasi-static pressure-volume curves in rabbits that received either CLSE or saline. Rabbits instilled with CLSE had normal lung mechanics. At this time PaO_2 values in these animals after breathing 100% O_2 for 15 min were 446 $\pm$ 31 vs. 197 $\pm$ 55 Torr in saline controls (mean $\pm$ 1 SEM; n = 8).

In summary, our results indicate that instillation of CLSE in rabbits exposed to hyperoxia prevents the decrease in lung volumes and the development of arterial hypoxemia. Increased levels of alveolar protein, present in the alveolar space due to hyperoxic injury to the alveolar epithelium [1, 2], interfere with normal surfactant function resulting in increased mini-

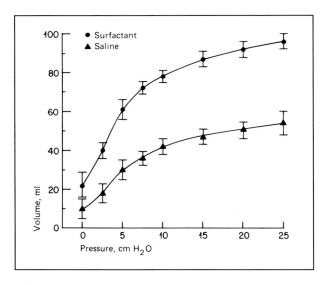

Fig. 1. Deflation limbs of quasi-static pressure-volume curves for surfactant (n = 7)-
and saline (n = 5)-instilled rabbits. Each curve was calculated by averaging the volume of
each animal at the indicated pressures. All rabbits breathed 100% O_2 for 64 h and air for
24 h. Values are means ± SE. From Matalon et al. [2], with permission.

mum surface tension. Instillation of CLSE increases lung phospholipid
levels, which helps overcome the inhibitory effects of proteins [3] and
returns minimum surface tension to its normal value.

References

1   Holm, B.A.; Notter, R.H.; Seigle, J.; Matalon, S.: Pulmonary physiological and sur-
     factant changes during injury and recovery from hyperoxia. J. appl. Physiol. *59:*
     1402–1409 (1985).
2   Matalon, S.; Holm, B.A.; Notter, R.H.: Mitigation of pulmonary hyperoxic injury by
     administration of exogenous surfactant. J. appl. Physiol. *62:* 756–761 (1987).
3   Holm, B.A.; Notter, R.H.; Finklestein, J.N.: Surface property changes from interac-
     tions of albumin with natural lung surfactant and extracted lung lipids. Chem. Phys.
     Lipids *38:* 287–298 (1985).

Sadis Matalon, PhD, Department of Anesthesiology,
University of Alabama at Birmingham, 619 South 19th Street,
Birmingham, AL 35233 (USA)

Wichert P von, Müller B (eds): Basic Research on Lung Surfactant.
Prog Respir Res. Basel, Karger, 1990, vol 25, pp 321–323

Acute Effects of Surfactant Therapy on Cerebral Circulation in Preterm Neonates

G. Jorch[a], *M. Garbe*[a], *H. Rabe*[a], *E. Michel*[a], *L. Gortner*[b]

Childrens Hospitals, Universities of [a] Münster and [b] Ulm, FRG

Since the introduction of natural surfactant into the treatment of hyaline membrane disease of the preterm infant there is a dramatic improvement in the course of this severe disease [1–3]. In particular, some studies have shown a positive effect on the neurological morbidity of preterm infants [4]. However, there are hardly any studies about the acute effects of intratracheal surfactant application on cerebral circulatory parameters.

Recently we described the application of transfontanellar pulsed Doppler ultrasonography to perform reproducible and noninvasive measurements of blood flow velocity in the internal carotid artery [5]. While participating in a multicentre clinical trial of the bovine surfactant SF-RI 1 [6], we used this method to assess the acute effects of intratracheally applied surfactant to the cerebral circulation in preterm infants. Since 'fluctuating pattern of blood flow velocity' has been regarded as a risk factor for intraventricular hemorrhage [7–9], the main aim of this study was to assess the variability of blood flow velocity following treatment. Additionally, we questioned whether there was a trend of blood flow velocity values in the first hour after treatment.

Material and Methods

In 10 preterm infants receiving prophylactic treatment with SF-RI 1, time-averaged maximum blood flow velocity of the right internal carotid artery (V_{max}), oscillometric mean arterial blood pressure (MABP; Dinamap, Criticon Tampa, Fla.), and transcutaneous pCO_2 ($tcpCO_2$) was measured before and every 10 min up to 60 min after intratracheal application of SF-RI 1. The study collective consisted of 10 preterm infants (3 male, 7 female) admitted to the neonatal intensive care unit of the University Children's Hospital of Münster between October 1987 and February 1988. Median gestational age was

27 weeks (range 24–29) and median birth weight was 848 g (range 710–1,380 g). Indications for surfactant treatment were gestatinal age 24–30 weeks, mechanical ventilation, and absence of congenital malformations.

The infants were intubated immediately after birth. SF-RI 1 is a liquid solution extracted from bovine lungs. 1.1 ml/kg of this solution (50 mg/kg BW dry substance) was instilled into the lower part of the trachea by a single bolus injection. Treatment was given within the first hour of life. All infants were monitored continuously by transcutaneous pO_2 and pCO_2 sensors (TCM 2/20 Radiometer, Copenhagen, Denmark), so that ventilatory settings could be adjusted precisely.

The study was approved by the ethical committee of our hospital. Informed consent was obtained from both parents prior to delivery.

Doppler-sonographic measurements of the right internal carotid artery were performed by using the range-gated pulsed Doppler ATL Mk 500 Duplex scanner (ATL, Bellevue, Wash.). We calculated the time-averaged maximum blood flow velocity (V_{max}) directly by dividing the area under the Doppler curve by the length of the cycle [5]. Previous studies showed that the variability of V_{max} of the internal carotid artery was dependent on the time interval between the measurements. CV was 7% with 15-second intervals [10] and 14% with 10-min intervals [unpubl. data]. Variability of the measured parameters during the first hour after treatment with SF-RI 1 was calculated as coefficient of variation (CV = SD/mean value · 100). CV was calculated for each patient. The mean CV was derived from these values. Analysis of variance (Friedman test) was used to test whether the observed changes of mean values were significant ($p < 0.05$). Mann-Whitney-Wilcoxon tests were performed for the $p < 0.05$ level where applicable.

Results

There was a great variability in the values for cerebral blood flow velocity. Mean CV of V_{max} within 60 min after intratracheal surfactant application was 19% (SD 7%, range 6–34%). Mainly due to this variability changes of V_{max} were not significant. However, there was a trend to an initial increase of this parameter (15.2–18.7 cm/s) and a subsequent decrease (to 13.4 cm/s) below the initial value. In an individual case V_{max} increased as much as 100% at 10 min after application. This was accompanied with an increase of pCO_2 of 14 mm Hg and of MABP of 14 mm Hg at the same time. In other cases a more stable pattern with changes of all parameters less than 10% was observed.

Testing of the variability of pCO_2 revealed a mean CV of 8% (SD ± 4%). Similar to V_{max}, an initial increase (47–50 mm Hg) was followed by a decrease below the initial level (43 mm Hg). This course was significant by the Friedman test for $p < 0.05$. CV of MABP was 9% (SD ± 3%). There was no significant course during the time of observation. Mean values were in the range of 34–38 mm Hg.

Conclusions

Surfactant-treated preterm infants recover more quickly from respiratory distres syndrome than those under conventional therapy. The increase of pCO_2 shortly after intratracheal application of natural surfactant is probably due to a transient increase of pulmonary resistance. We suppose that in single cases with improper adjustment of ventilatory parameters there may be the risk of inadvertent increase of pCO_2 and cerebral blood flow velocity. We recommend continuous transcutaneous measurements of pCO_2 to avoid this risk. Routine treatment regimes must be adjusted in order to deal with the changes in lung compliance and to improve the long-term outcome in cerebral morbidity.

References

1   Enhorning G: Surfactant can be supplemented before the neonate needs it. J Perinat Med 1987;15:479.

2   Notter RH, Egan EA, Kwong MS, et al: Lung surfactant replacement in premature lambs with extracted lipids from bovine lung lavage. Effects of dose, dispersion technique, and gestational age. Pediatr Res 1985;19:569.

3   Vidyasagar D, Shimada S: Pulmonary surfactant replacement in respiratory distress syndrome. Clin Perinatol 1987;14:991.

4   Notter H, Shapiro DL: Lung surfactants for replacement therapy. Biochemical, biophysical, and clinical aspects. Clin Perinatol 1987;14:433.

5   Jorch G, Pfannschmidt J, Rabe H: Die nicht invasive Untersuchung der intrazerebralen Zirkulation bei Früh- und Neugeborenen mit der gepulsten Dopplersonographie. Monatsschr. Kinderheilkd 1986;134:804.

6   Gortner L: Effects of bovine surfactant (SF-RI 1) on RDS in very premature infants. Special Ross Conference, Washington, Abstract book, 1988.

7   Van Bel F, van de Bor M, Stijnen T, et al: Cerebral blood flow velocity pattern in healthy and asphyxiated newborns. A controlled study. Eur J Pediatr 1987;146:461.

8   Hill A, Perlman JM, Volpe JJ: Relationship of pneumothorax to occurrence of intraventricular hemorrhage in the premature newborn. Pediatrics 1982;69:144.

9   Perlman JM, McMenamin JB, Volpe JJ: Fluctuating cerebral blood flow velocity in respiratory distress syndrome. Relationship to the development of intraventricular hemorrhage. N Engl J Med 1983;309:209.

10 Mirro R, Busija D, Green R, et al: Relationship between mean airway pressure, cardiac output, and organ blood flow with normal and decreased respiratory compliance. J Pediatr 1987;111:101.

Priv.-Doz. Dr. G. Jorch, Universitäts-Kinderklinik, Albert-Schweitzer-Strasse 33, D–4400 Münster (FRG)

Wichert P von, Müller B (eds): Basic Research on Lung Surfactant.
Prog Respir Res. Basel, Karger, 1990, vol 25, pp 324–328

Pulmonary Surfactant Enhances the Surface Phagocytosis of *Staphylococcus aureus* by Rat Alveolar Macrophages

J.F. van Iwaarden[a, b], *A. Welmers*[b], *J. Verhoef*[b], *L.M.G. van Golde*[a]

[a] Laboratory of Veterinary Biochemistry and [b] Laboratory of Microbiology, University of Utrecht, The Netherlands

Pulmonary surfactant forms the inner lining material of the alveoli. It consists of lipids, proteins and carbohydrates. One of its major functions is the prevention of lung collapse at end-expiration. Another important function of pulmonary surfactant may be a role in the host defense system of the lung.

In 1973, Laforce et al. [1] showed that pulmonary surfactant enhanced the killing of *Staphylococcus aureus* by alveolar macrophages; the cells which form the first line of defense of the lung against invading microorganisms [2, 3]. Since then several studies have evolved ascribing either a positive [4, 5], negative [6] or no effect [7] of surfactant on the function of alveolar macrophages. To reevaluate the role of surfactant in the phagocytosis of bacteria by alveolar macrophages, we studied the phagocytosis of bacteria *(S. aureus)* by rat alveolar macrophages on a surface in the presence or absence of rat lung surfactant, as the phagocytosis of bacteria by alveolar macrophages, in vivo, will probably take place on a surface.

Methods

Isolation of Alveolar Macrophages and Pulmonary Surfactant
Rat lungs (male, Wistar, 200 g) were lavaged via the trachea with sterile PBS (10 mM phosphate, 137 mM NaCl pH 7.4; 8 × 10 ml/rat). The lavage fluid was centrifuged (132 g; 15 min, 4 °C). The pellet contained predominantly macrophages (> 95%). The macro-

phages were resuspended in Hank's balanced salt solution supplemented with 0.1 % gelatine (w/v) (Gel HBBS). For the isolation of pulmonary surfactant the same procedure was used as for the isolation of the alveolar macrophages. After removal of the macrophages the supernatant of the lavage fluid was centrifuged at 50,000 g for 60 min at 4 °C. The pellets (pulmonary surfactant) derived from the lavage material of 40 rats were pooled and resuspended with PBS (final volume 13 ml).

Surface Phagocytosis Assay

S. aureus were metabolically labeled with ^{3}H-thymidine (overnight; 37 °C) and diluted to 2×10^8/ml Gel HBSS. The bacteria were transferred to a 96-well tissue culture plate (50 µl/well) and centrifuged for 1 min at 1,600 g and 4 °C. Alveolar macrophages (2×10^6/ml Gel HBSS) were added to the wells (50 µl/well) and centrifuged on top of the bacteria. The incubation of the bacteria with the macrophages was initiated by transferring the 96-well tissue plate to a 37 °C incubator. After the indicated time intervals the incubation was terminated by placing the 96-well tissue culture plate on ice and by adding 150 µl of ice-cold PBS to the wells. The bacteria and macrophages were removed from the bottom of the wells with ice-cold PBS containing 1 mM EDTA, and transferred to plastic vials (biovials). By centrifugation the macrophage-associated bacteria were separated from the free bacteria (4 ×, 132 g, 5 min, 4 °C). The pellet which consisted of macrophage-associated bacteria was counted in a β-counter. The nonspecific background was measured in each experiment and determined by the amount of radioactivity detected in incubations which contained only bacteria and no macrophages. The nonspecific background was subtracted from the total radioactivity of the incubations containing macrophages and bacteria. Phagocytosis was expressed as the number of bacteria found associated per macrophage.

Results

Effect of Surfactant on the Surface Phagocytosis of S. aureus by Alveolar Macrophages

Preincubation (opsonization) of the bacteria with rat serum is a prerequisite for phagocytosis of the bacteria by alveolar macrophages (fig. 1). No phagocytosis of the bacteria occurred when the bacteria were opsonized with either buffer or surfactant. However, preincubation of the alveolar macrophages with surfactant significantly enhanced the phagocytosis of opsonized bacteria, by approximately 70%, compared to the phagocytosis of opsonized bacteria by nonsurfactant-treated macrophages (fig. 1).

The phagocytosis of opsonized bacteria by macrophages is a time-dependent process (fig. 2). In the first 5 min of the incubation only a few bacteria are found associated with the macrophages. After 30 min the phagocytosis of bacteria by the macrophages has reached a maximum level and remains at that plateau even after 60 min of incubation. No cell-asso-

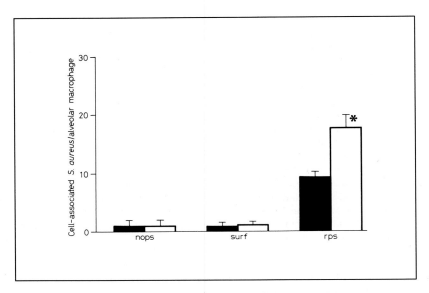

Fig. 1. The influence of opsonization of bacteria and preincubation of the alveolar macrophages on the surface phagocytosis. *S. aureus* (2×10^8/ml; 5 ml final volume) were opsonized (preincubated) for 30 min at 37 °C with buffer (nops), 10% (v/v) rat surfactant (surf) or 5% (v/v) rat pooled serum (rps) prior to the incubation (30 min, 37 °C) with alveolar macrophages. The alveolar macrophages were preincubated with buffer (black bars) or with 10% (v/v) surfactant (white bars) for 30 min at 37 °C (2×10^6/ml; 0.5 ml final volume) prior to the incubation with opsonized *S. aureus*. Significantly different from the control; * $p < 0.025$; n = 5.

ciated bacteria were detected when the incubation of the rat serum opsonized bacteria with the alveolar macrophages was performed at 0 °C (results not shown).

Discussion

This study demonstrates that *S. aureus* opsonized with rat serum can be readily phagocytozed by rat alveolar macrophages on a surface. Opsonization of bacteria with rat pooled serum is a prerequisite for phagocytosis of the bacteria by rat alveolar macrophages, as preincubation or 'opsonization' of bacteria with either buffer or surfactant does not lead to phagocytosis by alveolar macrophages. The phagocytosis of opsonized *S. aureus*

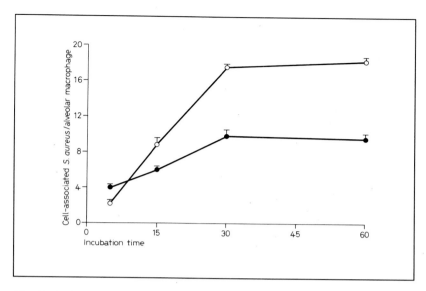

Fig. 2. Time course of the surface phagocytosis of rat serum opsonized *S. aureus* by alveolar macrophages. *S. aureus* opsonized with rat pooled serum were incubated with alveolar macrophages preincubated with buffer (•) or surfactant (o) (see legend to fig. 1) for the time indicated at 37 °C.

by alveolar macrophages is a time-dependent process reaching a maximum after 30 min of incubation of the bacteria with the macrophages. No phagocytosis of opsonized bacteria occurred when the bacteria were incubated with the alveolar macrophages at 0 °C (results not shown). This indicates that the opsonized bacteria found associated with the macrophages have been internalized by the macrophages and are not bacteria bound to the outer membrane of the macrophages, because cells do not take up substances at 0 °C. Preincubation of the alveolar macrophages with pulmonary surfactant enhances the phagocytosis of opsonized bacteria by macrophages, approximately 70%, compared to the phagocytosis of opsonized *S. aureus* by nonsurfactant-treated macrophages.

Pulmonary surfactant only enhanced the surface phagocytosis of *S. aureus* by alveolar macrophages when the bacteria are opsonized with rat serum, indicating that surfactant may influence the IgG or complement-mediated phagocytosis. Whether surfactant influences the affinity, number of IgG and complement receptors present on alveolar macrophages or acts on the ingestion process itself, is not clear at the moment. IgG and complement factors were reported to be present in lung lavage fluid [8], so opsonization of bacteria by these factors may occur in vivo. Hence, the surface phagocytosis of opsonized bacteria by alveolar macrophages, which can be further enhanced by pretreatment of the macrophages with surfactant, may play an important role in the host defense system of the lung.

Acknowledgement

This research was supported by the Dutch Asthma Foundation (Nederlands Astma Fonds).

References

1   LaForce, F.M.; Kelley, W.J.; Hubert, G.L.: Inactivation of staphylococci by alveolar macrophages with preliminary observations on the importance of alveolar lining material. Am. Rev. resp. Dis. *108:* 784–790 (1973).

2   Green, G.M.; Kass, E.H.: The role of the alveolar macrophage in the clearance of bacteria from the lung. J. exp. Med. *119:* 167–175 (1964).

3   Hocking, W.G.; Golde, D.W.: The pulmonary alveolar macrophage. New Engl. J. Med. *301:* 580–587 (1979).

4   Juers, J.A.; Rogers, R.M.; McCurdy, J.B.; Cook, W.W.: Enhancement of bacterial capacity of alveolar macrophages by human alveolar lining material. J. clin. Invest. *58:* 271–275 (1976).

5   O'Neill, S.; Lesperance, E.; Klass, J.: Rat lung lavage surfactant enhances bacterial phagocytosis and intracellular killing by alveolar macrophages. Am. Rev. resp. Dis. *130:* 225–230 (1984).

6   Coonrod, D.J.; Jarrells, M.C.; Yoneda, K.: Effect of rat surfactant lipids on complement and Fc receptors of macrophages. Infect. Immunity *54:* 371–378 (1986).

7   Jonsson, S.; Musher, D.M.; Goree, A.; Lawrence, E.C.: Human alveolar lining material and antibacterial defenses. Am. Rev. resp. Dis. *133:* 136–140 (1986).

8   Bell, D.Y.; Haseman, J.A.; Spock, A.; McLennan, G.; Hook, G.E.R.: Plasma proteins of the bronchoalveolar surface of the lungs of smokers and nonsmokers. Am. Rev. resp. Dis. *124:* 72–79 (1981).

J.F. van Iwaarden, PhD, Laboratory of Veterinary Biochemistry, PO Box 80.176, NL–3508 TD Utrecht (The Netherlands)

Wichert P von, Müller B (eds): Basic Research on Lung Surfactant.
Prog Respir Res. Basel, Karger, 1990, vol 25, pp 329–332

Ambroxol Reduces Paraquat Toxicity in the Rat[1]

M. Donnini[a], *M. Luisetti*[a], *L. Diomede*[b], *P.D. Piccioni*[a], *G. Gualtieri*[a], *E. Pozzi*[c], *M. Salmona*[b]

[a]Istituto di Tisiologia e Malattie Apparato Respiratorio IRCCS San Matteo, Università di Pavia; [b]Laboratorio di Enzimologia, Istituto di Ricerche Farmacologiche M. Negri, Milano;
[c]Cattedra di Fisiopatologia Respiratoria, Università di Torino, Italia

Paraquat (1,1'-dimethyl-4,4'-dipyridilium) is a widely used herbicide which causes acute (ARDS) and chronic (diffuse fibrosis) lung damage. There is evidence that the mechanism of lung injury may involve the generation of oxygen-free radicals [1] and that alveolar type 2 cells are specific targets of paraquat toxicity [2]. Reduced surfactant (a secretory product of alveolar type 2 cells) activity 24 h after injection of a sublethal dose of paraquat has been described [3]. This could be likely ascribed to peroxidation of type 2 cell membranes. Since ambroxol (a bromhexine derivative) is an agent able to induce the synthesis of surfactant from alveolar type 2 cells [4], we performed a preliminary study on its possible protective effect on the survival of rats injected with a lethal dose of paraquat.

Materials and Methods

In all experiments male CD-COBS rats (Charles River, Italy) with initial weight 175–200 g were used. Animals were administered paraquat dichloride (ICI, UK) i.p. at different dosages (see below). Ambroxol (Istituto De Angeli, Italy) was given orally at

[1] This work has been partially supported by a grant of the Italian National Research Council, contract number 87.01311.44.

Table 1. Mean survival time and death rate of rats treated with different doses of paraquat

Number of rats	Paraquat dosage mg/kg BW	Death rate %	Mean survival h
15	10	–	168
15	20	–	168
15	30	33.3	144.7
40	35	71.8	84.4
15	40	80	65.3

different dosages and schedules (see below). Controls received placebo at the same schedule. The survival rate (%) was checked daily. The mean survival (h) as well as the ambroxol-protected/control mean survival time ratio (T/C) were calculated.

Results

Table 1 shows the mean survival time and the death rate of paraquat-treated rats. On the basis of these data, we chose to treat the rats with paraquat 35 mg/kg BW single dose in protection experiments. Figure 1 summarizes the effects of ambroxol on paraquat-treated rats. The survival rate (fig. 1) was significantly enhanced by 3-day ambroxol pretreatment (50 mg/kg orally daily), ambroxol administered orally (50 mg/kg) within 30 min following paraquat, and ambroxol administered orally (50 mg/kg) within 2 h following paraquat. Ambroxol was continued for a further 7 days after paraquat treatment in all groups. The best mean survival time ratio (T/C) was observed in the 3-day ambroxol pretreatment group (164.6), followed by the group of rats treated with ambroxol within 30 min (143.7). The ambroxol administered within 2 h had no effect on this parameter (115.6).

Conclusions

On the basis of these results, ambroxol seems to display a protective effect on acute paraquat poisoning in rats. Each ambroxol administration protocol had a significant effect on survival rate, while the mean survival

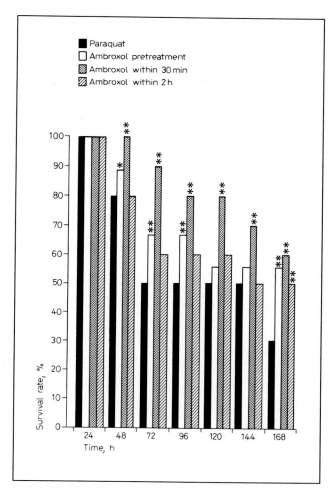

Fig. 1. Effect of ambroxol protection on the survival rate of paraquat-treated rats.
* p < 0.05; ** p < 0.01.

time ratio was significantly increased when ambroxol was administered beforehand or immediately following paraquat poisoning. It is likely to suppose that the protective effect of ambroxol could depend on the prevention of the pulmonary surfactant depletion induced by acute paraquat poisoning. Morphologic and biochemical studies are now in progress to support such a hypothesis.

References

1   Aldrich, T.K.; Fischer, A.B.; Cadens, E.; Chance, B.: Evidence for lipid peroxidation by paraquat in the perfused rat lung. J. Lab. clin. Med. *101:* 66–73 (1983).
2   Skillrud, D.M.; Martin, W.J., II: Paraquat-induced injury of type II alveolar cells. An in vitro model of oxidant injury. Am. Rev. resp. Dis. *129:* 995–999 (1984).
3   Lachmann, B.; Danzmann, E.: Adult respiratory distress syndrome; in Robertson, Van Golde, Batenburg, Pulmonary surfactant, pp. 505–548 (Elsevier, Amsterdam 1984).
4   Post, M.; Batenburg, J.J.; Schuurmans, E.A.J.M.; Oldenborg, V.; Molen, A.J. van der; Golde, L.G.M. van: The perfused rat lung as a model for studies on the formation of surfactant and the effect of ambroxol on this process. Lung *161:* 349–359 (1983).

M. Luisetti, MD, Istituto di Tisiologia, IRCCS San Matteo, via Taramelli 5, I-27100 Pavia (Italy)

Wichert P von, Müller B (eds): Basic Research on Lung Surfactant.
Prog Respir Res. Basel, Karger, 1990, vol 25, pp 333–337

Clearance of Vasoactive Intestinal Polypeptide by the Isolated, Ventilated and Perfused Rat Lung

W. Bernhard, P. v. Wichert

Medizinische Poliklinik of the Zentrum für Innere Medizin,
Philipps-Universität Marburg, BRD

Vasoactive intestinal peptide (VIP) has been shown to be degraded by the lung in vivo after intravenous application [Humphrey et al., 1979]. Receptors have been shown in the pulmonary capillary endothelium [Barrowcliffe et al., 1986]. It is unknown whether binding by these receptors is only a part of the degradation process or whether VIP exerts any biological effect on the tissue by the receptors. We investigated the influence of VIP concentration and of the lysosomal inhibitor chloroquine on the pulmonary clearance of VIP as a basis for the revelation of these putative effects.

Materials and Methods

Lungs of male Wistar rats of 320 g body weight were isolated and perfused as described elsewhere [Bernhard et al., 1989]. Portions of 18 kBq iodinated VIP (^{125}I-VIP) were supplemented with unlabeled VIP in siliconized glass vials to give the requested start concentrations of 10^{-12} to 10^{-7} mol/l in 100 ml recirculating perfusate. ^{125}I-VIP was added to the perfusate after 40 min equilibration time and recirculated for 2 min for equal distribution in the absence of lung perfusion. If chloroquine was used, it was added 30 min prior to ^{125}I-VIP.

Samples of 500 μl perfusate were taken from overflow and lung effluent at 1- to 30-min intervals and counted in a gamma-counter. ^{125}I-VIP was separated from iodinated degradation products by adding 500 μl of 10% trichloroacetic acid (TCA) and centrifugation for 15 min at 1,850 g at 4 °C [Bjoro et al., 1978]. ^{125}I-labeled degradation products in the supernatant were counted and the label subtracted from total ^{125}I radioactivity. The TCA precipitation method was calibrated using specific antibody binding of the ^{125}I label by VIP antibodies in excess.

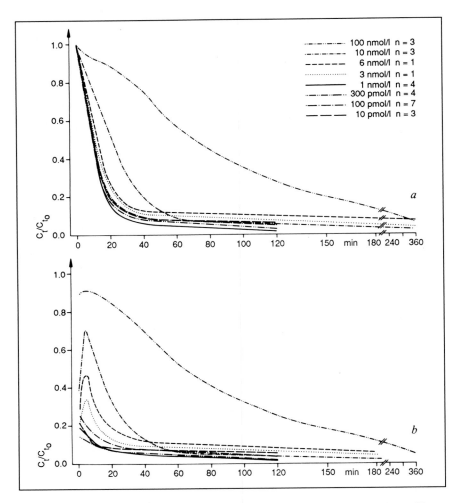

Fig. 1. Decrease of VIP concentration (C_t/C_{t_0}) in total perfusate (*a*) and lung effluent (*b*) as a function of time (t) and start concentration ($_{t_0}$). Lines for different C_{t_0} are explained in the upper part of the figure and are means of the advised numbers of experiments (n).

The time course of total ^{125}I label in the perfusate was calculated by dividing the label at any time through the initial label (cps_t/cps_{t_0}). VIP concentration was calculated as shown by the following equation: $[VIP]_t = [VIP]_{t_0} \times (cps_{ab+}/cps_{t_0})$; [VIP] = VIP concentration, cps_{ab+} = counts per second of ^{125}I specifically bound to anti-VIP antibodies). Statistics were done using the U test of Wilcoxon, Mann and Whitney.

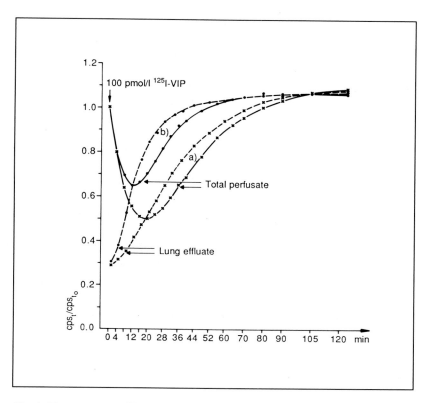

Fig. 2. Time course of ^{125}I radioactivity (^{125}I-VIP + ^{125}I-labeled degradation products) in total perfusate and lung effluent of experiments done in the presence (a; n=6) or absence b; n=7) of 70 µmol/l chloroquine. Lines are means of the experimental data. Time delay of the reincrease of ^{125}I-radioactivity in total perfusate of 12 min (without chloroquine) and 20 min (with chloroquine) was different with α=0.001.

Results

The concentration of VIP decreased quickly in the perfusate with a $t_{1/2}$ of 12 min at concentrations up to 1 nmol/l. 80% were cleared by a single lung passage. At higher concentrations more intact VIP left the lungs and $t_{1/2}$ increased to 70 min (fig. 1a, b). The degradation was accompanied by an uptake of ^{125}I-VIP by the lung, causing a decrease of the ^{125}I label of the perfusate, and a delayed extrusion of the degradation products into the perfusate, causing a reincrease of ^{125}I in the perfusate. Chloroquine retarded the extrusion of the degraded ^{125}I-VIP, which resulted in a stronger initial decrease of the ^{125}I label of the perfusate (fig. 2).

Discussion

The degradation of a molecule may be characterized by a three-step event: (1) binding; (2) cleavage; (3) release of cleavage products into the surrounding medium. Humphrey et al. [1979] could show that VIP accumulated in the perialveolar capillary endothelium but was not bound to endothelial cells of greater pulmonary vessels, implicating specific binding sites for VIP in these capillaries. The saturability of the extraction of VIP is in confidence with the existence of such specific receptors. The fact that the lungs were able to degrade the VIP of 100 ml of a 100-nM solution, i.e. 10^4-fold the physiological level, with a $t_{1/2}$ of 70 min may lead to the conclusion of a very high capacity of this mechanism. But quantitative transduction to in vivo conditions may be wrong. In VIPoma disease the lung is not able to extract all the VIP from the circulation to prevent a patient from systemic VIPergic effects. Nevertheless, at a concentration of 10 pmol/l VIP was quickly cleared from the perfusate implicating a very low dissociation constant of the endothelial receptors. The observations that (1) VIP is taken up by the lungs, (2) cleavage products are extruded into the perfusate with a time delay, and (3) the lysosomal inhibitor chloroquine [Lie and Schofield, 1973] elongated the delay time gives evidence that cleavage takes place within the cells and that lysosomes are involved in this process. So we conclude that VIP may be degraded by the lung capillary endothelial cells by a mechanism involving receptor-mediated endocytosis, lysosomal degradation and subsequent exocytosis of the cleavage products.

References

Barrowcliffe, M.P.; Morice, A.; Jones, J.G.; Sever, P.S.: Pulmonary clearance of vasoactive intestinal peptide. Thorax *41:* 88–93 (1986).

Bernhard, W.; Müller, B.; v. Wichert, P.: Effect of β-mimetic drugs on phospholipid synthesis in the isolated lung. Proc. 3rd Symp. Basic Research on Lung Surfactant, Marburg 1988 (Karger, Basel 1989).

Bjoro, T.; Wilk, P.; Opstad, P.K.; Gautvik, K.M.; Haug, E.: Binding and degradation of vasoactive intestinal peptide in prolactin-producing cultured rat pituitary tumor cells (GH$_4$C$_1$). Acta physiol. scand. *130:* 609–618 (1978).

Humphrey, C.S.; Murray, P.; Ebeid, A.M.; Fischer, J.E.: Hepatic and pulmonary clearance of vasoactive intestinal peptide in the rat. Gastroenterology *77:* 55–60 (1979).

Lie, S.O.; Schofield, B.: Inactivation of lysosomal function in normal cultured human fibroblasts by chloroquine. Biochem. Pharmac. *22:* 3109–3114 (1979).

Wolfgang Bernhard, Dr. rer. physiol., Medizinische Poliklinik,
Zentrum für Innere Medizin der Philipps-Universität Marburg, Baldingerstrasse,
D–3550 Marburg (FRG)

Wichert P von, Müller B (eds): Basic Research on Lung Surfactant.
Prog Respir Res. Basel, Karger, 1990, vol 25, pp 338–342

Biphasic Response of Dipalmitoyl Phosphatidylcholine Content to Respiratory Distress Induced by Oleic Acid[1]

C. Casals[a], *L. Herrera*[a], *E. Miguel*[a], *P. García-Barreno*[b], *A.M. Municio*[a]

[a] Department of Biochemistry, Faculty of Chemistry, Universidad Complutense;
[b] Department of Experimental Medicine, Hospital Gregarious Marañón,
Madrid, Spain

The present study was undertaken to determine the alteration of the acyl species composition of phosphatidylcholine (PC) in both lamellar bodies and purified surfactant (free of blood contamination) from rabbits under oleic acid-induced acute respiratory distress syndrome (ARDS). Since the turnover of surfactant is around 10 h in adult rabbits [1], the work used rabbits sacrificed at intervals of 12 and 24 h after oleic acid infusion. The data presented allow one to determine whether type II cells respond to injury altering the particular pattern of molecular species of surfactant PC.

Materials and Methods

Oleic Acid-Induced Injury

Prior to oleic acid administration rabbits were anesthetized with intramuscular injection of ketamine (20–30 mg/kg). Lung injury was induced by intravenous infusion of pure oleic acid via the ear vein. The oleic acid (100 μl/kg) was administered over 30 min with a constant flow of normal saline. The lungs were removed 12 or 24 h after oleic acid infusion. Arterial blood gases and pH were monitored (Gas Check 938 AVL) at the beginning and the end of the experiment. The animal group under ARDS induced by oleic acid showed hypoxemia, acidosis, congested and hyperemic lungs, and hemorrhagic bronchoalveolar lavage fluids.

[1] These investigations were supported by the Grant 0490-84 from Comisión Asesora de Investigación Científica y Técnica.

Isolation of Surfactant and Lamellar Bodies

Surfactant was obtained from rabbit lung lavage fluid as described in Schlame et al. [2]. To separate the surfactant from blood components transudated into the air spaces, the surfactant, resuspended in 16% NaBr, is placed beneath a two-layer discontinuous gradient of NaBr as described by Shelley et al. [3]. The surfactant has a density of about 1.085 at 4 °C which is lower than that of most of the blood-contaminating components.

Lamellar bodies were isolated from postlavage lung tissue as previously described [2].

Lipid Analysis

For analysis of the acyl species pattern, PC was isolated as described in Schlame et al. [2] and hydrolyzed to diacylglycerol (DG) using phospholipase C from *Clostridium welchii* [4]. DGs were derived with α-naphthylisocianate as previously described [4, 5]. Separation of molecular species of DG naphthylurethanes was carried out on ultrasphere ODS (5 μm) HPLC columns using a Beckman Model 163 equipped with a variable wavelength detector. The solvent system was a mixture of methanol-water (80:20) (solvent A) and methanol (solvent B). A convex gradient was run from 70%A to 15%A between 0 and 30 min. After that an isocratic elution with 15%A was run for 60 min. The flow rate was 0.7 ml/min and the column was set at 60 °C. Detection was achieved by UV absorption at 290 nm. HPLC peaks were identified by gas chromatography analysis of the fatty acid methyl esters using a Wcot capillary column (30 m × 0.25 mm ID) coated with Supelco-wax-10 stationary phase 0.25 μm thick. The column temperature was 185 °C for 32 min. Afterwards, a thermal gradient was used from 185 °C up to 200 °C at a rate of 2 °C/min.

Results and Discussion

The molecular species of PC from lamellar bodies and surfactant from control rabbits and rabbits under ARDS (12 and 24 h after oleic acid infusion) are listed in table 1. The resolution of the acyl species derived from PC is shown in the HPLC profile (fig. 1).

The species composition of PC from lamellar bodies and surfactant were very similar containing mainly 16:0 species which confirmed previous results from rat lung [2]. Dipalmitoyl PC (16:0/16:0-PC) responded to ARDS in two phases: (a) a marked increase of the 16:0/16:0 percentage of lamellar body and surfactant PC 12 h after ARDS induction, and (b) a reversion of this enhancement in lamellar bodies to a lower level than control 24 h after ARDS induction.

The increase of the 16:0/16:0 proportion of PC 12 h after injury induction might be due to the increase of de novo synthesis and/or acyl chain remodelling at the endoplasmic reticulum (ER). However, it is not well understood how PC is transferred from ER to the lamellar bodies. We have shown that lamellar bodies contain more 16:0/16:0-PC and less 16:0/18:1-

Table 1. Molecular acyl species of PC: molecular acyl species are arranged by increasing retention times of their respective diacylnaphthyl urethanes upon HPLC separation (% distribution given as mean of n experiments)

Molecular species	Peak No.	Lamellar bodies			Surfactant		
		control (n = 3)	12 h (n = 3)	24 h (n = 2)	control (n = 3)	12 h (n = 3)	24 h (n = 2)
14:0–16:1 14:0–16:0 18:0–20:4 18:2–22:6	1	1.65	1.74	3.18	2.12	1.76	3.39
18:2–28:2	2	3.76	4.62	1.22	6.04	4.63	2.90
16:0–16:1	3	12.06	8.36	9.96	11.26	10.44	10.30
16:0–22:6 16:0–20:4 16:0–18:3	4	1.70	0.69	2.30	1.69	1.09	1.30
16:0–18:2	5	13.73	7.07	14.57	13.62	9.82	9.25
18:1–18:2 16:0–22:5 16:0–20:2	6	4.98	3.10	4.10	6.07	2.28	1.82
16:0–16:0	7	34.06	53.22	27.14	34.71	44.11	40.79
16:0–18:1	8	15.94	13.79	26.00	15.20	14.75	18.53
18:1–18:1 18:0–22:6	9	2.18	1.35	3.12	2.10	1.22	1.26
18:0–20:4	10	3.83	1.56	3.19	3.55	1.79	1.02
18:0–18:2	11	0.37	0.78	0.23	0.32	0.50	0.12
18:0–22:4	12	0.42	0.56	0.34	0.34	0.95	0.13
18:0–20:2	13	2.21	1.71	2.03	1.50	1.97	1.10
16:0–18:0	14	0.76	0.49	1.47	0.65	1.39	0.52
18:0–18:1	15	1.55	1.02	0.50	0.19	1.19	0.70

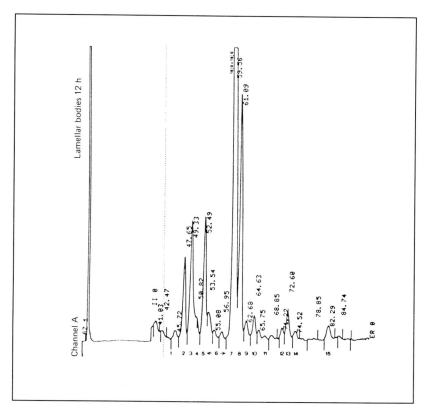

Fig. 1. Chromatogram of PC molecular species from lamellar bodies from rabbits under oleic acid-induced ARDS. Retention times (min) are given at the top of the peaks, which refer to species listed in table 1.

PC than other pneumocyte membranes in rat lung [2]. Therefore, the subcellular processing resulting in assembly of lamellar bodies was thought to carry out the acyl species selection in order to enrich 16:0/16:0 at the expense of 16:0/18:1. Lecerf et al. [6] reported the localization of a PC:ceramide choline phosphotransferase activity in lamellar bodies. This enzyme seems to have a substrate preference of 16:0/18:1-PC over 16:0/16:0-PC as the phosphocholine donor. We have found that the sphingomyelin percentage of total lamellar body phospholipids increases 2.5-fold 12 h after ARDS induction from 2.5 ± 0.6 in the control group (n = 11) to 6.3 ± 1.5

in the ARDS group (n = 5). Since the first period of ARDS induced by oleic acid is characterized by both an increase of 16:0/16:0-PC and of sphingo-myelin, this phosphocholine transferase enzyme activity might be involved in the cellular response to the pathological condition.

On the other hand, the late decrease of the 16:0/16:0 percentage of lamellar body PC (24 h after ARDS induction) could be the result of reup-take of nonsurfactant PC from edema material and cellular debris which fills the injured alveoli.

References

1   Baritussio, A.G.; Magoon, M.W.; Goerke, J.; Clements, J.A.: Precursor product rela-tionship between rabbit type II cell lamellar bodies and alveolar surface active mate-rial. Surfactant turnover time. Biochim. biophys Acta *666:* 382–393 (1981).
2   Schlame, M.; Casals, C.; Rustow, B.; Rabe, H.; Kunze, D.: Molecular species of phosphatidylcholine and phosphatidylglycerol in rat lung surfactant and different pools of pneumocytes type II. Biochem. J. *253:* 209–215 (1988).
3   Shelley, S.A.; Paciga, J.E.; Balis, J.U.: Purification of surfactant from lung washings and washings contaminated with blood constituents. Lipids *12:* 505–510 (1977).
4   Kruger, J.; Rabe, H.; Reichmann, G.; Rustow, B.: Separation and determination of diacylglycerols as their naphthylurethanes by high-performance liquid chromatogra-phy. J. Chromatogr. *307:* 387–392 (1984).
5   Schlame, M.; Rustow, B.; Kunze, D.; Rabe, H.; Reichmann, G.: Phosphatidylglyc-erol of rat lung. Intracellular sites of formation de novo and acyl species pattern in mitochondria, microsomes and surfactant. Biochem. J. *240:* 247–252 (1986).
6   Lecerf, J.; Fouilland, L.; Gagniarre, J.: Evidence for a high activity of sphingomyelin biosynthesis by phosphocholine transfer from phosphatidylcholine to ceramides in lung lamellar bodies. Biochim. biophys. Acta *919:* 48–59 (1987).

C. Casals, MD, Department of Biochemistry, Faculty of Chemistry,
Universidad Complutense, E–28040 Madrid (Spain)

Wichert P von, Müller B (eds): Basic Research on Lung Surfactant.
Prog Respir Res. Basel, Karger, 1990, vol 25, pp 343–346

Structure-Function Relationships of Surfactant Proteins SP-B and SP-C

A. Waring, B. Fan, T. Nguyen, J. Amirkhanian, W. Taeusch

Department of Pediatrics, King/Drew Medical Center, Los Angeles, Calif., USA

SP-B and SP-C are low molecular weight hydrophobic proteins that are unique to pulmonary surfactant [1]. While the sequences of these proteins have been identified, their secondary and tertiary structures, and the mechanism by which they interact with lipids are largely unknown. Hydrophobic surfactant proteins from bovine lung forms three bands on SDS-polyacrylamide gels under nonreducing conditions of 26, 14, and 5 kdaltons. Direct sequencing of these peptides after transfer to polyvinyline difluoride membranes show that the 26-kdalton peptide is SP-B, the 14-kdalton band is SP-C, and the 5-kdalton band is SP-C. These findings are in general agreement with Hawgood et al. [2] who carried out sequencing after electroeluting surfactant proteins from polyacrylamide gels. These findings are in contrast to earlier preliminary reports by us [3] and others [4] based on sequencing of fractions after column chromatography.

Computer-assisted analysis of amino acid sequences of these proteins reveals unique structural features. By using the hydrophobic moment plot computer program of Eisenberg [5], we find that SP-B resembles a membrane surface-seeking protein having several sequences with amphipathic helixes. In contrast, SP-C has a large segment that resembles a transmembrane anchoring protein, a domain similar to those for signal peptides in eucaryotic and procaryotic organisms [6]. Edmundson helical wheel representations of an amphipathic segment of SP-B and the transmembrane segment of SP-C emphasize the differences between these proteins (fig. 1).

The disposition and redox state of the cysteine residues in SP-B and SP-C may also have an important role in the function of these proteins. SP-B has cysteine residues over the entire length of the protein while SP-C

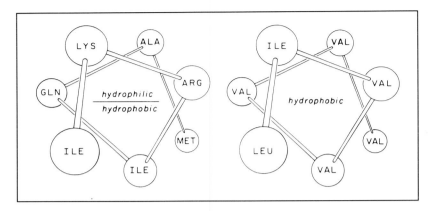

Fig. 1. Edmundson helical wheel diagrams of low molecular weight protein segments from SP-B and SP-C are shown. Amino acid residues 15–21 for the first amphipathic sequence of SP-B are represented in the left panel. The hydrophobic amino acid sequence (residues 20–26) of SP-C is diagrammed on the right.

has a vicinal pair of cysteine residues near its N-terminus. By titration with SH-sensitive reagents, we infer that alveolar fluid lacks reductants so that proteins in that environment are likely to be sulfhydryl linked. Exposing proteins to reducing agents causes a decrease in apparent molecular weights by two- to threefold. These observations coupled with results of experiments using redox-sensitive spin labels suggest that the peptides exist in situ predominantly as homologous dimers and trimers. Whether the hydrophobicity of these proteins is completely due to their primary amino acid composition or to covalent linkage to lipid as well is not known.

The interactions with lipids of the two proteins are also of interest. At the molecular level, lipid-protein interactions of SP-B, SP-C can be accurately monitored by using electron paramagnetic spectroscopy of nitroxide spin-labeled phospholipids (1-palmitoyl-2-[12-doxyl stearoyl]phosphatidylcholine) with liposomal dispersion of the proteins in phospholipids. At 30 °C (above the phospholipid gel to liquid crystal transition temperature), mixtures of phospholipids and SP-B and SP-C show large changes in molecular order compared to dispersions of phospholipid without protein, even when concentrations of protein are low. For example, a dispersion of dimyristoylphosphatidylcholine (DMPC) with 1% w/w SP-B, SP-C, 2:1

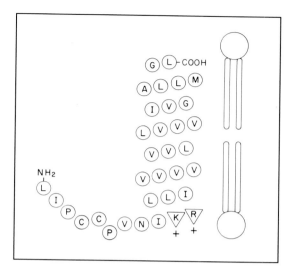

Fig. 2. An illustration of bovine SP-C is based on computer modeling. The hydrophobic transmembrane domain is depicted using a cylindrical plot projection. The non-helical 'hook' domain near the N-terminus is shown adjacent to the polar interface. Ball-stick figures represent phospholipids (ball = polar head group; stick = acyl chain) in a bilayer configuration. Scale is approximate.

shows a change in molecular order of 0.02 when compared with DMPC alone. These changes are replicated when we use SP-B and SP-C with mixtures of phospholipids [7].

Biophysical measurements using mixtures of synthesized and native bovine SP-B and SP-C with phospholipids have been carried out using a pulsating surfactometer, a Wilhelmy balance, and a King/Clements adsorption device. Despite presumably large differences in tertiary structure, it is surprising that SP-B or SP-C when mixed with surfactant phospholipids, each have large effects on minimal surface tension. In some experiments, SP-B and SP-C together have better activity than either alone. IgG or serum containing polyclonal antibody against both SP-B and SP-C does not inhibit (more than control IgG or serum) the ability of surfactants to reduce surface tension rapidly. Likewise, labelling SP-B and SP-C with lysine-linked spin labels does not interfere with activity. These results suggest that alteration of some domains on the hydrophilic face of the peptides does not interfere with their interaction with lipid.

In summary, several types of protein domains may be important for surfactant protein function. These domains include amphipathic surface domains such as found in SP-B that provide for interactions at the bulk aqueous-phospholipid polar head group interface. An apolar hydrophobic domain in SP-C exists adjacent to the phospholipid fatty acyl chains. The combination of these two types of domains suggests that the peptides may interact in a cooperative manner. Based on these observations and preliminary energy minimization structural analysis, we suggest that a model of SP-C in phospholipid dispersions would include a hydrophobic alpha-helical segment in the apolar domains with a bend near the polar residues adjacent to the N-terminus (fig. 2). A model of SP-B would include several amphipathic helical domains punctuated by proline residues and stabilized by intra- and intermolecular disulfide linkages. Oligomeric assemblies of the two molecules would act to provide a proper balance of amphipathic and hydrophobic domains that would maximize protein interaction with surfactant lipid.

References

1   Possmayer, F.: A proposed nomenclature for pulmonary surfactant-associated proteins. Am. Rev. resp. Dis. *138:* 990–998 (1988).

2   Hawgood, S.; Benson, B.; Schilling, J.; Damm, D.; Clements, J.; White, T.: Nucleotide and amino acid sequences of pulmonary surfactant protein SP 18 and evidence for cooperation between SP 18 and SP 28-36 in surfactant lipid adsorption. Proc. natn. Acad. Sci. USA *84:* 66–70 (1987).

3   Takahashi, A.; Smith, G.; Taeusch, W.: Separation of two pulmonary surfactant proteolipids in lung surfactant using HPLC (Abstract). Clin. Res. *36:* 245 (1988).

4   Yu, S.; Chung, W.; Olafson, R.; Harding, P.; Possmayer, F.: Characterization of the small hydrophobic proteins associated with pulmonary surfactant. Biochim. biophys. Acta *921:* 437–448 (1987).

5   Eisenberg, D.: Three dimensional structure of membrane and surface proteins. A. Rev. Biochem. *53:* 595–623 (1984).

6   Briggs, M.; Gierasch, L.: Molecular mechanisms of protein secretion. The role of the signal sequence. Adv. Protein Chem. *38:* 109–180 (1986).

7   Tanaka, Y.; Takei, T.; Aiba, T.; Masuda, K.; Kiuchi, A.; Fujiwara, T.: Development of synthetic lung surfactants. J. Lipid Res. *27:* 475–485 (1986).

H. William Taeusch, MD, Department of Pediatrics, King/Drew Medical Center, Los Angeles, CA 90059 (USA)

Subject Index